ALASKA'S CHANGING BOREAL FOREST

LONG-TERM ECOLOGICAL RESEARCH NETWORK SERIES
LTER Publications Committee

Grassland Dynamics: Long-Term Ecological Research in Tallgrass Prairie
Edited by
Alan K. Knapp, John M. Briggs,
David C. Hartnett, and Scott L. Collins

Standard Soil Methods for Long-Term Ecological Research
Edited by
G. Philip Robertson, David C. Coleman,
Caroline S. Bledsoe, and Phillip Sollins

Structure and Function of an Alpine Ecosystem: Niwot Ridge, Colorado
Edited by
William D. Bowman and Timothy R. Seastedt

*Climate Variability and Ecosystem Response
in Long-Term Ecological Research Sites*
Edited by
David Greenland, Douglas G. Goodin, and Raymond C. Smith

Biodiversity in Drylands: Toward a Unified Framework
Edited by
Moshe Shachak, James R. Gosz, Steward T. A. Pickett,
and Avi Perevolotsky

*Long-Term Dynamics of Lakes in the Landscape:
Long-Term Ecological Research on North Temperate Lakes*
Edited by
John J. Magnuson, Timothy K. Kratz, and Barbara J. Benson

Alaska's Changing Boreal Forest
Edited by
F. Stuart Chapin III, Mark W. Oswood, Keith Van Cleve,
Leslie A. Viereck, and David L. Verbyla

*Structure and Function of a Chihuahuan Desert Ecosystem:
The Jornada Basin Long-Term Ecological Research Site*
Edited by
Kris M. Havstad, Laura F. Huenneke, and William H. Schlesinger

ALASKA'S CHANGING BOREAL FOREST

Edited by

F. Stuart Chapin III

Mark W. Oswood

Keith Van Cleve

Leslie A. Viereck

David L. Verbyla

Illustration Editor

Melissa C. Chapin

2006

OXFORD
UNIVERSITY PRESS

Oxford University Press, Inc., publishes works that further
Oxford University's objective of excellence
in research, scholarship, and education.

Oxford   New York
Auckland   Cape Town   Dar es Salaam   Hong Kong   Karachi
Kuala Lumpur   Madrid   Melbourne   Mexico City   Nairobi
New Delhi   Shanghai   Taipei   Toronto

With offices in
Argentina   Austria   Brazil   Chile   Czech Republic   France   Greece
Guatemala   Hungary   Italy   Japan   Poland   Portugal   Singapore
South Korea   Switzerland   Thailand   Turkey   Ukraine   Vietnam

Copyright © 2006 by Oxford University Press, Inc.

Published by Oxford University Press, Inc.
198 Madison Avenue, New York, New York 10016

www.oup.com

Oxford is a registered trademark of Oxford University Press

All rights reserved. No part of this publication may be reproduced,
stored in a retrieval system, or transmitted, in any form or by any means,
electronic, mechanical, photocopying, recording, or otherwise,
without the prior permission of Oxford University Press.

Library of Congress Cataloging-in-Publication Data
Alaska's changing boreal forest / edited by F. Stuart Chapin III . . . [et al.] ;
illustration editor, Melissa C. Chapin.
 p. cm.—(Long-Term Ecological Research Network series)
ISBN-10: 0-19-515431-2
ISBN-13: 978-0-19-515431-3
1. Taiga ecology—Alaska.   I. Chapin, F. Stuart, III.   II. Series.
QH105.A4A743 2005
577.3'7'09798—dc22
                            2004030988

9 8 7 6 5 4 3 2 1

Printed in the United States of America
on acid-free paper

Preface

The boreal forest is the northern-most forested biome, whose organisms and dynamics have been shaped by low temperature and other features of a high-latitude environment. The Alaskan boreal forest is now warming as rapidly as any place on Earth, providing an unprecedented opportunity to examine a biome as it adjusts to change. This book is written to provide an understanding of the processes that have shaped the development and current dynamics of Alaska's boreal forest. These processes will constrain its response to future change. Therefore, understanding how Alaskan boreal ecosystems function is essential to guide sound management in a changing climate. We also hope to provide insights for those parts of the world where climate is changing more slowly.

The book is intended to provide a synthesis of understanding about Alaska's boreal forest. It also provides the most complete bibliography currently available for the Alaskan boreal forest and will be a useful guide for those who want to know additional details. It is written in a style that we hope is both scientifically accurate and accessible to general readers who have a curiosity about the North.

The first part of the book provides a geographic and historical context for understanding the boreal forest. In the second part, we describe the dynamics of plant and animal communities that inhabit this forest. The third part of the book delves into the biogeochemical processes that link these organisms. The fourth part of the book explores landscape phenomena that operate at larger temporal and spatial scales and integrates the processes described in earlier sections.

We are grateful to our co-investigators for the team effort that made this synthesis possible. We also thank the many graduate and undergraduate students, postdoctoral fellows, technicians, and colleagues who participated in the research that is synthesized in this book. We particularly thank the site managers (Phyllis Adams,

Wendy Davis, and Jamie Hollingsworth) and data managers (Darrell Blodgett, Phyllis Adams, Scott Miller, and Brian Riordan), whose dedication and hard work have kept the Bonanza Creek LTER site functioning over the long term. We also recognize the inspiration provided by Keith Van Cleve, Les Viereck, and Ted Dyrness, who initiated the Bonanza Creek LTER program and provided much of its current intellectual focus. We thank the National Science Foundation Long-Term Ecological Research program and the USDA Forest Service Pacific Northwest Experiment Station, which provided the major funding for this research. Finally, we thank the Alaskan boreal forest itself for providing such a magnificent place to work.

Contents

Contributors xi

Part I. Alaska's Past and Present Environment

1. The Conceptual Basis of LTER Studies in the Alaskan Boreal Forest 3
   F. Stuart Chapin III, John Yarie, Keith Van Cleve, and Leslie A. Viereck

2. Regional Overview of Interior Alaska 12
   James E. Begét, David Stone, and David L. Verbyla

3. State Factor Control of Soil Formation in Interior Alaska 21
   Chien-Lu Ping, Richard D. Boone, Marcus H. Clark, Edmond C. Packee, and David K. Swanson

4. Climate and Permafrost Dynamics of the Alaskan Boreal Forest 39
   Larry D. Hinzman, Leslie A. Viereck, Phyllis C. Adams, Vladimir E. Romanovsky, and Kenji Yoshikawa

5. Holocene Development of the Alaskan Boreal Forest 62
   Andrea H. Lloyd, Mary E. Edwards, Bruce P. Finney, Jason A. Lynch, Valerie Barber, and Nancy H. Bigelow

Part II. Forest Dynamics

6. Floristic Diversity and Vegetation Distribution in the Alaskan Boreal Forest   81
   F. Stuart Chapin III, Teresa Hollingsworth, David F. Murray, Leslie A. Viereck, and Marilyn D. Walker

7. Successional Processes in the Alaskan Boreal Forest   100
   F. Stuart Chapin III, Leslie A. Viereck, Phyllis C. Adams, Keith Van Cleve, Christopher L. Fastie, Robert A. Ott, Daniel Mann, and Jill F. Johnstone

8. Mammalian Herbivore Population Dynamics in the Alaskan Boreal Forest   121
   Eric Rexstad and Knut Kielland

9. Dynamics of Phytophagous Insects and Their Pathogens in Alaskan Boreal Forests   133
   Richard A. Werner, Kenneth F. Raffa, and Barbara L. Illman

10. Running Waters of the Alaskan Boreal Forest   147
    Mark W. Oswood, Nicholas F. Hughes, and Alexander M. Milner

Part III. Ecosystem Dynamics

11. Controls over Forest Production in Interior Alaska   171
    John Yarie and Keith Van Cleve

12. The Role of Fine Roots in the Functioning of Alaskan Boreal Forests   189
    Roger W. Ruess, Ronald L. Hendrick, Jason G. Vogel, and Bjartmar Sveinbjörnsson

13. Mammalian Herbivory, Ecosystem Engineering, and Ecological Cascades in Alaskan Boreal Forests   211
    Knut Kielland, John P. Bryant, and Roger W. Ruess

14. Microbial Processes in the Alaskan Boreal Forest   227
    Joshua P. Schimel and F. Stuart Chapin III

15. Patterns of Biogeochemistry in Alaskan Boreal Forests   241
    David W. Valentine, Knut Kielland, F. Stuart Chapin III, A. David McGuire, and Keith Van Cleve

Part IV. Changing Regional Processes

16. Watershed Hydrology and Chemistry in the Alaskan Boreal Forest:
    The Central Role of Permafrost   269
    Larry D. Hinzman, W. Robert Bolton, Kevin C. Petrone,
    Jeremy B. Jones, and Phyllis C. Adams

17. Fire Trends in the Alaskan Boreal Forest   285
    Eric S. Kasischke, T. Scott Rupp, and David L. Verbyla

18. Timber Harvest in Interior Alaska   302
    Tricia L. Wurtz, Robert A. Ott, and John C. Maisch

19. Climate Feedbacks in the Alaskan Boreal Forest   309
    A. David McGuire and F. Stuart Chapin III

20. Communication of Alaskan Boreal Science with
    Broader Communities   323
    Elena B. Sparrow, Janice C. Dawe, and F. Stuart Chapin III

21. Summary and Synthesis: Past and Future Changes
    in the Alaskan Boreal Forest   332
    F. Stuart Chapin III, A. David McGuire, Roger W. Ruess,
    Marilyn W. Walker, Richard D. Boone, Mary E. Edwards,
    Bruce P. Finney, Larry D. Hinzman, Jeremy B. Jones, Glenn P. Juday,
    Eric S. Kasischke, Knut Kielland, Andrea H. Lloyd, Mark W. Oswood,
    Chien-Lu Ping, Eric Rexstad, Vladimir E. Romanovsky,
    Joshua P. Schimel, Elena B. Sparrow, Bjartmar Sveinbjörnsson,
    David W. Valentine, Keith Van Cleve, David L. Verbyla, Leslie A.
    Viereck, Richard A. Werner, Tricia L. Wurtz, and John Yarie

Index   339

Contributors

Phyllis C. Adams
Forest Inventory and Analysis—PNW
　Research Station
620 SW Main St., Suite 400
Portland, OR 97205

Valerie Barber
University of Alaska Southeast Sitka
1332 Seward Ave.
Sitka, AK 99835

James E. Begét
Department of Geology and Geophysics
University of Alaska Fairbanks
Fairbanks, AK 99775

Nancy H. Bigelow
Alaska Quaternary Center
University of Alaska Fairbanks
Fairbanks, AK 99775

W. Robert Bolton
Water and Environmental Research Center
University of Alaska Fairbanks
Fairbanks, AK 99775

Richard D. Boone
Institute of Arctic Biology
University of Alaska Fairbanks
Fairbanks, AK 99775

John P. Bryant
Institute of Arctic Biology (retired)
Box 306
Cora, WY 82925

F. Stuart Chapin III
Institute of Arctic Biology
University of Alaska Fairbanks
Fairbanks, AK 99775

Marcus H. Clark
USDA—Natural Resources Conservation
　Service
800 W. Evergreen, Suite 100
Palmer, AK 99645

Janice C. Dawe
P.O. Box 82003
Fairbanks, AK 99708

Contributors

Mary E. Edwards
Department of Geography
University of Southampton
Highfield, Southampton SO17 1BJ UK
and Institute of Arctic Biology
University of Alaska Fairbanks
Fairbanks, AK 99775

Christopher L. Fastie
Biology Department
Middlebury College
Middlebury, VT 05753

Bruce P. Finney
Institute of Marine Science
University of Alaska Fairbanks
Fairbanks, AK 99775

Ronald L. Hendrick
School of Forest Resources
University of Georgia
Athens, GA 30602

Larry D. Hinzman
Institute of Environmental Engineering
and Environmental Quality Science
University of Alaska Fairbanks
Fairbanks, AK 99775

Teresa Hollingsworth
Boreal Ecology Cooperative Research Unit
USDA Forest Service, PNW Research Station
Fairbanks, AK 99775

Nicholas F. Hughes
School of Fisheries and Ocean Sciences
University of Alaska Fairbanks
Fairbanks, AK 99775

Barbara L. Illman
USDA Forest Service
Forest Products Laboratory
University of Wisconsin
Madison, WI 53705

Jill F. Johnstone
Department of Geography and
Environmental Studies
Carleton University
1125 Colonel By Drive
Ottawa, ON K1S 5B6 Canada

Jeremy B. Jones
Institute of Arctic Biology
University of Alaska Fairbanks
Fairbanks, AK 99775

Glenn P. Juday
Forest Sciences Department
University of Alaska Fairbanks
Fairbanks, AK 99775

Eric S. Kasischke
Department of Geography
University of Maryland
College Park, MD 20742

Knut Kielland
Institute of Arctic Biology
University of Alaska Fairbanks
Fairbanks, AK 99775

Andrea H. Lloyd
Department of Biology
Middlebury College
Middlebury, VT 05753

Jason A. Lynch
Department of Biology
North Central College
Naperville, IL 60566

John C. Maisch
Division of Forestry, Northern Region
Alaska Department of Natural Resources
Fairbanks, AK 99709-4699

Daniel Mann
Institute of Arctic Biology
University of Alaska Fairbanks
Fairbanks, AK 99775

A. David McGuire
U.S. Geological Survey
Alaska Cooperative Fish and Wildlife
  Research Unit
University of Alaska Fairbanks
Fairbanks, AK 99775

Alexander M. Milner
School of Geography, Earth and Environ-
  mental Sciences
University of Birmingham
Edgbaston, Birmingham B15 2TT, UK
and Institute of Arctic Biology
University of Alaska Fairbanks
Fairbanks, AK 99775

David F. Murray
University of Alaska Museum of the North
University of Alaska Fairbanks
Fairbanks, AK 99775

Mark W. Oswood
Biology and Wildlife Department (retired)
University of Alaska Fairbanks
2390 Jeffrey Court
Wenatchee, WA 98801

Robert A. Ott
Forestry Program
Tanana Chiefs Conference
122 First Ave., Suite 600
Fairbanks, AK 99701

Edmond C. Packee
Forest Sciences Department
University of Alaska Fairbanks
Fairbanks, AK 99775

Kevin C. Petrone
Institute of Arctic Biology
University of Alaska Fairbanks
Fairbanks, AK 99775

Chien-Lu Ping
Agricultural and Forestry Experiment
  Station
533 E. Fireweed
Palmer, AK 99645

Kenneth F. Raffa
Department of Entomology
University of Wisconsin
Madison, WI 53706

Eric Rexstad
Institute of Arctic Biology
University of Alaska Fairbanks
Fairbanks, AK 99775

Vladimir E. Romanovsky
Geophysical Institute
University of Alaska Fairbanks
Fairbanks, AK 99775

Roger W. Ruess
Institute of Arctic Biology
University of Alaska Fairbanks
Fairbanks, AK 99775

T. Scott Rupp
Forest Sciences Department
University of Alaska Fairbanks
Fairbanks, AK 99775

Joshua P. Schimel
Department of Ecology, Evolution &
  Marine Biology
University of California
Santa Barbara, CA 93106-9610

Elena B. Sparrow
School of Natural Resources and Agricul-
  tural Sciences
University of Alaska Fairbanks
Fairbanks, AK 99775

David Stone
Department of Geology and Geophysics
University of Alaska Fairbanks
Fairbanks, AK 99775

Bjartmar Sveinbjörnsson
Department of Biological Sciences
University of Alaska Anchorage
Anchorage, AK 99508

David K. Swanson
USDA Forest Service
1550 Dewey Ave.
Baker City, OR 97814

David W. Valentine
Forest Sciences Department
University of Alaska Fairbanks
Fairbanks, AK 99775

Keith Van Cleve
279 Kanaka Bay Road
Friday Harbor, WA 98250

David L. Verbyla
Forest Sciences Department
University of Alaska Fairbanks
Fairbanks, AK 99775

Leslie A. Viereck
Boreal Ecology Cooperative Research
  Unit
University of Alaska Fairbanks
Fairbanks, AK 99775

Jason G. Vogel
Department of Botany
University of Florida
Gainesville, FL 32611-8526

Marilyn D. Walker
Boreal Ecology Cooperative Research
  Unit (retired)
548 W. Cactus Ct.
Loiusville, CO 80027

Richard A. Werner
Pacific Northwest Research Station
Corvallis Forestry Sciences Laboratory
Corvallis, OR 97330

Tricia L. Wurtz
Boreal Ecology Cooperative Research
  Unit
USDA Forest Service, PNW Research
  Station
Fairbanks, AK 99775

John Yarie
Forest Sciences Department
University of Alaska Fairbanks
Fairbanks, AK 99775

Kenji Yoshikawa
Water and Environmental Research Center
University of Alaska Fairbanks
Fairbanks, AK 99775

Part I

Alaska's Past and Present Environment

1

The Conceptual Basis of LTER Studies in the Alaskan Boreal Forest

F. Stuart Chapin III
John Yarie
Keith Van Cleve
Leslie A. Viereck

Introduction

The boreal forest occupies 10% of the ice-free terrestrial surface and is the second most extensive terrestrial biome on Earth, after tropical forests (Saugier et al. 2001). It is a land of extremes: low temperature and precipitation, low diversity of dominant plant species, dramatic population fluctuations of important insects and mammals, and a generally sparse human population.

The boreal forest is also a land poised for change. During the last third of the twentieth century, many areas of the boreal forest, such as western North America and northern Eurasia, warmed more rapidly than any other region on Earth (Serreze et al. 2000). This pattern of warming is consistent with projections of general circulation models. These models project that human-induced increases in greenhouse gases, such as carbon dioxide, will cause the global climate to warm and that the warming will occur most rapidly at high latitudes (Ramaswamy et al. 2001). If we accept the projections of these models, the climate of many parts of the boreal forest will likely continue to warm even more rapidly than it has in the past.

The ecological characteristics of the boreal forest render it vulnerable to warming and other global changes. Because the boreal forest is the coldest forested biome on Earth, organisms are adapted to low temperatures, and many of its physical and biological processes are molded by low temperature. Permafrost (permanently frozen ground) is widespread and governs the soil temperature and moisture regime of a large proportion of the boreal forest. Yet permafrost temperatures are close to the freezing point throughout much of interior Alaska (Osterkamp and Romanovsky 1999), so only a slight warming of soils could greatly reduce the extent of permafrost. Low temperature and anaerobic soil conditions associated with permafrost-impeded drainage constrain decomposition rate, leading to thick layers of soil organic matter. Consequently, boreal soils account for about a third of the readily decomposable soil organic matter on Earth (McGuire et al. 1995). This represents a quantity of carbon similar to that in the atmosphere. Fire is another process that could rapidly return this undecomposed carbon from the organic layers of the soil to the atmosphere. Although most boreal organisms are adapted to a cold environment and frequent disturbance, organisms and properties of the boreal forest are sensitive to changes in climate, permafrost, and fire regime, so climatically induced changes are likely to propagate through most ecological processes.

Changes in the boreal forest will certainly affect its human residents through modifications to the biota, soil stability, hydrologic regime, forest productivity, fire regime, insect outbreaks, recreational opportunities, and many ecosystem goods and services. These changes could also affect nonboreal residents through changes in Earth's climate system. The boreal forest provides most of the freshwater input to the Arctic Ocean. This input influences the stability of thermohaline circulation, the conveyor belt that drives global heat transport in the world's oceans (Peterson et al. 2002). Release of the large stores of readily decomposable soil organic matter in boreal forest soils could alter the warming potential of the atmosphere and therefore the rate at which climate warms globally (McGuire et al. 1995). Finally, changes in solar energy absorbed by the boreal forest and transferred to the atmosphere could alter climate well beyond the limits of boreal forest (Bonan et al. 1995).

Long-term studies of the boreal forest are the most effective way to document the past changes and predict the future responses of this biome to global change. The Bonanza Creek Long-Term Ecological Research (LTER) site was established in 1987, as part of the United States LTER network, to develop a process-based understanding of the structure and functioning of the boreal forests of interior Alaska—that is, the region north of the Alaska Range and south of the Brooks Range (Fig. 1.1). This book summarizes our current understanding, which is based on this and other boreal research. We next describe two conceptual paradigms (state factors and interactive controls) that underlie the Bonanza Creek LTER research and are therefore threads that weave together many of the chapters in this book.

The State Factor Concept

Dokuchaev (1879) and Jenny (1941) proposed that five independent *state factors* govern the properties of soils and terrestrial ecosystems. These state factors are climate, parent material (i.e., the rocks that give rise to soils), topography, potential biota (i.e., the organisms present in the region that could potentially occupy a site), and time (Fig. 1.2; Jenny 1941, Amundson and Jenny 1997). Together these five factors define the major characteristics of terrestrial ecosystems. Over thousands to millions of years, these state factors change, as climate responds to changes in solar input, as new terranes are rafted across oceans and collide with continents, as mountains rise and erode, and as biota migrate in response to glacial cycles or human introductions. However, over the decadal-to-century time scales studied by ecologists, these state factors, as put forth by Jenny, can usually be treated as con-

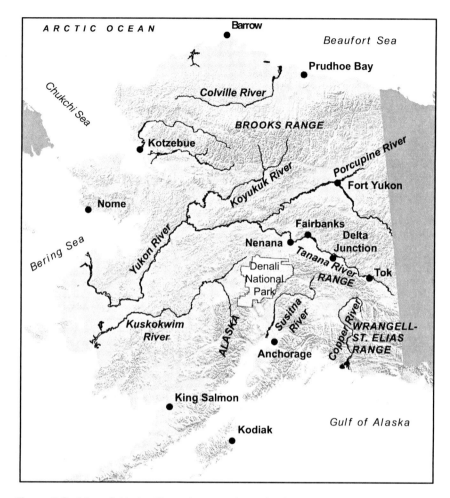

Figure 1.1. Map of Alaska. Ecoregions are shown in Fig. 2.3.

stants. Within interior Alaska, climate and potential biota are relatively homogeneous, so the state factors that account for most regional variation in ecosystem processes in interior Alaska are parent material, topography, and time. Much of the research in the Bonanza Creek LTER site has therefore focused on ecosystem variation due to these three state factors.

Over broad geographic scales, larger than interior Alaska, *climate* is the state factor that most strongly determines the global patterns of ecosystem structure and functioning. Climate is ultimately the factor that differentiates the boreal forest from other biomes. Within interior Alaska, climate is a consequence of the northerly latitude of the region and the presence of the Alaska and Brooks Ranges, which block maritime and polar influences, respectively, yielding a dry continental climate with cold winters and warm summers (Chapters 2 and 4).

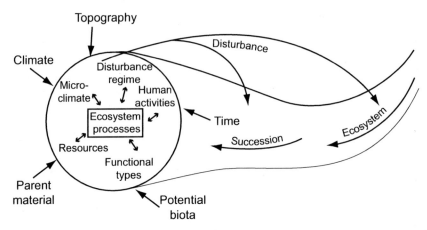

Figure 1.2. The relationship between state factors (outside the circle), interactive controls (inside the circle), and ecosystem processes (inside the box). The circle represents the boundary of the ecosystem, whose structure and functioning respond to and affect interactive controls, which are ultimately governed by state factors.

The *potential biota* of interior Alaska is the result of landscape filters (Tonn 1990) that operate to winnow the world biota to a much smaller regional biota. The coarsest filter operates at the continental level, eliminating species that have not dispersed to Alaska from ancient centers of origin (e.g., marsupials from Australia). Finer filters, operating within continents and regions, eliminate more organisms. For many North American organisms, the rigorous climate of interior Alaska exceeds their physiological tolerances (e.g., warm water fishes such as the sunfishes). The potential biota of the region has not changed radically for the past 5,000 years, by which time most species in interior Alaska had established their current distributions. However, introductions of exotic species, such as lodgepole pine and Siberian larch, could significantly alter the future community composition of the Alaskan landscape. Local and regional variations in the distributions of plants, animals, and microbes reflect local heterogeneity in environment, disturbance regime, and ecological processes, rather than regional variation in availability of organisms to colonize a site.

The *parent material* at a site reflects the underlying bedrock and materials transported from adjacent areas. A partially metamorphosed shist underlies about 75% of interior Alaska, including most of the upland sites studied by the Bonanza Creek LTER program. Floodplains, in contrast, develop primarily on alluvium that is transported from the Alaska Range by glacial rivers. Most lowlands of interior Alaska have never been glaciated, so, over millions of years, large quantities of wind-blown loess have been transported from the floodplains to adjacent uplands, forming a loess cap that is frequently many meters in depth. Much of the regional variation in parent material in the LTER study sites near Fairbanks reflects variation in the quantity of wind-blown loess that accumulated above the schist bedrock during the Pleistocene (Chapters 2 and 3; Bejét 1990).

Topography influences ecosystem processes through its effects on slope steepness, aspect (direction that a slope faces), and elevation. Because of the low sun angle at high latitudes, aspect plays a critical role in influencing energy input, soil temperature, and occurrence of permafrost (Chapter 4). At a landscape level, floodplains differ strikingly from uplands due to frequent flooding near rivers and the potential supply of moisture from groundwater not available in upland locations. Comparisons among ecosystems in different topographic positions have enabled LTER researchers to document the role of topography and associated differences in microclimate and parent material on ecosystem processes. Topography is determined by bedrock geology and by the weathering and fluvial processes that have operated over geologic time. In our LTER studies, topography is essentially constant, with the exception of alluvial terraces, which may be newly deposited or eroded, changing their elevation above the river.

Time influences ecosystems at many scales. The deposition and erosion of loess shape regional topography and parent material over decadal to millennial time scales (Fig. 1.3). The biota that dominate the boreal forest migrated to their current locations over thousands of years. The cycles of disturbance and succession that account for much of the regional variation in ecosystem processes occur over decades to a few centuries. The physiological and ecosystem processes that interact with disturbances and succession happen over time scales of seconds to centuries. As ecologists, we can directly observe only those changes that occur over brief time periods. In the LTER program, we emphasize those temporal changes that occur over seasons to decades, and we use retrospective analyses and comparative studies to extend our understanding back in time and to project into the future.

Interactive Controls

In addition to state factors, which are independent of processes occurring in ecosystems, there are a multitude of ecological factors, termed *interactive controls*, that both affect and respond to ecosystem processes at the level of individual stands (Fig. 1.2; Van Cleve et al. 1991, Field et al. 1992, Chapin et al. 1996). These interactive controls include the functional types of organisms that actually occupy the ecosystem (as contrasted with the potential biota present in the region); resources (e.g., light, water, nutrients, oxygen) that are used by plants, animals, and decomposers to grow and reproduce; and microclimate, which varies with topography and vegetation. All of these factors can be dramatically altered by disturbances (such as fire, flooding, and human activities) and the subsequent recovery of ecosystems through succession. These controls over ecosystem processes operate hierarchically, with state factors providing the ultimate bounds on ecosystem processes, interactive controls (e.g., soil resources, microclimate, and functional types of organisms) mediating the effects of these state factors at the local scale, and disturbance and succession explaining the temporal variation in ecosystem processes (Fig. 1.2). Much of the research in the Bonanza Creek LTER program has focused on elucidating the nature of interactions and feedbacks associated with interactive controls (i.e., the mechanisms by which state factors influence ecosystem processes).

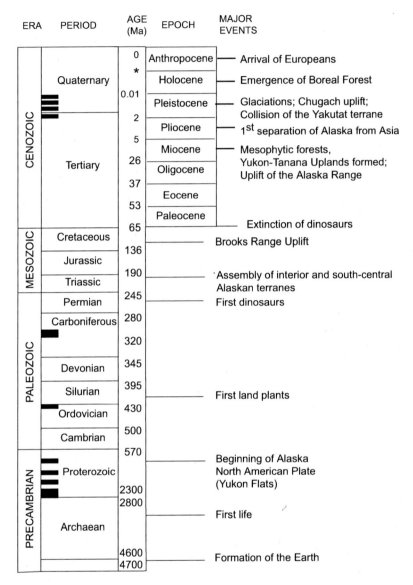

Figure 1.3. Geological time periods in the earth's history showing major glacial (solid bars) and other events that have shaped the current functioning of Alaska's boreal forest. Note the changes in time scale (Ma = millions of years; * the Anthropocene began in about 1750). Based on information from Chapters 2 and 5 and from Thorson (1985).

Resources are the energy and materials in the environment that are used by organisms to support their growth and maintenance. Resource supply is governed by state factors such as climate, parent material, and topography. It is also sensitive to processes that occur within the ecosystem. Light availability, for example, depends on climatic elements, such as cloudiness and on topographic position, but also changes during succession due to changes in shading by vegetation. Similarly, soil fertility depends on parent material and climate but is also sensitive to ecosystem processes, such as the inputs of nitrogen from nitrogen-fixing organisms and successional changes in litter quality. An important discovery of the Bonanza Creek LTER program has been the critical role that vegetation-induced changes in nutrient availability play in governing the productivity and nutrient cycling of the boreal forest (Chapters 12–15).

Functional types are groups of species that are similar to one another in their effects on ecosystem processes. Examples include ectomycorrhizal fungi, nitrogen-fixing shrubs, and large mammalian herbivores. The relative abundance and the nature of interactions among functional types can influence ecosystem processes just as strongly as do large differences in climate or parent material. A key discovery of the LTER program has been the strength and rapidity with which changes in the dominant species induce changes in environment and biogeochemistry that alter most ecosystem processes. These vegetation shifts define *turning points*, which are threshold changes in most of the interactive controls that govern ecosystem processes (Chapter 7).

Microclimate is the climatic variation that occurs within and among ecosystems within a region. It depends on regional climate (a state factor), but it is also sensitive to topography (another state factor) and vegetation development during succession. Soil temperature is particularly sensitive to canopy cover of plants, to the density and types of mosses in the understory, and to the thickness of the forest floor.

Many other factors such as soil pH change with succession. Soil acidity is sensitive to species composition and parent material and is a key factor governing resource availability and species composition.

Landscape-scale *disturbances* by fire, wind, floods, insect outbreaks, ice storms, and human activities are critical determinants of the natural structure and process rates in ecosystems (Pickett and White 1985, Sousa 1985). As with other interactive controls, the probability of disturbance and the rate and pattern of succession after disturbance depend on both state factors and ecosystem processes. Seasonal variation in climate, for example, directly affects fire probability and spread, but geographic variation in climate also influences the types and quantity of plants present in an ecosystem and therefore the fuel load and flammability of vegetation. Deposition and erosion during floods shape river channels and influence the probability of future floods. The LTER program has shown that vegetation-induced changes in disturbance probability are key factors in governing the landscape dynamics of the boreal forest (Chapter 17). An important challenge for the future is to manage interactive controls in such a way that the most valuable attributes of the boreal forest, both economically and aesthetically, are sustained in the future (Chapter 18).

Roadmap

This book begins with a brief description of the past and present environment of the boreal forest of interior Alaska, including its geologic and vegetation history (Chapters 2 and 5) and the major ecoregions (Chapter 2). We then describe the soils (Chapter 3) and climate (Chapter 4). The second section of the book describes the dynamics of the boreal forest, including the spatial and temporal variation in vegetation (Chapters 6 and 7) and the population dynamics of mammals (Chapter 8) and herbivorous insects (Chapter 9). The ecology of streams and rivers is described in Chapter 10. The third section of the book describes the ecosystem processes that underlie forest dynamics. These processes include primary production (Chapters 11 and 12), herbivory (Chapter 13), and biogeochemical and hydrologic cycling (Chapters 14–16). The fourth section of the book, on landscape and regional processes, addresses the role of fire and forest harvest (Chapters 17 and 18), and the climatic feedbacks from changes in the boreal forest (Chapter 19). Finally, we consider the critical role of communicating our understanding with other scientists, students, and the general public (Chapter 20) and summarize key findings that determine the present future state of the boreal forest (Chapter 21). The companion website provides a more visual view of the boreal forest, including photographs of key processes and changes and model simulations of likely future changes. The interested reader will find a wealth of additional information in specific studies that are cited by individual chapters. In addition, there are excellent syntheses of the history of the Alaskan boreal forest (Aigner et al. 1986) and its patterns of biogeochemistry (Van Cleve et al. 1986). Two key syntheses of research in the Canadian boreal forest focus on trophic dynamics (Krebs et al. 2001) and natural history (Henry 2002).

References

Aigner, J. S., R. D. Guthrie, M. L. Guthrie, R. K. Nelson, W. S. Schneider, and R. M. Thorson, editors. 1986. Interior Alaska: A Journey through Time. Alaska Geographic Society, Anchorage.

Amundson, R., and H. Jenny. 1997. On a state factor model of ecosystems. BioScience 47:536–543.

Bejét, J. 1990. Mid-Wisconsin climate fluctuations recorded in central Alaskan loess. Geographie Physique et Quaternaire 544:3–13.

Bonan, G. B., F. S. Chapin III, and S. L. Thompson. 1995. Boreal forest and tundra ecosystems as components of the climate system. Climatic Change 29:145–167.

Chapin, F. S. III, M. S. Torn, and M. Tateno. 1996. Principles of ecosystem sustainability. American Naturalist 148:1016–1037.

Dokuchaev, V. V. 1879. Abridged historical account and critical examination of the principal soil classifications existing. Transactions of the Petersburg Society of Naturalists 1:64–67.

Field, C., F. S. Chapin III, P. A. Matson, and H. A. Mooney. 1992. Responses of terrestrial ecosystems to the changing atmosphere: A resource-based approach. Annual Review of Ecology and Systematics 23:201–235.

Henry, J. D. 2002. Canada's Boreal Forest. Smithsonian Institution of Washington, Washington, DC.

Jenny, H. 1941. Factors of Soil Formation. McGraw-Hill, New York.

Krebs, C. J., S. Boutin, and R. Boonstra, editors. 2001. Ecosystem Dynamics of the Boreal Forest: The Kluane Project. Oxford University Press, New York.

McGuire, A. D., J. W. Melillo, D. W. Kicklighter, and L. A. Joyce. 1995. Equilibrium responses of soil carbon to climate change: Empirical and process-based estimates. Journal of Biogeography 22:785–796.

Osterkamp, T. E., and V. E. Romanovsky. 1999. Evidence for warming and thawing of discontinuous permafrost in Alaska. Permafrost and Periglacial Processes 10:17–37.

Peterson, B. J., R. M. Holmes, J. W. McClelland, C. J. Vörösmarty, R. B. Lammers, A. I. Shiklomanov, I. A. Shiklomanov, and S. Rahmstorf. 2002. Increasing arctic river discharge: Responses and feedbacks to global climate change. Science 298:2171–2173.

Pickett, S. T. A., and P. S. White. 1985. The Ecology of Natural Disturbance as Patch Dynamics. Academic Press, New York.

Ramaswamy, V., O. Boucher, J. Haigh, D. Hauglustaine, J. Haywood, G. Myhre, T. Nakajima, G. Y. Shi, and S. Solomon. 2001. Radiative forcing of climate change. Pages 349–416 *in* J. T. Houghton, Y. Ding, D. J. Griggs, M. Noguer, P. J. van der Linden, X. Dai, K. Maskell, and C. A. Johnson, editors. Climate Change 2001: The Scientific Basis. Cambridge University Press, Cambridge.

Saugier, B., J. Roy, and H. A. Mooney. 2001. Estimations of global terrestrial productivity: Converging toward a single number? Pages 543–557 *in* J. Roy, B. Saugier, and H. A. Mooney, editors. Terrestrial Global Productivity. Academic Press, San Diego.

Serreze, M. C., J. E. Walsh, F. S. Chapin III, T. Osterkamp, M. Dyurgerov, V. Romanovsky, W. C. Oechel, J. Morison, T. Zhang, and R. G. Barry. 2000. Observational evidence of recent change in the northern high-latitude environment. Climatic Change 46:159–207.

Sousa, W. P. 1985. The role of disturbance in natural communities. Annual Review of Ecology and Systematics 15:353–391.

Thorson, R. M. 1986. The ceaseless contest. Pages 1–51 *in* J. S. Aigner, R. D. Guthrie, M. L. Guthrie, R. K. Nelson, S. W. Schneider, and R. M. Thorson, editors. Interior Alaska: A Journey through Time. Alaska Geographic Society, Anchorage.

Tonn, W. M. 1990. Climate change and fish communities: A conceptual framework. Transactions of the American Fisheries Society 119:337–352.

Van Cleve, K., F. S. Chapin III, C. T. Dyrness, and L. A. Viereck. 1991. Element cycling in taiga forest: State-factor control. BioScience 41:78–88.

Van Cleve, K., F. S. Chapin, III, P. W. Flanagan, L. A. Viereck, and C. T. Dyrness, editors. 1986. Forest Ecosystems in the Alaskan Taiga: A Synthesis of Structure and Function. Springer-Verlag, New York.

2

Regional Overview of Interior Alaska

James E. Begét
David Stone
David L. Verbyla

Introduction

From continental macroclimate to microalluvial salt crusts, geology is a dominant factor that influences patterns and processes in the Alaskan boreal forest (Fig. 2.1). In this chapter, we outline important geologic processes as a foundation for subsequent chapters that discuss the soil, hydrology, climate, and biota of the Alaskan boreal forest. We conclude the chapter with a discussion of interior Alaska from a regional perspective.

Physiographic Regions

Alaska can be divided into four major physiographic regions (Fig. 2.2). The arctic coastal plain is part of the Interior Plains physiographic division of North America, analogous to the great plains east of the Rocky Mountains. The arctic coastal plain is predominantly alluvium underlaid by hundreds of meters of permafrost, resulting in many thaw lakes and ice wedges. South of the arctic coastal plain lies the Northern Cordillera, an extension of the Rocky Mountain system dominated by the Arctic Foothills, Brooks Range, Baird Mountains, and Delong Mountains. These mountains were glaciated during the Pleistocene. South of the Brooks Range lies interior Alaska, which is an intermontane plateau region analogous to the Great Basin/Colorado Plateau regions. This extensive region is characterized by wide alluvium-covered lowlands such as the Yukon Flats, Tanana Valley, and Yukon Delta, as well as moderate upland hills, domes, and mountains. Largely unglaciated, this region served as a refugium for biota during glacial periods. With the Northern

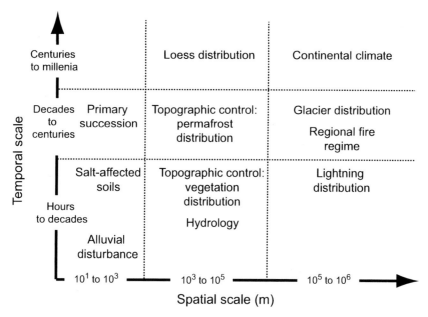

Figure 2.1. Temporal and spatial scales of selected ecosystem controls. In general, the spatial and temporal scales tend to be positively correlated, although some factors operate at disjunct spatial-temporal scales.

and Southern Cordilleras acting as barriers, the major rivers of this region have long, meandering paths to the Bering Sea. The Southern Cordillera is composed of two mountain ranges: the Alaska Range to the north and the Kenai/Chugach/Wrangell-St. Elias Mountains to the south. The lowland belt between these mountains includes the Susitna and Copper River lowlands. The entire Southern Cordillera was glaciated during the Pleistocene and today has extensive mountain glaciers.

Tectonic Geology

Much of Alaska is made up of multiple geologic fragments that have been rafted together by the movements of the major plates called tectonic terranes (Thorson 1986, Connor and O'Haire 1988). Plate-tectonic theory explains such observations as the changing distribution of fossils with geologic time, the deep Aleutian Trench, high Alaskan mountain barriers, and mountain glaciers. The plate movements have transported fragments of oceanic plateaus, slivers of continental margins, and island arcs. When plates collide, the denser oceanic plate subducts beneath the lighter continental plate, and the material being rafted by them accumulates. In the case of the North Pacific, the eastern edges of the plates slide north, tearing off pieces of the North America continental plate.

In Alaska, the region south of the Brooks Range has been assembled by a combination of simple rafting and the dragging of continental margin fragments parallel to

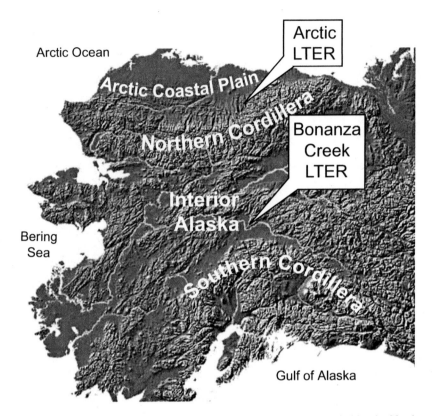

Figure 2.2. Physiographic regions of Alaska. Interior Alaska is bounded by the Northern and Southern Cordilleras. The locations of the two Alaskan LTER sites are shown.

the boundary between the two plates. The tectonic models for the origins of arctic Alaska, including the Brooks Range, are more controversial. The most accepted model involves the initiation of the ancestral Arctic Ocean with a spreading center, splitting arctic Alaska and Siberia and rotating them away from the Canadian arctic.

These terrane collisions leave their mark on the landscape. The only autochthonous (original) part of Alaska that has remained part of North America since the Precambrian is the far eastern corner of the Yukon Flats area in the eastern interior. The Brooks Range resulted from a terrane pushed from the north. Southeast of interior Alaska, the Wrangell–St.Elias Range is the result of the northward-moving Pacific plate. A consequence is the dramatic uplift of the coastal ranges and the high level of earthquake activity in the area. Three of the ten largest earthquakes in the world since 1900 have occurred in Alaska. The northern margin of the rafted terranes is marked by the prominent Alaska Range, as well as other topographic features that form an arcuate chain from the Bering Sea margin across to southeast Alaska and British Columbia. The central parts of the Alaska range, south of Fairbanks, were low-lying about 6 million years ago and have grown to their present height since then and are still growing today.

Interior Alaska lies between southern rafted terranes and a composite of terranes to the north. This region is made up of large slivers taken from the continental margin of North America, which have been dragged north. Some of these slivers may have themselves been rafted in and accreted to the continental margin of North America in much earlier times, but, from the perspective of the formation of Alaska, they have been translated only a relatively short distance.

Quaternary Glaciation and Loess Deposition

The modern landscapes and soils of the Alaskan boreal forest are influenced by large-scale, relict geologic features that formed under a different environment during the Pleistocene Ice Ages. Many of these geologic features are important to the ecology and include salt-affected soils of the Tanana River (Van Cleve et al. 1993); deep silt soils in some uplands; dry, fire-prone outwash areas; and current glacial hydrology of many interior Alaskan rivers.

Continental glaciation played a surprisingly small role in modifying the landscapes of interior Alaska. While Pleistocene ice caps, ice fields, and valley glaciers episodically covered Alaska south of the Alaska Range, interior Alaska remained mostly glacier-free. North of the Alaska Range, the maximum advances of the valley glaciers are delineated by moraines and erratic fields that are only tens of kilometers beyond the termini of modern glaciers. Some of the most extensive ice advances progressed down the Delta and Nenana River drainages, where ice lobes from multiple valley glaciers of the Alaska Range and ice fields centered south of the Alaska Range coalesced and flowed north through gaps in the Alaska Range (Hopkins et al. 1982, Begét and Keskinen 2003).

The small extent of Pleistocene glacier advances north of the Alaska Range was due to relatively low snowfall in interior Alaska during the Ice Ages, The precipitation decrease was probably due to (1) a colder climate that resulted in decreased evaporation from the ice-covered northern Pacific Ocean, (2) an increase in the extent and elevation of mountain glaciers that acted as barriers to maritime air masses, and (3) reduction of the area of the Bering Sea to the west. Global sea level was 30–150 m lower during the Ice Ages. This would have eliminated what is currently a major source of precipitation for interior Alaska (Hopkins et al. 1982; Chapter 4).

Although moraines are restricted in extent, Pleistocene glaciofluvial deposits are very significant to the landscape of interior Alaska. Coarse-grained alluvial terraces can be traced down-valley from the Alaska Range moraines for hundreds of kilometers along the Tanana and Yukon Rivers. More discontinuous fragments of older terrace systems are also locally preserved along river systems, with an extensive high terrace system preserved east of Fairbanks along the Tanana River (Begét 2000, Begét and Keskinen 2003). Extensive dune fields, covering thousands of hectares, as well as smaller sand deposits, occur near rivers, particularly near the towns of Nenana and Delta. These dune fields are vegetated today but were active during the last glaciation, when the sand supply in local rivers was greater due to increased glaciation, higher wind speed, and less continuous vegetation cover.

The most widespread and important surficial sediments found in the unglaciated areas of interior Alaska are loess (wind-deposited silt) deposits that deeply bury many low-lying areas, with increasingly thicker loess sections found on progressively older terraces and on bedrock surfaces adjacent to river systems. The thickest loess deposits, found near Fairbanks, are more than 100 m deep and contain loess more than 2 million years old (Westgate et al. 1990). Thinner and younger deposits of loess are ubiquitous in the lowlands. Loess deposits typically become thinner with increasing elevation and are essentially absent at elevations above 500 m. The altitudinal constraints on loess distribution reflect (1) low concentrations of dust initially carried by winds to high elevations, and (2) subsequent erosion of loess deposits from higher elevations during subsequent interglacial periods and Ice Ages.

Loess in interior Alaska accumulated very slowly, typically taking 10,000 years or more for a meter of deposition (Begét 1988, 1990). Many loess deposits in interior Alaska are perennially frozen and contain a unique record of the chronology and pattern of climate change in interior Alaska. Loess stratigraphy can produce long, continuous paleoclimatic records similar to those found in marine and other environments (Begét 1991, 1996). Massive ice bodies and ice wedge casts are commonly found in loess deposited during glacial periods, while loess sections dating to interglacial warm episodes are characterized by buried paleosols (ancient soils). Some fossils are frozen in excellent condition in permafrost, providing important information on the paleoecology and paleoenvironments of high-latitude areas (Guthrie 1990). Deposits containing frozen peat and wood layers, including perfectly preserved logs, plants, and leaves provide a direct record of ancient high latitude vegetation during previous interglacials (Edwards and McDowell 1990, Hamilton and Brigham-Grette 1991, Begét et al. 1991, Muhs et al. 2001). The chronology of ash layers preserved in the loess and in glacial sediments show that loess accumulated during both past glacial and interglacial periods. The last several major glaciations in interior Alaska were in phase with major cold periods recorded in the loess record and with the periods of global glaciation recorded from marine sediments (Begét 2000, Begét and Keskinen 2003).

Most modern loess in interior Alaska is derived from glacial rivers, which transport glacial silt for hundreds of kilometers downstream. Loess in the Bonanza Creek LTER site is derived from the glacier-fed Tanana River. Significant dust storms (and concomitant loess deposition) occur only within the five- to six-month snow-free period during spring, summer, and early fall, although some river bars may be exposed to the wind in winter and dust may accumulate during winter in the snowpack before melting out in the spring (Pewe 1955). Some river floodplains produce dust clouds regularly, while others may do so only in unusually dry, windy conditions.

Quaternary loess deposits in interior Alaska appear to have been formed by similar processes to those associated with modern loess deposits, although coarser grain sizes and other factors suggest that wind intensities and storminess were greater during glacial episodes of the Pleistocene. The loess deposition began in latest Pliocene time (Westgate et al. 1990, Begét 2000). Loess deposition in most midlatitude regions, such as the central United States and China, slowed or completely stopped during warm interglacial and interstadial intervals (Pye 1987). In contrast, some central Alaskan loess deposition appears to be essentially continuous through the last glacial cycle, reflecting the continued presence of mountain glaciers (Begét and Hawkins 1989, Begét 1996, Chlachula et al. 1997).

Interior Alaska: A Regional Perspective

The mountains of the Northern and Southern Cordilleras are major barriers, defining the region of interior Alaska (Fig. 2.2). These ranges delimit the drainages of major interior rivers (e.g., the Tanana, Yukon, Koyukuk, and Kuskokwim Rivers), which meander to the Bering Sea. Glacial meltwater from these mountains produces a classic glacial hydrology—clear water beneath ice-covered rivers in the winter and milky waters in summer, turbid with glacial flour (Chapter 10). The Northern and Southern Cordilleras are climatic gates, partially isolating interior Alaska from polar air to the north and from warm, moist air from the south, thus influencing the continental climate and fire regime of interior Alaska. The severe climate at high elevations is also a barrier to dispersal and restricts the ranges of organisms.

Because of the high latitude of the region, the solar elevation is relatively low, and therefore topography can strongly control microclimate (Chapter 4), permafrost distribution and associated hydrology (Chapter 16), and vegetation patterns (Chapter 6). The interaction of topography, disturbance, biota, and climate result in a heterogeneous landscape at many spatial scales.

Ecoregions of Interior Alaska

Four major ecoregions (Gallant et al. 1996) dominate interior Alaska (Fig. 2.3). The Interior Forested Lowlands and Uplands Ecoregion is characterized by a lack of Pleistocene glaciation, a continental climate, a mantling of undifferentiated alluvium and slope deposits, a mosaic of forests stands dominated by white and black spruce and by birch/aspen, and a very high frequency of lightning fires. The two major study areas of the Bonanza Creek LTER site (Bonanza Creek Experimental Forest [BCEF] and Caribou-Poker Creek Research Watershed [CPCRW]) fall primarily within this ecoregion (Fig. 2.3). The Interior Highlands Ecoregion is characterized by a highland with dwarf scrub vegetation and open spruce stands. Mountains in most parts of this region rise to at least 1,200 m, and many are taller than 1,500 m. Most of the higher peaks were glaciated during the Pleistocene. The higher elevations within CPCRW have vegetation typical of this ecoregion. The Interior Bottomlands Ecoregion is characterized by floodplains along large rivers such as the Yukon, Porcupine, Tanana, and Koyukuk Rivers. The Tanana River floodplain within BCEF is representative of this ecoregion. Soils are poorly drained and shallow, often over permafrost. Oxbows and thaw ponds dominate this landscape, and fluvial deposition and erosion are annual disturbance events. The Yukon Flats Ecoregion is similar to the Interior Bottomlands Ecoregion, except that it is a continuous region of extreme continental climate and has the highest fire frequency of any ecoregion in interior Alaska.

18 Alaska's Past and Present Environment

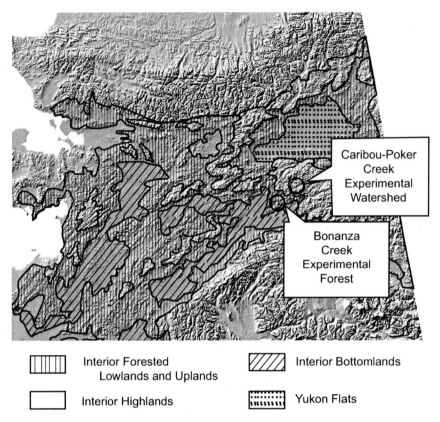

Figure 2.3. Ecoregions of interior Alaska. Climate and topography are the dominant state factors that control the regional distribution of ecoregions in Alaska. The locations of the two intensive study sites of the Bonanza Creek LTER are shown.

Interior Alaska has few roads and a population of less than 100,000 (Fig. 2.4). This has important consequences in terms of fire suppression (Chapter 17) and human disturbance (Chapter 18). With some of the nation's largest national parks and national wildlife refuges, no dams, and a population density of approximately one person per 1,000 ha, interior Alaska is a region where natural processes dominate. However, it is also a region of changing climate and permafrost dynamics (Chapter 4), which influence the forest dynamics and ecosystem and regional processes discussed in subsequent chapters.

Conclusions

Alaska is a collage of geologic fragments that came from both north and south and attached to the northwestern tip of the North American plate. The resulting collisions gave rise to mountain ranges whose glaciers produced silt that blew from river valleys onto surrounding uplands and that acted as barriers to isolate interior Alaska from the

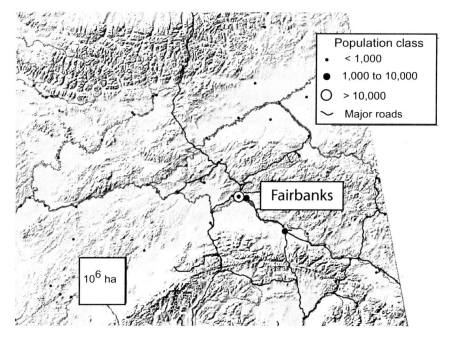

Figure 2.4. Interior Alaska's population and transportation infrastructure. Fairbanks is Alaska's second largest city, with a 2000 census of 30,224.

climates and biota to the north and south. The resulting mosaic of uplands and lowlands in interior Alaska is the substrate on which Alaska's boreal forest has developed.

References

Begét, J. 1988. Tephras and sedimentology of frozen loess. Pages 672–677 *in* I. K. Senneset, editor. Fifth International Permafrost Conference Proceedings. Tapir, Trondheim, Norway.
Begét, J. 1990. Mid-Wisconsinan climate fluctuations recorded in central Alaskan loess. Geographie Physique et Quaternaire 544:3–13.
Begét, J. 1991. Paleoclimatic significance of high latitude loess deposits. Pages 594–598 *in* G. Weller, editor. International Conference on the Role of the Polar Regions in Global Change II, Proceedings. Geophysical Institute, Fairbanks.
Begét, J. 1996. Tephrochronology and paleoclimatology of the last interglacial-glacial cycle recorded in Alaskan loess deposits. Quaternary International 34–36:121–126.
Begét, J. 2000. Continuous late Quaternary proxy climate records from loess in Beringia. Quaternary Science Reviews 20:63–71.
Begét, J., and D. Hawkins. 1989. Influence of orbital parameters on Pleistocene loess deposition in central Alaska. Nature 337:151–153.
Begét, J., and M. Keskinen. 2003. Trace element geochemistry of individual glass shards of the Old Crow Tephra and the age of the Delta Glaciation, central Alaska. Quaternary Research 60:63–69.

Begét, J., M. Edwards, D. Hopkins, M. Keskinen, and G. Kukla. 1991. Old crow tephra found at the Palisades of the Yukon. Quaternary Research 34:291–297.

Chlachula, J., N. W. Rutter, and M. E. Evans 1997. A late Quaternary loess-paleosol record at Kurtak, southern Siberia. Canadian Journal of Earth Sciences 34:679–686.

Connor, C., and D. O'Haire. 1988. Roadside Geology of Alaska. Mountain Press Publishing Co., Missoula, MT.

Edwards, M., and P. McDowell. 1990. Interglacial deposits at Birch Creek, northeast interior Alaska. Quaternary Research 35:41–52.

Gallant, A. L., E. F. Binnian, J. M. Omernik, and M. B. Shasby. 1996. Ecoregions of Alaska: U.S. Geological Survey Professional Paper 1567.

Guthrie, R. D. 1990. Frozen Fauna of the Mammoth Steppe: The Story of Blue Babe. University of Chicago, Chicago, IL.

Hamilton, T., and J. Brigham-Grette. 1991. The last interglaciation in Alaska: Stratigraphy and paleoecology of potential sites. Quaternary International 10–12:49–71.

Hopkins, D. M., J. V. Matthews, C. D. Schweger, and S. B. Young, editors. 1982. Paleoecology of Beringia. Academic Press, New York.

Muhs, D. R., T. A. Ager, and J. Begét. 2001. Vegetation and paleoclimate of the last interglacial period, central Alaska. Quaternary Science Reviews 20:41–61.

Pewe, T. 1955. Origin of the upland silt near Fairbanks, Alaska. Geological Society of America Bulletin 66:699–724.

Pye, K. 1987. Aeolian Dust and Dust Deposits. Academic Press, London.

Thorson, R. M. 1986. The ceaseless context: Landscapes in the making. Pages 1–51 *in* J. S. Aigner, R. D. Guthrie, M. L. Guthrie, R. K. Nelson, W. S. Schneider, and R. M. Thorson, editors. Interior Alaska: A Journey through Time. Alaska Geographic Society, Anchorage.

Van Cleve, K., L. A. Viereck, and G. M. Marion. 1993. Introduction and overview of a study dealing with the role of salt-affected soils in primary succession on the Tanana River floodplain, interior Alaska. Canadian Journal of Forest Resources 23:879–888.

Westgate, J., B. Stemper, and T. Péwé. 1990. A 3 m.y. record of Pliocene-Pleistocene loess in interior Alaska. Geology 18:858–861.

3

State Factor Control of Soil Formation in Interior Alaska

Chien-Lu Ping
Richard D. Boone
Marcus H. Clark
Edmond C. Packee
David K. Swanson

Introduction

The most striking feature of Alaska's boreal soils, compared to those from most other biomes, is the lack of significant soil development even though much of interior Alaska has not been glaciated for millions of years. Soils in the boreal region support forest ecosystems that account for nearly half of the land area of Alaska. Worldwide, boreal forests store nearly a third of the terrestrial carbon (Apps et al. 1993). Hence, changes in boreal soils could greatly impact the global carbon balance (Chapter 19). Another striking feature of boreal soils is their great local variation due to slope and aspect. Compared to the well-developed, colorful soils of the temperate and tropical regions, boreal soils generally have ochric (yellowish brown) colors in uplands and dark-colors in lowlands due to organic matter accumulations. These lowlands account for 85% of the wetland inventory in the United States (Bridgham et al. 2001).

Soil is a mixture of geological parent material and organic matter altered by weathering and the action of living organisms and conditioned by topography over time. Jenny (1941) defined the major factors that influence soil formation as parent material, organisms, topography, climate, and time. When only one of these factors varies, soil characteristics can be treated as a function of that factor (Chapter 1). Soils are also responsive to anthropogenic changes, such as agriculture and forestry, that alter interactive controls at the local scale (Moore and Ping 1989).

Interior Alaska consists of several broad, nearly level lowlands with elevations mostly below 500 m and rounded mountains with elevations up to about 2000 m (Wahrhaftig 1965). The Interior Highlands Ecoregion (Chapter 2) includes the Kuskokwim Highlands and the Interior Highlands. The Interior Bottomlands

Ecoregion includes the Koyukuk-Innoko Lowland, the Kanuti Flats, and the Tanana-Kuskokwim Lowlands (Rieger et al. 1979). The Copper River Basin south of the Alaska Range has a climate and vegetation similar to that of interior Alaska. The Copper River Basin has fine-textured subsoils of lacustrine origin. Soils from the Alaska Range, the Kuskokwim, and the Interior Highlands generally form in glacial deposits and residual materials and have abundant rock fragments. The lowlands have widespread bogs and organic soils. However, despite these variations, soils in these regions share similar soil-forming factors. The objective of this chapter is to discuss the factors that govern soil formation and soil distribution in interior Alaska, relying on available literature and ongoing soils research and inventories. The few soils studies that have been conducted in interior Alaska are concentrated near Fairbanks. Information presented in this chapter without citation is based on current United States Department of Agriculture Natural Resources Conservation Service (NRCS) soil survey projects and Rieger et al. (1979).

Soil-Forming Factors and Pedogenesis

Soils develop through an interaction of four processes: (1) the addition of materials, such as leaf litter, to the soil, (2) transformation of these materials to new forms within the system, (3) vertical transfer up or down the soil profile, and (4) loss of materials from the system (Simonson 1959). The most important transformations that contribute to soil formation are weathering, which converts original minerals to smaller, more stable forms such as oxides and clays, and decomposition, which converts dead organic material to CO_2 and small, recalcitrant organic molecules. The products of weathering and decomposition can be transferred downward by leaching or upward by capillary rise. Physical mixing can also occur by freeze-thaw action (the process of cryoturbation), burrowing by animals, and growth of plant roots. Additions of organic matter to the soil surface and transformations that yield small mobile particles or compounds, followed by their downward leaching, lead to the formation of soil profiles with characteristic horizons. The uppermost organic horizon, or O horizon, consists of organic material that accumulates above the mineral soil (Fig. 3.1). The O horizon can be further subdivided into a relatively undecomposed litter layer (Oi), a partially decomposed organic layer (Oe), and a fully decomposed humus layer (Oa). These horizons are designated as L, F, and H, respectively, in the Canadian soil classification system (Agriculture Canada Expert Committee on Soil Survey 1998). The A horizon is the uppermost mineral horizon. It has a dark color from the organics that originate from dead roots and material mixed in from the O horizon. The B horizon, beneath the A horizon, is the zone of alteration such as oxidation/reduction and accumulation of minerals leached from the A horizon. Beneath the B horizon is the C horizon, which is relatively unmodified by soil-forming processes. These horizons differ in many important chemical and physical properties. For example, the abundance of negative charges on soil organic matter and clay minerals enables a soil to retain cations. Therefore, the more organic- or clay-rich soils typically have a high cation exchange capacity (CEC). CEC is the capacity of a soil to hold exchangeable cations, such as calcium,

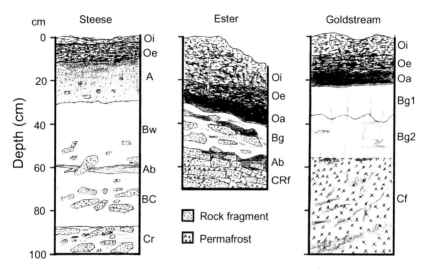

Figure 3.1. Block diagram showing the profiles of three soils common to interior Alaska. The Steese series formed on south-facing slopes, the Ester series formed on north-facing slopes, and the Goldstream series formed on bottomlands. Both the Steese and Ester series formed in weathered bedrock with an admixture of loess in the A and B horizons. The effect of slope movement is reflected by the buried A horizon (Ab) and by streaks of organic matter in the Bg horizon of the Ester series. The Goldstream series formed on a deep reworked loess deposit, and the permafrost table is generally within 60 cm of the surface. Streaks of organics in the Bg and Cf horizons are caused by slope movement. The lower-case letter "w" stands for weathered, "b" for buried, "f" for frozen (permafrost), "r" for abundance of fractured bedrock, and "g" for gleying. CR stands for C horizon dominated by fractured bedrock.

magnesium, potassium, and sodium. These cations are essential plant nutrients. Decomposed organic matter produces organic acids that give the soil a lower pH value. In soils formed on uplands, the organic matter content decreases and, thus, the soil pH increases with depth because there are not enough acids to neutralize all the salts. Soils rich in organic matter generally have low bulk density, and, as the organic matter decreases with depth, the bulk density increases. A horizons generally have granular structures that hold water and also provide good aeration for plant growth. B horizons generally have a platy or blocky structure and higher bulk densities. C horizons generally lack structure.

Although most of interior Alaska north of the Alaska Range has never been glaciated, and although soil-forming processes have operated uninterruptedly for millions of years, there are no well-developed soils such as the reddish soils of the tropics or the highly leached or clay-rich soils of the humid temperate regions. Also lacking are well-developed paleosols (ancient soils). The buried paleosols in the loess deposits in interior Alaska were weakly developed because of the syngenetic nature (i.e., deposition on the surface) of loess deposits (Gubin 1997). The current and past cold climates of interior Alaska limit the rates of key soil-forming processes. For example, decomposition is extremely slow in cold, wet soils (Chapters

14 and 15), and chemical weathering to form clay minerals occurs at a negligible rate in cold dry climates, with most of the clays being inherent in the parent material (Dement 1962). At the same time, other processes that counteract soil profile development are particularly active in Alaska. These include the input of new unmodified mineral material to the soil surface and the physical mixing of soils by cryoturbation. A consideration of the five soil-forming factors gives further insight into the patterns of variation in Alaska's boreal soils.

Parent Material

Parent material is the material from which soil forms. The dominant parent materials in interior Alaska are residuum (weathered bedrock), colluvium (slope deposits moved by gravity), lacustrine (lake sediments), glacial deposits, eolian (wind deposits), and alluvium (stream sediments). Parent material has a direct effect on soil particle size distribution, which, in turn, affects the physical and hydrological properties of the soils. Upon weathering, parent material also contributes to the biochemical and chemical properties of soils.

Residuum and Colluvium

The Yukon-Tanana metamorphic complex (Precambrian metamorphosed sedimentary rocks formerly called the Birch Creek Formation) underlies approximately 75% of the Yukon-Tanana Uplands. The complex consists of greenschist facies dominated by chloritic and quartz-mica schists with some micaceous quartzites, garnet-mica schists, phyllites, and possibly greenstone or impure marbles (Chapman et al. 1971). Superimposed on the bedrock is a discontinuous blanket of loess. Soils formed from residuum and colluvium generally have loamy textures (i.e., a mixture of sand, silt, and clay) and contain rock fragments. These soils are commonly found on steep slopes and ridge tops.

Eolian Deposits

Parent materials of eolian origin are wind-transported materials and vary in particle size depending on the force of the wind and the distance from sources. Loess consists mainly of silt and very fine sand that are transported from exposed sediment deposits of braided rivers. During Illinoian and Wisconsin glaciations that occurred during the Pleistocene (550–400 kA and 80–10 kA, respectively), silt from the Yukon and Tanana Rivers and their tributaries was carried by southerly winds and deposited on the Yukon-Tanana Uplands, with the thickest deposits occurring on south-facing slopes and in valley bottoms immediately adjacent to the rivers (Péwé 1975). Thickness of upland loess deposits can exceed 3 m adjacent to rivers and decreases gradually over 30–50 km from the rivers (Mulligan 2005). Most uplands in the Copper River basin are mantled with a layer of calcareous loess that ranges from a few meters thick on lacustrine terraces adjacent to the Copper River to a few centimeters thick on glacial and mountain landforms more distant from the river (Clark and Kautz 1998). Thus, most uplands in interior Alaska have only a thin cap of loess, normally less than 20 cm, overlying residual or colluvial materials. Soils formed in loess generally have a uniform, very fine sandy loam to silt loam texture. One reason for weak soil development in boreal Alaska is the intermittent or continuous loess deposition. While soil is forming at the existing loess surface, new loess is deposited; this process, called syngenesis, deters the development and differentiation of soil horizons.

In several areas of interior Alaska, large sand dunes were produced by strong winds during the Late Pleistocene and early Holocene (Péwé 1975, Dijkmans et al. 1986). Tephra (volcanic ash), another type of eolian deposit produced mainly from volcanoes in the Wrangell Mountains, blankets the uplands of northeastern Alaska (Lerbekmo and Campbell 1969). On the northern flanks of the Alaska Range, pockets of thin tephra caps (<5 cm) overlay outwash deposits (Ping 1987b). Tephra in eastern Alaska usually has a fine sandy loam texture. On stabilized sand dunes, this thin, volcanic ash cap supports dense forest because it has a higher water-holding capacity than the sand below (Ping and Packee 2003). Extensive areas of sand dune deposits occur between the Yukon and Tanana Rivers (Brannen and Swanson 2001, Furbush et al. 1980). On most of these dunes, a 10–15 cm thick loess cap supports mixed forest cover that stabilizes the land against erosion.

Alluvium

The Bonanza Creek LTER study area on the Tanana River is a good example of alluvium deposited by rivers and streams on floodplains and terraces (Van Cleve et al. 1993). Depending on transport energy, particle size varies from silt to extremely gravelly sand. Many alluvial soils have a loamy cap on top of sandy or gravelly substrata. The thickness of the loamy layer tends to increase with surface age, a result of silt deposition during seasonal flooding (Viereck 1970). These soils are generally well drained but are subject to seasonal inundations (Shaw et al. 2001, Van Cleve et al. 1992).

Glacial Deposits

Glacial deposits include glacial outwash, glacial till, moraines, kames, and eskers. Large areas of outwash plain formed during the glacial retreat at the end of the Wisconsin glaciation, about 12,000–9,000 years ago, and are found in the Tanana River basin from the Canadian border to Delta, in the Nenana Valley, and on both sides of the Alaska Range. Péwé (1975, 1977) identified glacial moraine and kames in the northern foot slopes of the Alaska Range between Delta and Tok. Outwash deposits have a gravelly or cobbly sand texture and low water-holding capacity. In the Tok and Delta areas, a loess cap over outwash has a thickness ranging from 10 to 100 cm (Ping and Packee 2003, Schoephorster 1973). The loess cap holds most of the moisture that supports plant growth. Permafrost generally forms under lowland soils with more than a 50-centimeter loess cap. The gravelly substratum in these areas permits complete thawing of permafrost after land clearing and thus makes agriculture possible. Glacial deposits are also present in the Brooks and Alaska Ranges, in the Wrangell Mountains, and at high elevations in interior Alaskan uplands (Péwé 1975). These deposits are associated with hills, plains, and lower mountain slopes. Deposits include sand and gravelly outwash and gravelly loamy moraine.

Lacustrine Deposits

Lacustrine deposits are lake sediments. Because of the low energy of transport, only very fine particles (mainly clay and silt) are sorted, carried, and precipitated in lakes. Extensive areas of lacustrine deposits are a unique feature of the Copper River Basin. There, clayey and loamy calcareous materials were deposited in a large proglacial lake that occupied the basin during the Pleistocene from about 35,000–9,000 years ago (Ferrians et al. 1983). The carbonate-rich parent materials, dominantly chlorite, are well flocculated (precipitated) and express strong granular or blocky structure and moderate permeability (Clark and Kautz 1998). Most of the soils formed in the lacustrine sediments of the Copper River Basin have a thin loess cap (<20 cm), a nearly neutral reaction, and a medium texture. The subsoils are slightly alkaline (pH 7.5) with a fine texture but low CEC because of the dominance of chlorite in the clay fraction.

Organisms

Vegetation affects soil formation both physically and biochemically. Physically, it shades and covers the soil surface, alters the thermal and moisture regimes, protects the soil from erosion, and maintains productivity (Chapter 7). In interior Alaska, especially the region of discontinuous permafrost, vegetation moderates soil temperature by providing both shade and insulation (Van Cleve et al. 1992). An important biochemical function of vegetation is its addition of litter to the soil where soil microorganisms and fauna transform a portion into humus that is the major fraction of soil organic matter (SOM). SOM decomposes slowly in interior Alaska because of the cold climate. As a result, thick organic layers accumulate under spruce forests, and organic soils form in depressions and bottomlands where saturation causes anaerobic conditions that retard decomposition. More than 95% of the peatlands of the United States are in Alaska, and most are in the boreal region (Bridgham et al. 2001). SOM is the reactant for many biogeochemical processes. It forms aggregates with mineral particles that stabilize soils, acts as a nutrient reserve for nutrient cycling, yields organic acids that dissolve primary minerals and release metal ions and other inorganic components, and serves as an energy source for soil organisms. Organo-metal and organo-clay complexes give soil horizons their characteristic color (Schwertmann 1993). When transported downward, organic acids and their metal complexes enhance the biogeochemical weathering of the underlying horizons. However, leaching is weak in interior Alaska due to low precipitation, so there is minimal downward transport of such complexes. Instead, they tend to accumulate in surface horizons, mostly as a thin, dark A horizon and a weakly developed, pale brown B horizon (Fig. 3.1).

Microorganisms transform SOM ultimately to CO_2, release mineral nutrients from SOM during decomposition, facilitate plant uptake of nutrients, and act as a binding agent for soil particles (Chapter 14). Soil fauna contribute not only to the transformation of SOM but also to its transport, especially along the south-facing river bluffs, through the creation of channels or burrows.

Nowhere else in the world can human activities impact soil formation as in the boreal and arctic regions. The reason? Once the surface vegetation is removed due to farming, urban development, human-caused fire, or any other disturbance, the soil loses its insulation and warms. As a result of warming, the underlying permafrost thaws, soil drainage improves, biogeochemical processes are accelerated, and the soils are no longer classified as Gelisols (permafrost-dominated soils). The effects are similar to those caused by natural fire. Few other soils in the world change so radically due to fire or human disturbance that they are reclassified into another soil order (Moore and Ping 1989).

Climate

Permafrost

The cold climate retards most soil-forming process (except for organic-matter accumulation) and leads to the formation of permafrost. Interior Alaska is in the zone of discontinuous permafrost (Péwé 1975, Ferrians 1965). In the Tanana-Yukon and Kuskokwim Highlands, permafrost underlies most of the north slopes and some foot slopes and most toe slopes of south-facing slopes (Chapter 4). In the lowlands, permafrost underlies most of the landscape except Holocene-age terraces, alluvial fans, and active floodplains. In the Copper River Basin, except on floodplains and under lakes, permafrost underlies most of the plateau. Ground ice occurs as finely segregated ice crystals, thin discontinuous ice lenses, and large masses of ice. In clayey and silty deposits, ice content accounts for an average 30–60% of the dry weight of the soil (Nichols 1956). In the Copper River Basin, massive ground ice formation and thermokarst (subsidence due to thaw of ground ice) occur locally but are not extensive. In the interior, massive ground ice and thermokarst are common (Péwé 1982).

Ice-rich permafrost affects soil formation in several ways. First, it serves as a barrier to roots and water percolation, thus promoting anaerobic or reducing conditions and hydrophilic vegetation. Most biogeochemical turnover is in the active layer (the layer of soil subject to the annual freeze-thaw cycle); however, recent studies indicate that biochemical processes can continue in soils at subfreezing temperatures (Hobbie et al. 2000, Michaelson and Ping 2003). Second, due to ice-lens formation during the freeze cycle, a granular structure usually forms in the A horizons of soils associated with earth hummocks (Gubin 1988), and lenticular and reticulate structures form in the subsoils (French 1996). Earth hummocks are pillow-shaped earth mounds with diameters of 1–2 m caused by the freezing and squeezing upward of soil materials. Commonly, an ice-rich layer occurs just above the permafrost table (Hoefle and Ping 1996). Third, permafrost has a profound impact on soil morphology due to cryoturbation. Cryoturbation occurs commonly in both permafrost-affected areas and in the alpine zone, where soils typically lack permafrost and are also subject to strong freeze-thaw cycles. Cryoturbation results in warped soil horizons, mixing of horizons, and, most important, frost-churning that moves surface organic matter downward into upper permafrost where carbon is sequestered. In boreal forest soils affected by permafrost, nearly 50% of the total organic carbon is stored in the upper permafrost (Ping et al. 2002).

Fire

The vegetation mosaic and soil morphology in interior Alaska strongly reflect the interactive effects of fire, permafrost, and topography. Charcoal particles from past fires commonly occur in soil profiles and are found especially at the base of O horizons. Combustion and consolidation of the insulating organic mat result in warming of underlying soils, lowering of the permafrost table, and, often, a transition to a well-drained and permafrost-free state before eventual return to pre-burn conditions (Ping et al. 1992). For example, following the 1981 Wilson Camp fire near Glennallen, thaw depth increased to 80 cm in the burn after one year to 100 cm after two years, compared to 40 cm below the mineral surface in the unburned stand (Clark and Kautz 1998). Removal of vegetation and some or all of the surface organic horizons deepens the active layer, increases effective overland flow, decreases infiltration, and produces warmer, drier soils. These changes may, in turn, change the soil classification from a poorly drained, permafrost-affected (Gelisol) to a well-drained, permafrost-free soil (Inceptisol; Viereck and Dyrness 1979, Dyrness and Viereck 1982, Moore and Ping 1989). The amount of soil thaw after fire is typically greatest on sites of intermediate wetness and can be minimal on very wet soils with thick organic surface layers (Swanson 1996a). Soil reaction in the surface organic horizons ranges from moderately alkaline (pH 7.9–8.4) immediately after fire to slightly acid (pH 6.1–6.5) several years later.

The broad effect of fire on soils (increased temperature and a more mesic moisture regime) is to accelerate biogeochemical processes and the leaching of inorganic nutrients and dissolved organic matter through the soil profile. After fire, exposed mineral soils on steeper slopes can have increased mass movement and erosion. However, on convex slopes, leaching regimes are altered by combinations of higher surface runoff, less infiltration, and limited capillary transport of soluble salts. After fire, increased oxidative weathering of primary minerals may produce reddening on the surface of exposed mineral soils. Enhanced decomposition after fire causes a pulse of available nutrients, such as base cations and phosphate, previously complexed with SOM and nitrites. Total pools of C and N decrease through combustive oxidation of SOM to CO_2 and nitrogen oxides. CEC of the surface horizons declines due to loss of SOM and the formation of black carbon (charcoal), the most recalcitrant and biogeochemically inactive form of SOM.

The location of soils in a watershed dictates their response to fire. The morphology of south-facing, upland soils, which are drier and permafrost-free, is little affected by fire because they already have shallow O horizons and are pedogenically in a relatively steady state. Soils in the coldest and wettest regimes (north aspect and floodplains) are not severely affected by fire because most of the saturated organic mat and permafrost persist after fire. Marginal soils with permafrost and those located in potentially warmer, drier topographic positions, such as east- and south-aspect toe slopes, are most likely to show major changes in moisture and temperature regimes after fire (Swanson 1996a).

Topography

Greater variation in soil properties occurs among topographic locations within a broad landscape unit than between the northern and southern extremes of interior Alaska because of differences in microclimates (Chapter 4; Swanson 1996b, Krause et al. 1959).

North-Aspect Slopes

North-facing slopes are cooler and wetter than south-facing slopes because they receive less solar energy, evaporate less water, and exhibit delayed snowmelt (Chapter 4). On north slopes, the cooler temperatures favor the establishment of black spruce and an ericaceous shrub and moss understory that further insulate the soil. Permafrost is typically present at shallow depths, usually less than 1 m. The permafrost layer perches water and creates a saturated zone, generally on slopes <20%, and results in reducing conditions in the soil. North-aspect soil mineral horizons tend to be thin, with high overall organic contents and irregular horizonation due to frost action. The cold, wet conditions result in thicker organic accumulations and weak weathering in the subsoil. Thick (25–45 cm) organic horizons overlie thin mineral horizons that are mixed with fractured bedrock. Textures of the organic horizons grade from discernible moss fiber composition on the surface to highly decomposed muck just above the mineral contact. Anaerobiosis due to poor drainage and relatively cold soils limits rooting depths of most shrub species to the lower boundary of the organic horizons, whereas black spruce is shallowly rooted regardless of soil conditions (Reiger et al. 1972, Furbush and Schoephorster 1977).

The mineral horizons are dominated by a platy structure, with thin plate-like layers caused by persistent seasonal frost in active layers and fluctuating permafrost tables in lower zones. Organic and mineral soil materials are commonly mixed together as a result of slope movement caused by the annual freeze-thaw cycle. The entire process is referred to as gelifluction. Ground movement on north slopes approaching 30° occurs entirely during the early summer when soils are saturated and gelifluction is common (Wu 1984). The thick organic horizons increase the water-holding capacity of north-slope soils and further insulate the mineral matrix and permafrost from outside influences. Water content ranges from 550% to 725% of the dry weight for organic horizons and from 160% to 220% for mineral horizons. This is three to five times higher than water content in comparable horizons on south aspects. Soils of the north aspect generally have lower pH, higher organic carbon content, higher CEC, and lower base (Ca and Mg) status than soils of other slopes. The pH values are much lower than those found on south slopes and range from extremely acidic (4) in the surface organic horizons to very strongly acidic (5) in the underlying mineral horizons.

South-Aspect Slopes

The well-drained and relatively warm soils of south-aspect slopes are generally deeper and more mineral-dominated than those on north-aspect slopes. Soils below mixed deciduous-conifer forests usually have thinner (<15 cm) organic horizons (Oi and Oe). The Oa horizon is generally lacking because of rapid SOM decomposition in the warmer conditions. Underlying mineral soils horizons typically have an upper grayish-brown B horizon that indicates limited organic matter accumulation in a well-drained environment; below this is a yellowish-brown BC or C horizon. The yellowish-brown color is the result of brunification, a process by which Fe is released as oxides that then coat the mineral grains.

Typically, rooting zones are limited to the O, A, and B horizons, where maximum available water and nutrients are found. In areas with moderate to high loess influences, rooting zones are shallow on erosional landscape positions such as ridge tops and shoulder slopes and deeper on relatively stable landscape positions such as back and foot slopes (Furbush and Schoephorster 1977). Deeper rooting zones contribute to increased productivity, belowground biomass, and humus storage. In areas with minimal loess influence, soils have sandy loam textures and increasing rock content with depth. Back and foot slope profiles contain abundant rock fragments due to long-term erosion (Ping et al. 2005). Irregular horizon boundaries and truncated soil horizon sequences, along with high rock fragment content, indicate that episodic slope failures and mass movements interrupted pedogenic development.

Soils on south slopes are generally slightly acidic (pH 6) to strongly acidic (pH 5) on the foot slopes. Low pH values in A horizons (4.2–4.6) indicate the influence of organic acids. In subsoils (BC or C horizons), pH values commonly increase with depth to 5.6–6.1, indicative of decreasing leaching at depth as a result of the low precipitation in interior Alaska. These soils also have lower CEC values due to lower organic matter content compared to soils on north-facing slopes but have higher base (Ca and Mg) status due to lower leaching losses and increased nutrient cycling associated with the hardwood species.

Lowlands

In lowlands and depressions, soil drainage is restricted by landscape position and presence of permafrost. The wet and cold conditions slow decomposition rate, favoring accumulation of organic matter (Chapter 15). Black spruce and occasionally tamarack dominate these poorly drained areas; the organic horizon is typically 20–40 cm thick. The poor to very poorly drained organic soils (with organic layers >40 cm), known locally as muskeg, generally occur in bogs, fens, and depressions. The dominant soil-forming process in poorly to very poorly drained mineral soils is reduction and in somewhat poorly drained soils is an alternation of reduction and oxidation. In a reduced (oxygen-depleted) environment, microbes decompose (oxidize) organic matter by getting oxygen from the iron oxides that coat the mineral grains. Thus, the iron is reduced and the soil loses its brownish color and turns gray in color, a soil condition referred to as gleyed. When the water table drops later in the growing season, reduced soil may become oxidized again in small, localized areas and result in mottles of mixed gray and rusty color.

Time

Even though most of the boreal region of interior Alaska has not been glaciated for millions of years, time has left only weak marks on the soil landscape. Most soils are considered to be of Holocene age, quite recent in geologic terms, because of the intermittent nature of loess deposition, slope movement due to gelifluction, and erosion that either mixed soil horizons or removed surface horizons on uplands. Deeper loess deposits in the Fairbanks area are more than 2 million years old (Péwé 1975). However, the early soils (paleosols) formed in older materials, including ancient loess and loess deposits of Holocene age, were buried. In the Nenana Valley, modern soils, formed in loess on the upper terrace, date to less than 400 years (Bigelow 1991, Ping et al. 1992). The stratification of silt with thin lenses of organic rich layers in soils of the foothills of the Fairbanks and Caribou-Poker Creek areas reflects repeated episodes of deposition (Mulligan 2005, Ping et al. 2005). Soils are morphologically young across interior Alaska because of the abundance of young parent materials, the continual addition to the surface of parent materials such as loess and alluvium, and the slow rate of weathering because of the cold climate. There are, however, some old soils in interior Alaska. In the Tanana River Basin, soils formed in Pleistocene alluvium are found on the older (higher) terraces, and those formed in Holocene (recent) alluvium are on the younger (lower) terraces (Mulligan 2005).

Floodplain soils follow a predictable chronological sequence after the sediments are initially deposited by the river or stream (Chapter 7). The LTER Tanana River study sites provide a classic example of soil formation (Viereck et al. 1983). The factor that controls soil formation is time; the youngest soils formed in the most recent deposits show no horizonation. On the floodplains of the glacially fed Nenana and Tanana Rivers, most soils have a sandy texture and sparse vegetation.

Soils on the second terrace generally have a finer texture with shrub vegetation. There is minimal horizon development except for a thin A horizon, where some organic matter is incorporated into the surface mineral horizon. The saturated subsurface horizons experience reducing conditions that produce a bluish gray color of gleyed soil. Along the Tanana River, saline soils are found in a narrow belt on the second terrace due to capillary concentration of soluble salts on the soil surface during the dry periods (Chapter 7; Van Cleve et al. 1993).

On the third terrace, where there are deciduous trees, soils have a thin litter layer (Oi) over a grayish A horizon. The underlying B horizon has a yellowish or yellowish gray color that indicates an early stage of mineral weathering. There are multiple organic layers separated by alluvial deposits of mostly fine sand and silt deposited by occasional floods. The strong mottles and iron oxide linings around root channels in the subsoils indicate fluctuating water tables during the growing season.

On the fourth terrace, white spruce dominates the stands, and the O horizons have thickened to 7–20 cm. The soils are moderately to well drained and have a grayish brown color with a rooting zone >40 cm. In later stages dominated by black spruce, the ground temperature becomes colder due to shading from the mature spruce (Chapter 7). A permafrost table appears and rises to <60 cm below the surface, and the soils become poorly drained. The rooting zone is mostly limited to the O horizon, and the underlying mineral soils are gleyed due to shallow permafrost table. The slightly higher soil pH values in the foot slopes and floodplains reflect limited leaching and conservation of base cations.

Soil Classification

Soil classification provides a tool for comparing soils within Alaska and elsewhere. Alaskan boreal soils have been classified according to the distinct classification systems of the United States (Soil Survey Staff 1999), Canada (Agriculture Canada Expert Committee on Soil Survey 1998), Russia (Shishov et al. 2001), and the World Reference Base (WRB) (Food and Agriculture Organization [FAO] 1998, Table 3.1).

Weakly developed soils without permafrost on well-drained south-facing slopes are classified within the Inceptisol order. The Cryepts suborder includes soils having a mean annual soil temperature (MAST) at 50 cm of 0–8°C. The Gelepts suborder includes soils having a MAST <0°C but without permafrost. The Dystrocryepts and Dystrogelepts great groups includes those Cryepts with more than 40% of their bases (mainly Ca and Mg) depleted in the A and B horizons (measured by dividing the sum of cations by CEC). At the subgroup level, Typic Dystrocryepts are found on uplands and Aquic Dystrocryepts on the lower slopes because of fluctuating water tables within 25 cm of the mineral surface. Depending on the depth of gleyed horizons, these soils are classified as Cambisols or Gleysols according to the Canadian and the FAO-WRB classification systems, respectively.

Soils on north-facing slopes and bottomlands have permafrost tables within 1 m and are thus classified in the Gelisol order (permafrost-affected soils). Soils with cryoturbation features are classified in the Turbel suborder. Turbels experiencing saturation that results in reducing conditions are classified in the Aquiturbels great group, and those with discontinuous organic horizons, such as those under tussocks, are in the Ruptic-Histic Aquiturbels subgroup. Depending on the thickness of organic horizons, these soils are classified in the Canadian and FAO-WRB systems as either Cryosols or Histosols. In the Russian soil classification system, the concept of Cambisols and Gleysols is similar to that of the Canadian and WRB systems except that the category Cryosol is limited to soils with cryoturbation. Both the Russian and the WRB systems emphasize soil material over presence of permafrost. Therefore, soils formed in deep organic matter and with permafrost are classified as Histosols (organic soils). In the U.S. and Canadian systems, however, the presence of permafrost precedes soil material, so organic soils underlain by permafrost are classified as belonging to the Histic suborder of Gelisols or Cryosols, respectively.

Table 3.1. Common boreal soils and comparison of their classification according to different systems.

Landform	Drainage class	Soil texture	Organic layer, cm	Depth to permafrost, cm	Soil color moist	Soil taxonomy (U.S.A.)	Canadian	FAO-WRB	Russian
Floodplain	poor	sandy to loamy	0–3	>200	gray	Typic Cryofluvents	Orthic Regosols	Fluvisols	Floodplain soils
Terrace	poor	silty to loamy	8–20	<100	gray to bluish gray	Typic Aquorthels Typic Aquiturbels	Static Cryosols Turbic Cryosols	Stagnic Cryosols Turbic Cryosols	Taiga gley peaty soils
Terrace	well	sandy to loamy	2–5	>200	pale brown	Typic Cryorthents Typic Dystrocryepts	Orthic Regosols Dystric Brunisols	Haplic Arenosols Dystric Cambisols	Taiga burozemic soils
Valley or Bottomland	very poor	peaty or mucky	>40	<50	reddish brown to black	Typic Fiberistels Fluvaquentic Hemistels	Organic Cryosols	Cryic Histosols	Oligotrophic peat wetland soils
Valley or Bottomland	poor	peaty or mucky	20–40	<80	reddish brown to black	Histic Aquorthels Histic Aquiturbels	Gleysolic Cryosols	Histic Cryosols	Taiga gley peaty-mucky and peaty soils
Ridge top, shoulder slope	well	silty or gravelly, loamy	2–5	>200	pale brown	Typic Eutrocryepts	Eutric Brunisols	Eutric Cambisols	Taiga burozemic soils
Back slope (S aspect)	well	silty or loamy	3–7	>200	pale brown	Typic Eutrocryepts	Eutric Brunisols	Eutric Cambisols	Taiga burozemic soils

(*continued*)

Table 3.1. (Continued)

Landform	Drainage class	Soil texture	Organic layer, cm	Depth to permafrost, cm	Soil color moist	Soil taxonomy (U.S.A.)	Canadian	FAO-WRB	Russian
								Soil classification	
Foot slope (S aspect)	mod. well	silty or loamy	5–18	<200	grayish brown	Typic Eutrocryepts	Eutric Brunisols	Eutric Cambisols	Taiga burozemic soils
Toe slope	poor	mucky or loamy	15–30	<100	black to gray	Histic Aquorthels	Gleysolic Cryosols	Histic Cryosols	Taiga gley peaty soils
Back slope (N aspect)	poor	peaty, gravelly loam	30–50	<100	gray to dark brown	Typic Fibristels, Histic Aquiturbels	Organic Cryosols, Gleysolic Cryosols	Cryic Histosols, Histic Cryosols	Oligotrophic peat wetland soils
Foot slope (N aspect)	poor	peaty to mucky	40–60	<60	dark brown	Typic Hemistels, Histic Aquorthels	Organic Cryosols	Cryic Histosols	Oligotrophic peat wetland soils
Toe slope (N aspect)	very poor	peaty to mucky	40–60	<50	dark brown, gray	Sphagnic Fibristels, Typic Hemistels	Organic Cryosols	Cryic Histosols	Oligotrophic peat wetland soils
Sand dune	excessive	sandy	0–3	>200	gray to brown	Typic Cryopsamments	Orthic Regosols	Regosols	Sands

34

Conclusions

Most soils of boreal Alaska are weakly developed, even though most of the region was not glaciated during the Quaternary, a period of 1.8 million of years. The most notable soil features are the combined effects of variations in soil temperature and hydrology due to topographic position. The dominant soil development on the well-drained south slopes is limited to leaching of carbonates and soluble salts below the rooting zone and brunification due to iron mineral release from weathering. On floodplains, plant succession and geomorphic age of the surface are the controlling factors in soil development, leading to a cooler, wetter soil with thickened organic horizons that ultimately can lead to the formation of permafrost. On north aspect and lowland soils, the presence of permafrost, poor drainage, and low soil temperature favors the formation of a thick surface organic horizon that often become mixed into the underlying mineral horizons by cryoturbation or gelifluction. The most unique feature of Alaskan boreal soils is the lack of correlation between soil development and landscape age due to cryoturbation and/or intermittent loess deposition.

References

Agriculture Canada Expert Committee on Soil Survey. 1998. The Canadian System of Soil Classification. 3rd ed. NRC Research Press, Ottawa.

Apps, M. J., W. A. Kurz, R. J. Luxmoore, L. O. Nilsson, R. J. Sedjo, R. Schmidt, L. G. Simpson, and T. Vinson. 1993. The changing role of circumpolar Boreal forest and tundra in global C cycle. Water, Air, and Soil Pollution 70:39–53.

Bigelow, N. H. 1991. Analysis of late quaternary soils and sediments in the Nenana Valley, central Alaska. M.A. Thesis. University of Alaska Fairbanks.

Brannen, K. M., and D. K. Swanson. 2001. Soil Survey of Kantishna Area, Alaska. U.S. Dept. of Agriculture, Natural Resources Conservation Service. Palmer, AK.

Bridgham, S. D., C. L. Ping, J. L. Richardson, and K. Updegraff. 2001. Soils of northern peatlands: Histosols and Gelisols. Pages 371–382 *in* J. L. Richardson and M. J. Vepraska, editors. Wetland Soils: Genesis, Hydrology, Landscapes and Classification. CRC Press, Boca Raton, FL.

Chapman, R. M, F. R. Weber, and B. Taber. 1971. Preliminary geologic map of Livengood quadrangle, Scale 1:250,000. U.S. Dept. of the Interior. Geological Survey Open File Report 483.

Clark, M. H., and D. R. Kautz. 1998. Soil survey of Copper River Area, Alaska. U.S. Dept. of Agriculture, Natural Resources Conservation Service. U.S. Government Printing Office, Washington, DC.

Dement, J. A. 1962. The morphology and genesis of the subarctic brown forest soils of central Alaska. Ph.D. Dissertation, Cornell University, Ithaca, NY.

Dijkmans, J. W. A., E. A. Koster, J. P. Galloway, and W. G. Mook. 1986. Characteristics and origin of calcretes in a subarctic environment, Great Kobuk sand dunes, northwestern Alaska, USA. Arctic and Alpine Research 18:377–387.

Dyrness, C. T., and L. A. Viereck. 1982. Control of depth to permafrost and soil temperature by the forest floor in black spruce/feathermoss communities. U.S. Department of Agriculture, Forest Service, Research Note PNW–396.

Ferrians, O. J., Jr. 1965. Permafrost map of Alaska: Scale 1:2,500,000. U.S. Dept. of the Interior, Geological Survey. Miscellaneous Geological Investigation Map I–445.

Ferrians, O. J., Jr., D. R. Nichols, and J. R. Williams. 1983. Copper River basin, resume of Quaternary Geology. Pages 137–175 *in* T. L. Péwé and R. D. Reger, editors. Guidebook to Permafrost and Quaternary Geology, Richardson and Glenn Highways, Alaska. Vol. 1. Alaska Department of Natural Resources, Fairbanks.

Food and Agriculture Organization. 1998. World Reference Base for Soil Resources. Food and Agriculture Organization of the United Nations, Rome.

French, H. M. 1996. The Periglacial Environment. 2nd ed. Longman Scientific and Technical, White Plains, NY.

Furbush, C. E., B. E. Koepke, and D. B. Schoephorster. 1980. Soil Survey of Totchaket Area, Alaska. U.S. Dept. of Agriculture, Soil Conservation Service. U.S. Government Printing Office, Washington, DC.

Furbush, C. E., and D. B. Schoephorster. 1977. Soil Survey of Goldstream-Nenana Area, Alaska, with Maps. U.S. Dept. of Agriculture, Soil Conservation Service. U.S. Government Printing Office, Washington, DC.

Gubin, S. 1997. Paleopedological analysis of late-Pleistocene deposits in Beringia. Pages 53–56 *in* M. E. Edwards, A. Sher, and R. D. Guthrie, editors. Terrestrial Paleoenvironmental Studies in Beringia, Proceedings of a Joint Russian-American Workshop, Fairbanks, Alaska. June 1991. Alaska Quaternary Center, University of Alaska Fairbanks.

Gubin, S. V. 1988. Soil formation paleogeographic aspects in Yakutiya. Pages 758–763 *in* K. Sennese, editor. Permafrost—Fifth International Conference Proceedings. Vol. 1. August 2–5. Tapir, Trondheim, Norway.

Hobbie, S. E., J. P. Schimel, S. E. Trumbore, and J. R. Randerson. 2000. Controls over carbon storage and turnover in high-latitude soils. Global Change Biology 6(Suppl. 1):196–210.

Hoefle, C. M., and C. L. Ping. 1996. Properties and soil development of late-Pleistocene paleosols from Seward Peninsula, northwest Alaska. Geoderma 71:219–243.

Jenny, H. 1941. Factors of Soil Formation. McGraw-Hill, New York.

Krause, H. H., S. Rieger, and S. A. Wilde. 1959. Soils and forest growth on different aspects in the Tanana watershed of interior Alaska. Ecology 40:492–495.

Lerbekmo, J. F., and F. A. Campbell. 1969. Distribution, composition, and source of the White River Ash, Yukon Territory. Canadian Journal of Earth Sciences 6:109–116.

Michaelson, G. J., and C. L. Ping. 2003. Soil organic carbon and CO_2 respiration at subzero temperature in soils of arctic Alaska. Journal of Geophysical Research 108(D2), 8164 ALT 5-1–5-10.

Moore, J. P., and C. L. Ping. 1989. Classification of permafrost soils. Soil Survey Horizons 30:98–104.

Mulligan, D. K. 2005. Soil Survey of the Greater Fairbanks Area. U.S. Dept. of Agriculture, Natural Resources Conservation Service, Fairbanks.

Nichols, D. R. 1956. Permafrost and ground-water conditions in the Glennallen area, Alaska. U.S. Dept. of the Interior, Geological Survey Open-File Report 392.

Péwé, T. L. 1975. Quaternary geology of Alaska. U.S. Geological Survey Professional Paper 835.

Péwé, T. L. 1977. Guidebook to the Quaternary Geology, Central and South-central Alaska. INQUA VIIth Congress, 1965. Reprint by Alaska State Department of Natural Resources, College, AK.

Péwé, T. L. 1982. Geologic hazards of the Fairbanks area, Alaska. Alaska Division of Geological and Geophysical Surveys Special Report 15, College, AK.

Ping, C. L. 1987a. Soil temperature profiles of two Alaskan soils. Soil Science Society of America Journal 51:1010–1018.
Ping, C. L. 1987b. Characterization and fertility status of soils and mine spoils in Healy, Alaska. Alaska Agricultural and Forestry Experiment Station Bulletin 66. School of Agriculture and Land Resources Management, University of Alaska Fairbanks.
Ping, C. L., and E. C. Packee. 2003. Guidebook for Alaska Soil Geography Field Study NRM–F489/689–F01. Boreal forest soils field trip, Fairbanks to Northway. Aug. 11–15, 2003. Agricultural and Forestry Experiment Station, School of Natural Resources and Agricultural Sciences, University of Alaska Fairbanks.
Ping, C. L., Y. L. Shur, and G. J. Michaelson. 1992. Pedological properties of the Dry Creek archaeological site in the Nenana Valley, Alaska. Page 101 *in* Proceedings, 43rd Arctic Science Conference, Valdez, Alaska. University of Alaska Fairbanks.
Ping, C. L., G. J. Michaelson, X. Y. Dai, J. M. Kimble, and L. Everett. 2002. Organic carbon stores in tundra soils of Alaska. Pages 485–494 *in* J. M. Kimble, R. Lal, and R. F. Follett, editors. Agricultural Practices and Policies for Carbon Sequestration in Soils. Lewis Publishers, Boca Raton, FL.
Ping, C. L., G. J. Michaelson, E. C. Packee, C. A. Stiles, D. K. Swanson, and K. Yoshikawa. 2005. Characterization and formation of soils in the Caribou-Poker Creek Research Watershed, Alaska. Soil Science of America Journal 69:1761–1772.
Rieger, S., D. B. Schoephorster, and C. E. Furbush. 1979. Exploratory soil survey of Alaska. U.S. Dept. of Agriculture, Soil Conservation Service. U.S. Government Printing Office, Washington, DC.
Rieger, S., C. E. Furbush, D. B. Schoephorster, H. Summerfield Jr., and L. C. Geiger. 1972. Soils of the Caribou-Poker Creeks Research Watershed, Interior Alaska with Maps. U.S. Army Cold Regions Research and Engineering Laboratory Tech. Rep. 236. Hanover, NH.
Schoephorster, D. B. 1973. Soil survey of Salcha-Big Delta area, Alaska. U.S. Dept. of Agriculture, Soil Conservation Service. U.S. Government Printing Office, Washington, DC.
Schwertmann, U. 1993. Relations between iron oxides, soil color, and soil formation. Pages 51–69 *in* J. M. Bigham and E. J. Ciolkosz, editors. Soil Color. SSSA Special Publication No. 31. Soil Science Society of America, Madison, WI.
Shaw, J. D., E. C. Packee, and C. L. Ping. 2001. Growth of balsam poplar and black cottonwood in Alaska in relation to landform and soils. Canadian Journal of Forest Research 31:1793–1804.
Shishov, L. L., V. D.Tonkonogov, I. I. Lebedeva, and M. I. Gerasimova. 2001. Russian Soil Classification System. Dokuchaev Soil Institute, Moscow.
Simonson, R. W. 1959. Outline of a generalized theory of soil genesis. Soil Science Society of America Proceedings 23:152–156.
Soil Survey Staff. 1999. Soil Taxonomy. 2nd ed. U.S. Dept. of Agriculture Handbook No. 436. U.S. Government Printing Office, Washington, DC.
Swanson, D. K. 1996a. Susceptibility of permafrost soils to deep thaw after forest fires in interior Alaska, USA, and some ecological implications. Arctic and Alpine Research 28:217–227.
Swanson, D. K. 1996b. Soil geomorphology on bedrock and colluvial terrain with permafrost in central Alaska, USA. Geoderma 71:157–172.
Van Cleve, K., F. S. Chapin, III, C. T. Dyrness, and L. A. Viereck. 1992. Elemental cycling in taiga forest: State factor control. Bioscience 42:78–88.
Van Cleve, K., C. T. Dyrness, G. M. Marion, and R. Erickson. 1993. Control of soil development on the Tanana River floodplain, interior Alaska. Canadian Journal of Forest Research 23:941–955.

Viereck, L. A. 1970. Forest succession and soil development adjacent to the Chena River in Interior Alaska. Arctic and Alpine Research 2:1–26.

Viereck, L. A., and C. T. Dyrness. 1979. Ecological effects of the Wickersham Dome fire near Fairbanks, Alaska. U.S. Dept. of Agriculture, Forest Service, General Technical Report PNW-90.

Viereck, L. A., C. T. Dyrness, K. Van Cleve, and M. J. Foote. 1983 Vegetation, soils, and forest productivity in selected forest types in interior Alaska. Canadian Journal of Forest Research 13:703–720.

Wahrhaftig, C. 1965. Physiographic divisions of Alaska. U.S. Geological Survey Professional Paper 482.

Wu, T. H. 1984. Soil movements on permafrost slopes near Fairbanks, Alaska. Canadian Geotechnical Journal 21:699–709.

4

Climate and Permafrost Dynamics of the Alaskan Boreal Forest

Larry D. Hinzman
Leslie A. Viereck
Phyllis C. Adams
Vladimir E. Romanovsky
Kenji Yoshikawa

Climate Distribution in the Circumboreal Region

There are large climatic differences among the boreal regions of the world. The extreme continental climates of central Siberia, with a mean annual temperature of −11°C or colder and precipitation of only 150 mm, for example, contrasts strikingly with the semicoastal climate of Newfoundland, with a mean annual temperature of +5°C and precipitation of 1400 mm. Yet both are considered boreal. This wide range in mean annual temperatures translates into large variation in the soil thermal conditions. Although much of the northern region of the boreal forest is underlain by continuous and discontinuous permafrost, southern regions are entirely permafrost-free.

Climates of the Canadian Boreal Forest

Boreal Canada has been classified into four major ecoclimatic provinces (Ecoregions Working Group 1989). The Subarctic Ecoclimatic Province extends from treeline in northern Canada south to the border with continuous stands of closed spruce. It ranges from the highly continental areas of northern Yukon Territory to the wetter and somewhat warmer regions of the Labrador Peninsula. The Boreal Ecoclimatic Province includes the main body of the boreal forests of Canada from the Mackenzie River east to Newfoundland. It is a complicated province that has been divided into High, Mid-, and Low Boreal, with a wide range of climate conditions. The Subarctic Cordilleran Ecoclimatic Province occurs only at higher elevations in western Canada. Forested areas in this region are usually restricted to valley bottoms or low, south-facing slopes. The Cordilleran Ecoclimatic Province includes the mountain ranges along the west coast and the continental divide from Montana to Alaska and from the Yukon River south to the boundary with the coastal forests. The boreal portion of this province has climates similar to that of the eastern section of the Interior Highland Ecoregion of Alaska (Fig. 2.3, Gallant et al. 1995).

Alaskan Climate Distribution

Alaska does not fit well into these Canadian ecoclimatic provinces because of differences in elevation, the effects of the two east-west-oriented mountain ranges (the Alaska and Brooks Ranges), and the coastal influences of the Bering Sea to the west and Cook Inlet to the south (Fig. 1.1; Hopkins 1959, Hare and Ritchie 1972). Hammond and Yarie (1996) separated Alaska into 35 ecoclimatic regions, of which nine include areas of boreal forest. These range from the cold northern interior regions along the Koyukuk River and the south slopes of the Brooks Range, with mean annual temperatures as low as –8°C (National Weather Service) and mean annual precipitation around 200 mm, to regions in the eastern Yukon and Tanana River valleys, with mean annual temperatures ranging from –5°C to –7°C and mean annual precipitation between 215 and 300 mm. To the west, closer to the influence of the Bering Sea, temperatures are warmer, –1.5°C to –4.4°C, and precipitation is higher, between 400 and 480 mm. The warmest and wettest regions of the Alaskan boreal forest are in the Cook Inlet and the western Kenai Peninsula areas, south of the Alaska Range, where mean annual temperature is above freezing (around +1°C), and precipitation is between 400 and 560 mm. Gallant et al. (1995) divided the state into 20 ecoregions based on climate, terrain (including physiography, geology, glaciations, permafrost and hydrological features), soils, and vegetation. Five of these 20 ecoregions (including two south of the Alaska Range) contain extensive areas of boreal forest (Fig. 2.3). Their climates are summarized in Table 4.1.

Climate of Interior Alaska

Most of the research described in this book has taken place in interior Alaska in the Interior Forested Lowlands and Uplands and the Interior Bottomlands Ecoregions, so we focus our discussion of the Alaskan boreal climate on this region, relying largely on information from the Fairbanks Airport and from LTER climate stations at the Bonanza Creek Experimental Forest (BCEF) and the Caribou-Poker Creeks Research Watershed (CPCRW). The boreal forest of Alaska is wedged between three mountain ranges: the Alaska Range to the south, the Brooks Range to the north, and the McKenzie Mountains (an extension of the Brooks Range/Rocky Mountain Range) to the east. These mountain ranges provide effective barriers to coastal air masses. As a result, the climate is strongly continental, with cold winters and warm, relatively dry summers. These forests are characterized by drastic seasonal fluctuation in day length (nearly 22 hours on June 21 and less than 4 hours on December 21). Temperature ranges from extremes of –50°C in January to more than +33°C in July, with a short growing season (135 days or less from early May to mid-September). Montane rain shadows and distance from the Bering Sea minimize precipitation from coastal storms, and the average annual precipitation is only 287 mm

Table 4.1. Climate data for the forested ecoregions of boreal Alaska, including the Cook Inlet and Copper River Ecoregions, which are south of the Alaska Range.

Ecoregion	Precipitation		Air temperature			Permafrost
	annual total (mm)	snowfall accumulation (cm)	mean winter min (°C)	mean summer max (°C)	mean annual (°C)	
(104) Interior Forested Lowlands and Uplands	250–550	125–205	−35 to −18	17 to 22	−2.5 to −5	East: discontinuous West: continuous
(106) Interior Bottomlands	250–400	95–205	−33 to −26	22	−3 to −6	Discontinuous with isolated masses
(107) Yukon Flats	170	115	−34	22	−3	Discontinuous but widespread
(115) Cook Inlet	380–680	160–225	−15	18	1 to 2	Free
(117) Copper River Plateau	250–460	100–190	−27	21	−2.5 to −3.5	Discontinuous thin to thick

Source: from Gallant et al. (1995).

in Fairbanks, 35% of which falls as snow (Slaughter and Viereck 1986). Snow covers the ground from mid-October until mid- to late April, and maximum accumulation averages 75–100 cm (Viereck et al. 1993). Soil temperatures are consistently low. In interior Alaska, the permafrost distribution and active layer thickness (portion of soil profile above permafrost that thaws and refreezes annually) are closely related to the topographic conditions of slope, aspect, drainage, soil moisture content, thermal properties of the parent material, and vegetation.

Temperature

The continental climate of interior Alaska has a wide range of air temperature between summer and winter and large fluctuations around the seasonal means. Mean annual temperatures in the Tanana Valley average –3.1°C at the Fairbanks International Airport, with a range during the period of record (1917–2000) of +0.3° C in 1926 to –5.9° C in 1956. The warmest month, July, averages 16.3°C, whereas in January the average is a cold –23.5°C. These averages do not present a good picture of either the summer or winter air temperatures. In the Tanana Valley, periods of extreme cold ranging in the vicinity of –40°C to –45°C can occur at any time from late November through February. In contrast, daily maximum temperatures occasionally reach 35°C to 37°C in June and July, often with only modest night cooling because of persistent daylight.

Degree-Day Concept

Cumulative degree-day sums tend to give a more biologically meaningful picture of seasonal temperatures than do monthly averages. Growing degree-days (GDD) is a concept developed in Finland (Sarvas 1970) to predict flowering and seed maturation time in northern trees. In our Alaskan studies, we have used 5°C (approximately 40°F) as a threshold for growing degree-day temperatures. This system simply accumulates the average daily air temperatures above 5°C through the growing season. A day with an average temperature of 10°C would contribute 5 degrees toward the accumulated summer total. In the Tanana Valley, the average accumulated GDD for the summer months is approximately 1140 GDD. Extremes since 1917 have run from 796, in 1922, to 1427, in 1995.

The soil degree-day (SDD) is a concept developed in Alaska to compare soil temperatures between stands and years (Viereck and Van Cleve 1984). This system is based on 0°C and on daily average soil temperature at a given depth. Depths of 5, 10, and 20 cm have commonly been used in our studies. These figures range from 3000 SDD at 5 cm depth in a south-facing aspen stand to 530 SDD in a north-facing black spruce stand, with 1225 SDD in a white spruce stand on a south-facing slope being a representative intermediate value.

Freezing and thawing degree-days are useful for comparing years and sites. They again use an average daily air temperature and sum the degrees above or below 0°C for the year. Freezing degree-days begin in October and accumulate through September each year. The 1917–2000 average for freezing degree-day accumulation in the Tanana Valley is about 3000 FDD (Fig. 4.1), with extremes during those

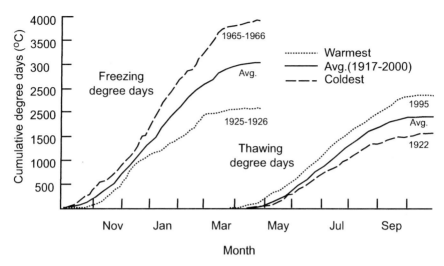

Figure 4.1. Minimum, average (1917–2000), and maximum accumulated sums of freezing (below 0°C) and thawing (above 0°C) air temperature degree-days in Fairbanks.

same years of 3918, in 1965–1966, and 2070, in 1925–1926. Thawing accumulated DD are somewhat fewer, 1950 TDD, with extremes of 2342, in 1995, and 1556, in 1922. Trends in GDD and FDD show that an abrupt warming in summer temperatures occurred in the mid-1960s, whereas an abrupt winter warming did not occur until the mid-1970s (Fig. 4.2). FDD and TDD of a specific site are strongly influenced by topography. If FDD (calculated using ground surface temperature) is consistently greater than TDD for many years, permafrost will be maintained at that site.

Precipitation

The mountain ranges that surround interior Alaska protect it from large frontal systems from the Arctic Ocean and the Gulf of Alaska. Consequently, annual precipitation is low and decreases from west to east, with a 50-year average for Fairbanks of 287 mm and a range from 142 mm, in 1957, to 478 mm, in 1990. Most summer and winter precipitation is generated from major frontal systems that cross the State, but convective storms add significantly to the summer precipitation. Precipitation events in early summer (May, June, and early July) are typically light and showery, with high spatial variability. The relatively dry summer conditions are replaced by the fall rain events, which can be heavy and sustained. On average, precipitation increases through the summer (Fig. 4.3). There is considerable variability in annual precipitation in Alaska, with low precipitation years, such as 1957, generating frequent wildfires and high-precipitation years, such as 1967, often resulting in flooding. Although precipitation during the growing season may be low, evaporation rates are also low because of the relative short growing season and cool temperatures. Nonetheless, as much as 76–100% of the summer precipitation may be lost as evapotranspiration (Dingman 1966, Gieck and Kane 1986). Thus, much of

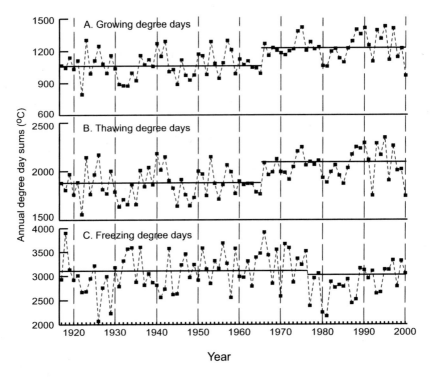

Figure 4.2. Growing (GDD), thawing (TDD), and freezing (FDD) degree-days from 1917 to 2000 for the Fairbanks airport. Horizontal lines show mean degree-day sums before and after the step change in regional climate.

the summer precipitation probably derives from recycling of water that evaporated from land (Serreze and Etringer 2003).

Snowpack

Snow is an important climate and ecological factor in the boreal forest. The ground is usually snow-covered from mid-October until early to mid-April, although in some years the snow period can be considerably longer. In 1992, for example, snow remained on the ground until the last week of May and returned permanently on 10 September, giving only a three-month snow-free period. Although snow accumulates during the entire winter period, maximum snowpack is generally shallow in interior Alaska, with an average maximum snowpack depth in mid-March to early April of 75 cm depth, with a water equivalent of 110 mm (at the Fairbanks International Airport). There are occasional heavy snow years, such as 1970–1971, 1990–1991, and 1992–1993, when maximum snow accumulations may reach 2 meters. There are also low snow years, such as 1969–1970, 1979–1980, 1980–1981, 1985–1986, 1986–1987, and 1998–1999, when less than 40 cm may accumulate throughout the entire winter. Precipitation data from the Fairbanks Airport (1948–2000)

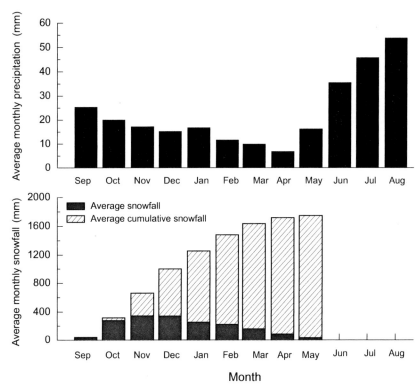

Figure 4.3. Average monthly rainfall and snowfall (1929–1999) and snowpack accumulation (1952–1999) in Fairbanks for the period of record.

indicate that snowfall accounts for about 35% of the yearly total precipitation (range 13–77%, standard deviation = 13%, Fig. 4.3). Little winter snowmelt occurs due to typically below-freezing temperature throughout the winter. During the snowmelt period (generally late April), snow is released as streamflow over a relatively short period, making snowmelt the major hydrological event of the year (Chapter 16). Incoming solar radiation (supplied by long daylight hours) is the major factor governing snowmelt as the albedo of the snow decreases from approximately 0.75–0.90 (fresh snow) to 0.40–0.70 (old snow; Campbell and Norman 1998, Woo 1986). Melting of snow from the ground surface is negligible as ground temperatures are <0°C. During snowmelt, south-facing slopes typically have less snow than north-facing slopes (more interception by vegetation and greater sublimation losses), and the snowpack melts one to two weeks earlier (Carey and Woo 1998). Spring snowmelt is the primary source of groundwater recharge in interior Alaska (Kane and Stein 1983), with infiltration occurring through frozen ground, primarily in areas absent of ice-rich permafrost (Kane et al. 1978).

The snowpack of interior Alaska is an excellent thermal insulator, despite its shallow depth, because winds are calm in winter and winter rain-on-snow events are rare. The snowpack therefore has a lower density and thermal conductivity than does arctic or temperate snow (Adams and Viereck 1997). Most of the interannual variation in the insulative properties of snow reflect the timing of the first major snow event. In the winter of 1995–1996, for example, snow depths at the end of January were only 30 cm at BCEF, resulting in very cold soil temperatures (–5°C at 5 cm). In 1992, however, an early deep snowpack caused soil temperatures at 5 cm to reach only –1.8°C.

Radiation

Low sun angles in both summer and winter limit the solar radiation that reaches the surface in interior Alaska. It is, of course, the low solar input that accounts for the overall low annual temperatures in the boreal zone. At the latitude of 65° N, the maximum solar angle is only 1.5° at the winter solstice and 48.5° at the summer solstice. There is also an extreme variation in day length, only 3 hours and 42 minutes of sunlight at the winter solstice as compared with 21 hours and 50 minutes on the longest day of the summer. This results in average daily solar radiation of 231 KJ m^{-2} d^{-1} in December and 22,375 KJ m^{-2} d^{-1} in June, nearly a 100-fold difference (Fig. 4.4). Disappearance of snow in April or May and its reappearance in September or October causes a sharp change in albedo (ratio of reflected to incoming shortwave radiation) and therefore in both radiation balance and local climate.

The low solar angle creates a striking contrast between north-facing and south-facing slopes. A south-facing slope of 10° receives 33% (133 vs. 100 W m^{-2}) greater radiation annually than does a comparable north-facing slope and 14% greater radiation than does a flat surface (estimated based on Rouse 1990). This results in warmer dryer soils on south-facing than on north-facing slopes, which in turn results in significant differences in distribution of forest types on adjacent slopes (Chapter 6).

Interaction of Climate and Topography

Effects of Slope and Aspect

The large differences in solar input between north- and south-facing slopes results in significant differences in ground temperatures but surprisingly little difference in air temperature, because convection effectively mixes air within the canopy with the bulk atmosphere. At the middle of a north-facing slope in CPCRW at 480 m elevation, the mean annual air temperature is only 1°C colder (–2.7 vs. –1.6°C) than at the same elevation of the south-facing slope. However, soil temperatures at these sites differ by 4°C (–1.2 and +2.7°C on north- vs. south-facing slopes, respectively). These large differences in soil temperature reflect ecosystem differences in radiation absorption and insulation by the forest floor (Slaughter and Long 1975).

The differences in solar input and soil temperature between north- and south-facing slopes of the same valley are greater than the differences between the northern and southern latitudinal extremes of boreal forest in interior Alaska due to differences in "equivalent latitude," that is, the latitude of a horizontal surface that would receive

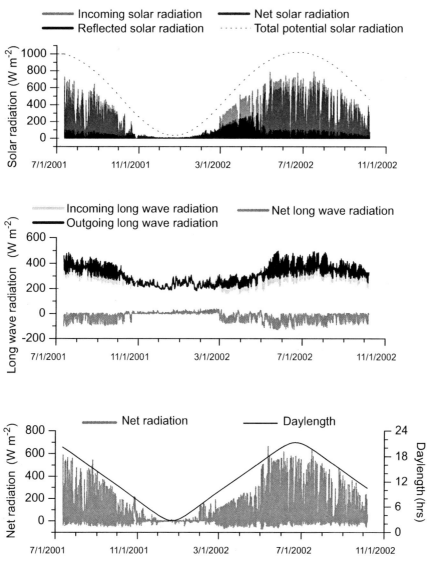

Figure 4.4. Radiation balance in CPCRW near Fairbanks. Total potential solar radiation is the theoretical maximum potential at that location, assuming no absorption or reflection in the atmosphere.

the same potential solar insolation (Lees 1964). Thus, a steep (37°) south-facing slope in interior Alaska has an equivalent latitude of 28° (central Florida), whereas that of a north-facing slope might be as high as 82° (northern Greenland; Dingman and Koutz 1974). The moderately rolling topography in interior Alaska produces equivalent latitude figures ranging between these two extremes. This contrasts with the actual latitudinal range of boreal forest in interior Alaska of 63–67°N.

Equivalent latitude is useful in characterizing landscapes with respect to their potential vegetation cover, which in turn is a useful indicator of permafrost distribution (Dingman and Koutz 1974). Morrissey and Strong (1986) mapped permafrost distribution using Thematic Mapper satellite imagery and a GIS database of vegetation distribution and equivalent latitude. Their mapping of permafrost had 78% accuracy and compared well with the Rieger et al. (1972) classification of permafrost distribution. Temperature profiles obtained from deep boreholes provide a good estimate of long-term mean annual ground surface temperature (Collins et al. 1988). A comparison of borehole temperatures from different equivalent latitudes shows that the permafrost boundary occurs at approximately 63° equivalent latitude, just south of the major LTER study sites at Bonanza Creek and CPCRW.

Effects of Elevation

Topography also generates important gradients in local climate as a result of strong temperature inversions (coldest air temperatures in valley bottoms and increasing temperature with elevation) during the calm winter cold periods. In the Fairbanks area, strong inversions occur as much as 80% of the time during cold spells and produce gradients of up to 21°C per 100 meters (Benson and Rizzo 1979). These inversions are occasionally broken up by winds from the Alaska Range but can persist for weeks during winter. Temperature inversions are most pronounced during cold high-pressure events in midwinter (Fig. 4.5) but also occur frequently at night during summer. During the day in summer, inversions are replaced by a normal adiabatic rate of cooling in which air temperature decreases with increased elevation (somewhat less than 1°C per 100 m; Slaughter and Viereck 1986). Air temperature inversions play a major role in maintaining permafrost in valley bottoms. Average annual temperatures in a valley bottom can be 3° colder than the average annual temperature of the neighboring hilltops (Caribou Peak, 1995–1999). In regions where the average annual temperature

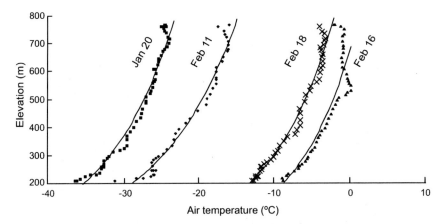

Figure 4.5. Measured air temperature inversions in CPCRW during the winter 1999. Redrawn with permission from the *Journal of Glaciology and Geocryology* (Yoshikawa et al. 2002).

is close to 0°, the significantly colder temperatures in the lower elevations preserve permafrost, while atop slightly warmer hilltops, permafrost is absent. Other factors also impact the topographic influence on permafrost distribution. Drier hilltops have less thermal inertia associated with less ground ice. Additionally, the wetter valley bottoms foster thick layers of moss and organic soil, which serve as good insulation for the underlying permafrost.

Terrain Influence on Permafrost Distribution

Two major climatic agents control the natural formation and degradation of permafrost: (1) solar radiation determines energy input to a site, and (2) temperature inversions control FDD during wintertime by determining the temperature gradient between the ground surface and the atmosphere. As a result, the distribution of the permafrost is strongly influenced by topography, being located mostly on north-facing slopes and valley bottoms. However, thermal and hydrological properties of the snow and ground surface also influence permafrost distribution, particularly when the permafrost temperature is close to phase change. Wetter valley bottom soils experience more evaporative cooling than drier hilltop sites, resulting in an additional heat loss mechanism. Topographic variation in soil texture also influences heat transport and therefore permafrost occurrence. Areas of exposed and/or weathering bedrock (usually on hilltops) have highly permeable soils and therefore low water content. Dry soils have a lower heat capacity than wet soils, so summer heat inputs cause greater increase in soil temperatures in dry than in wet areas. Ice-rich silt deposits resist penetration of heat by conduction, so the colluvially transported silt deposits at the base of south-facing slopes usually correspond with the boundary of the permafrost between uplands and lowlands.

Soil and Permafrost Climate

Most of the Alaskan boreal forest is underlain by discontinuous permafrost. Only the northernmost foothills of the Brooks Range are within the continuous permafrost zone (Pewe 1983, Ferrians 1965, Brown and Pewe 1973, Brown et al. 1997). This contrasts strikingly with the East Siberian boreal forest, most of which lies within the continuous permafrost zone, or with the West Siberian and Russian European North taiga, where permafrost is largely absent. The discontinuous character of permafrost distribution creates a complex interrelationship between permafrost and vegetation in the Alaskan boreal forest.

Permafrost thickness varies in this area from 0 to 200 meters (Ferrians 1965, Brown et al. 1997) with the most typical values being 25 to 90 meters. Lateral continuity of the perennially frozen layer changes from practically continuous (90–95%) in the southern foothills of the Brooks Range to discontinuous (50–90%) in interior Alaska and in most of the Alaska Range to sporadic (10–50%) and isolated patches (0–10%) south of the Alaska Range, Talkeetna, and the Wrangell Mountains (Brown et al. 1997).

Permafrost temperature is the most important indicator of permafrost stability. The closer the temperature is to 0°C, the more susceptible permafrost is to climate warming or surface disturbances. In interior Alaska, permafrost temperature varies seasonally within the upper 10 to 15 meters, explaining why the temperature at the depth of zero seasonal amplitude (typically 10 to 15 m) is usually used to describe the permafrost thermal state. Permafrost temperatures in the Alaskan boreal forest are 0 to –4°C and typically warmer than –2°C (Osterkamp and Romanovsky 1999), reflecting the mean annual air temperatures of the region (–2°C to –5°C). The thickness, thermal properties, and duration of the snow cover also influence permafrost temperature (Brown and Pewe 1973, Romanovsky and Osterkamp 1995). The mean annual ground surface temperatures are usually 3–6°C warmer than the mean annual air temperatures (Table 4.2). As a result of relatively warm air temperatures and the effect of snow cover in reducing winter heat loss from soil, the mean annual ground surface temperatures in the Alaskan boreal forest often exceed 0°C and can be as high as 4°C (Table 4.2; Osterkamp and Romanovsky 1999). The warm temperature and sensitivity to surface properties make permafrost in the Alaskan boreal forest vulnerable to climate change and disturbance.

Although the mean annual ground surface temperature may be above 0°C for a long time (a decade or longer), permafrost may stably exist below the active layer. The natural phenomenon that is responsible for permafrost preservation in this situation is called "thermal offset," that is, the difference between the mean annual temperature at the bottom of the active layer and that at the ground surface (Kudryavtsev et al. 1974, Goodrich 1978, Burn and Smith 1988, Romanovsky and Osterkamp 1995, 1997, 2000). The thermal offset occurs because the thermal conductivity of ice is greater than that of water, so soils conduct heat more effectively in winter than in summer. The magnitude of the thermal offset depends on soil ther-

Table 4.2. Mean annual air, ground surface, and permafrost-surface temperatures averaged over period of measurements (all temperatures in °C).

Sites and period of measurements	Mean annual air temperature	Mean annual ground surface temperature	Mean annual temp at bottom of the active layer
Permafrost sites			
Council, Seward Pen. Woodland, 1999-2000	–4.55	–0.5	–1.2
Fairbanks, LTER BNZ, Floodplain, 1995–1998	–4.26	0.26	–1.13
Fairbanks, LTER CPCRW 1999-2000	–4.19	1.55	–0.24
Fairbanks, Smith Lake #1 1997–1998	–3.73	0.82	–0.72
Fairbanks, Smith Lake #3 1997–1998		2.16	–0.02
Fairbanks, GI #1 1990–1995	–3.14	0.07	
Fairbanks, GI #4 1988–1994	2.72	0.27	–0.02
Denali Park, 1996–1997		1.4	–0.55
Nonpermafrost sites			
Council, Seward Pen. forest, 1999-2000	–4.06	0.52	0.76 (1 m depth)
Council, Seward Pen. shrubland, 1999-2000	–4.62	1.51	0.82 (1 m depth)
Fairbanks, UAF Farm, 1989–1997	–2.79	3.86	2.86 (1 m depth)

mal properties and is largest in organic sites with poor drainage (–1°C to –2.5°C). At dry sites, where moss and peat layers are absent, the thermal offset is minimal (usually less than –1°C; Romanovsky and Osterkamp 1995).

Numerical modeling suggests that permafrost temperatures have changed considerably during the twentieth century. The near-surface active layer temperatures were about 0°C in the late 1930s and 1940s (Fig. 4.6). This was followed by 19 years of below –0°C temperatures to 1971 and 16 years of temperatures varying about 0°C. Since 1987, active-layer temperatures have remained warmer than 0°C. The ground surface warmed by about 3°C over 30 years for an average rate of 0.1°C yr^{-1}. Permafrost survives at the Bonanza Creek site only because of the insulating effect of the organic mat at the ground surface and the related thermal offset in the active layer (Osterkamp and Romanovsky 1999). Near-surface permafrost temperatures paralleled the patterns of active-layer temperature. A warming trend began in the late 1960s and totaled about 2°C over about 30 years for an average rate of 0.07°C yr^{-1}.

Assuming that temperature trends of the past 30 years continue, model projections of active-layer and upper permafrost temperatures suggest that extensive permafrost instability and degradation will begin sometime between 2015 to 2025 (Romanovsky et al. 2001). If this occurs, thermokarst (the subsidence of the ground due to melting of ice-rich permafrost) may strongly affect Alaskan ecosystems and infrastructures, perhaps causing significant areas of boreal forest to degrade and creating wetlands in lowlands with poor drainage.

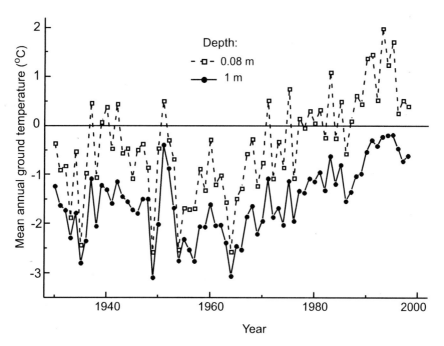

Figure 4.6. Calculated mean annual ground temperature dynamics within the active layer at Bonanza Creek (1930–1998).

Climatic Control over Ecosystem Processes

Climate ultimately regulates many ecosystem processes through control of interacting environmental and biotic factors. Some of these, including fire, flooding, and mortality due to snow and ice storms, result from stochastic events. Climate variability also influences boreal forest ecosystems indirectly through biotic processes such as seed production (Zackrisson et al. 1995), insect infestations, and productivity (Barber et al. 2000).

Fire

Climate strongly influences fire severity and frequency, with the greatest aerial extent of burning in the hottest, driest years (Chapter 17). For example, the average summer (June, July, and August) air temperatures in the large Alaska fire years of 1940 and 1957 were 15.5 and 16.3°C, respectively, well above the long-term average of 14.8°C for these months, while the total annual precipitation in 1957 was only 142 mm, about half of the long-term average. The GDD for 1940 and 1957 were 1270 and 1298, respectively, somewhat above the average 1139. Most of the Alaskan boreal forest is shaped by stand-replacing fires (Viereck 1973), and it has been hypothesized that landscape patterns of species distribution shift with increased fire frequency. Black spruce, with semiserotinous cones, is well adapted to fire, although increases in fire severity might alter seedbed conditions sufficiently to favor the establishment of early successional trees, such as birch and aspen. White spruce reproduction is synchronized with the climate-controlled fire regime. Seed production is enhanced by hot dry weather, a condition that also promotes wildfires. Thus, in the season following a year of high fire probability, white spruce seeds are more likely to be available to disperse to adjacent burn scars.

Productivity

Radial growth of interior Alaskan white spruce trees has decreased with increasing temperature (Barber et al. 2000). This temperature-induced drought stress limits carbon uptake in much of the North American boreal forest (Chapter 19). Ecosystem carbon balance may be sensitive to moisture even in boreal bogs, where carbon loss occurs during extended periods of summer drought (Alm et al. 1999) or in black spruce forests, which show a negative carbon balance in warm dry years (Goulden el al. 1998). On the other hand, there is a strong correlation between annual soil temperature sums and forest productivity in interior Alaska (Viereck and Van Cleve 1984), suggesting that a warmer climate would allow more productive forest types, such as white spruce and birch, to expand to colder sites in Alaska.

Snow

The amount and timing of snow accumulation during the winter is critical to many ecological processes, including forest productivity, tree mortality, and winter survival of animals. Damage from snow and ice can cause major changes to the forest structure and age distribution. Snow accumulation on branches results in tree mortality due either directly to top breakage and removal of the foliage or indirectly as a result of insect populations that infest heavily damaged forest stands. Heavy snow loads on white spruce trees during the winter of 1990–1991 resulted in extensive top breakage and subsequent mortality throughout the Tanana Valley. In contrast, deciduous trees sustained little damage as a result of the heavy snowfall because the leafless branches did not accumulate snow loads. A different scenario occurred with an early wet snowfall in mid-September 1992, before the leaves had fallen or changed color. Many birch trees were bowed over by this snow load and eventually died. Additionally, the leaf litter that accumulated when the leaves finally fell was high in nitrogen because nitrogen resorption had not occurred prior to leaf fall.

Interaction of Changing Climate and Disturbance Regimes

The boreal forests in northern North America and Eurasia are experiencing a rapidly changing climate (Serreze et al. 2000). The most dramatic changes appear to be significant winter warming (~0.5 to 2.0°C per decade from 1966 to 1995) throughout Eurasia and western North America, with only slight increase or no change in Eastern Canada (Fig. 4.7). Spring warming is less pronounced (~0.25 to 1.0°C per decade between 1966 and 1995) but follows similar spatial trends, resulting in somewhat earlier spring melt and increasing the probability of rain-on-snow events. The ecological significance of such events includes increased foraging difficulties for caribou and decreased oxygenation for small mammals dwelling beneath the snowpack. In general, earlier spring melt results in less harsh winter stress, a longer growing season, and longer ice-free periods on ponds and lakes (Magnuson et al. 2000). Changes in precipitation are less consistent. There has been a significant ($p = 0.05$) increase in precipitation in Fairbanks, Alaska, and Yakutsk, Russia, over the past century. This increase in precipitation may, however, be offset by higher rates of evapotranspiration resulting from the warmer temperatures, leading to drier soils (Oechel et al. 2000).

The observed changes in climate directly impact the stability and distribution of permafrost. An analysis of the change in the extent of discontinuous permafrost in CPCRW during the twentieth century shows a slight decrease in the total area underlain by permafrost and a substantial increase in the proportion of permafrost that is unstable (currently thawing; Fig. 4.8). In these analyses, the primary factor that determines the mean annual surface temperature is the winter surface temperature, because summer temperature (thawing degree days at the ground surface) and snow depth do not vary greatly from south-facing to north-facing slopes. The permafrost condition is therefore effectively controlled by freezing degree days. Using this method, we determined that 37.5% of CPCRW has unstable or thawing permafrost. At least 2.1% of the permafrost in this watershed has retreated in the past 90 years due to climate warming. It appears that 1.2% of the permafrost in the watershed did not recover after the forest fires early in the twenty-first century. Permafrost

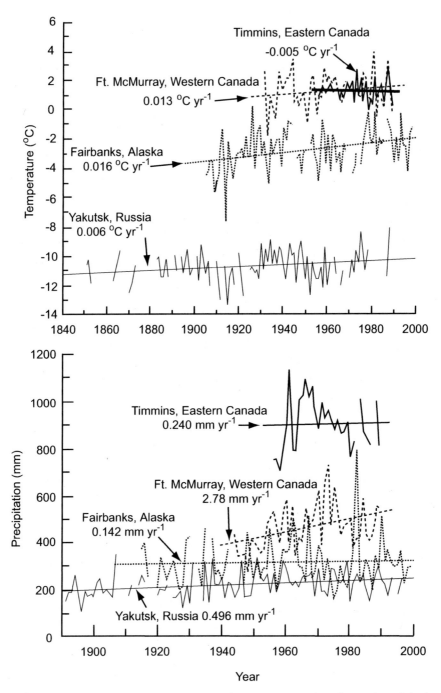

Figure 4.7. Recent changes in mean annual air temperature and total annual precipitation at four stations in the circumarctic boreal forest.

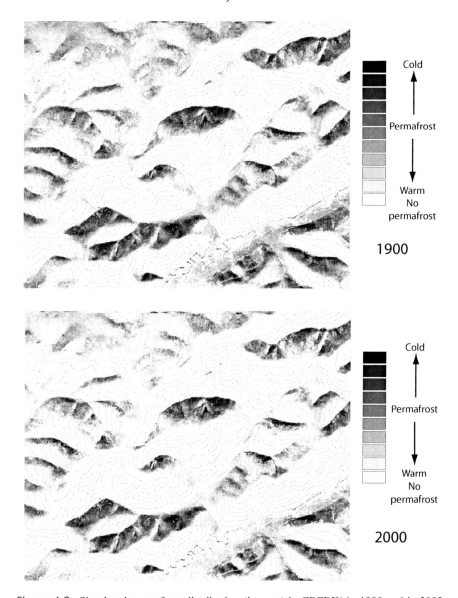

Figure 4.8. Simulated permafrost distribution (in gray) in CPCRW in 1900 and in 2002.

degradation can cause changes in surface hydrology, particularly soil moisture levels, slumping of frozen stream banks, increased erosion, and myriad ecological impacts as the system adjusts to these disturbances.

Continued climatic warming in the boreal forest may initiate a complex set of interdependent impacts in the physical environment that affect ecosystem responses and feedbacks (Weber and Flannigan 1997). Warm permafrost is quite sensitive to any disturbance, including changes in annual temperatures, hydrologic regime, surface configuration, mining, or wildfires. Permafrost may be ice-rich or ice-poor, depending upon soil type, terrain, and location (Chapter 3). Massive ice wedges developed across broad regions of the circumpolar north during the Pleistocene and early Holocene period (Hamilton et al. 1988), with most ice wedges forming after the last interglacial. Upon thawing, these ice wedges may leave huge voids allowing subsidence of the surface and, depending on terrain, formation of thaw ponds (Fig. 4.9). This process is called thermokarst.

Air temperatures have increased in most of the circumpolar boreal forest, while changes in precipitation have been more variable (Serreze et al. 2000). Drought, which results from high temperature and/or decreased precipitation, is the direct cause of increased fire frequency through reduction in fuel moisture (Chapter 17). The probability of fire is also enhanced by insect outbreaks, which increase in frequency in drought-stressed trees. Increased insect outbreaks associated with climate warming (Malmstrom and Raffa 2000) may contribute to future increases in fire frequency. Increases in precipitation may not be enough to offset increases in evapotranspiration that occur with warmer air temperature. The resulting drying of soils (Oechel et al. 2000) raises the potential and severity of forest fires. The increase in air temperature in western North America over the past 30 years coincides with a doubling of the annual area burned (Murphy et al. 2000), and the number of fires reported and the total area burned in North America have increased markedly over the past five decades (Chapter 17).

Fire may play an even more important role in shaping community dynamics, carbon balance, and surface energy exchanges if climate continues to warm. The immediate impact of fire is to reduce albedo, as blackened carbon accumulates at the surface, resulting in a change in the surface energy balance. During the summer, the surface temperature increases as the reduced albedo causes increased adsorption of solar radiation. At the same time, net radiation actually decreases because,

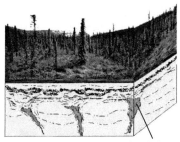

Figure 4.9. Thermokarst ponds develop as relict ice wedges melt. This has important ecological implications as forests in discontinuous permafrost warm due to changing climate.

as the temperature of the ground increases, the thermal (longwave) radiation emitted from the surface increases more than the increased adsorption of solar (shortwave) radiation (Chambers and Chapin 2003, Yoshikawa et al. 2003). As vegetation recovers, albedo increases above values typical of unburned spruce stands, again reducing net radiation. Immediately after a fire, soils become wetter because of the cessation of water loss through transpiration (Fig. 4.10). As the active layer becomes thicker, the soil drainage improves, and near-surface soils become drier. It may thaw to a depth greater than that which will completely freeze the following winter, forming a talik, or a layer of unfrozen soil above the permafrost and below the seasonally frozen ground (Fig. 4.10).

The short- and long-term impacts of forest fire depend on fire severity (Fig. 4.10). After a light or moderate burn, much of the organic soil remains intact and retains its insulative capacity, so the underlying permafrost is not strongly affected. After a more severe fire, however, the organic soil may be completely consumed, exposing the underlying mineral soil and resulting in rapid degradation of permafrost (Yoshikawa et al. 2003). As the active layer becomes thicker, subsurface drainage improves, leading to surface drying. If the permafrost does not recover after accumulation of a surface organic layer, the soil will remain drier, and the postrecovery vegetation may be different from the prefire vegetation (Chapter 7). The point at which soils tend to become drier may act as a second, more subtle threshold event.

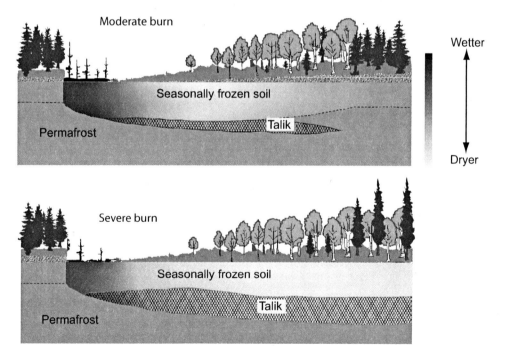

Figure 4.10. Successional changes in permafrost after fire. Redrawn from Hinzman et al. (2004).

Conclusions

Climate varies over space and time, and these differences, along with changes in parent material, have largely shaped the mosaic that is the boreal forest. The ecosystem not only responds and develops under the prevailing microclimatic conditions but also affects the local climate through impacts on the surface energy balance. Developing a greater understanding of the dynamic relationship among climatic and ecosystem processes increases our ability to predict ecosystem responses to a changing climate, project changes in the disturbance regime, and anticipate biological responses to changes in the physical regime.

References

Adams, P. C., and L. A. Viereck. 1997. Soil temperature and seasonal thaw: Control and interactions in floodplain stands along the Tanana river, interior Alaska. Pages 105–111 *in* I. K. Iskandar, E. A. Wright, J. K. Radke, B. S. Sharratt, P. H. Groenvelt, and L. D. Hinzman, editors. International Symposium on Physics, Chemistry, and Ecology of Seasonally Frozen Soils. U.S. Army Cold Regions Research and Engineering Laboratory, Hanover, NH.

Alm, J., L. Schulman, J. Walden, H. Nykanen, P. J. Martikainen, and J. Silvola. 1999. Carbon balance of a boreal bog during a year with an exceptional dry summer. Ecology 80:161–174.

Barber, V.A., G. P. Juday, and B. P. Finney. 2000. Reduced growth of Alaskan white spruce in the twentieth century from temperature-induced drought stress. Nature 405:668–673.

Benson, C. S., and K. R. Rizzo. 1979. Air pollution in Alaska. Alaska's Weather and Climate. UAGR-269, 94-113. Geophysical Institute, University of Alaska Fairbanks.

Brown, J., O. J. Ferrians, J. A. Heginbottom, and E. S. Melnikov. 1997. Circum-Arctic map of permafrost and ground-ice conditions, U.S. Geological Survey, scale 1:10,000,000.

Brown, R. J. E., and T. L. Pewe. 1973. Distribution of permafrost in North America and its relationship to the environment, a review, 1963–1973. Pages 71–100 *in* Proceedings of the Second International Conference on Permafrost, Washington, DC.

Burn, C. R., and C. A. S. Smith. 1988. Observations of the "thermal offset" in near-surface mean annual ground temperatures at several sites near Mayo, Yukon Territory, Canada. Arctic 41:99–104.

Campbell, G. S., and J. M. Norman. 1998. An Introduction to Environmental Biophysics. 2nd ed. Springer-Verlag, New York.

Carey, S. K., and M. K. Woo. 1998. A case study of active layer thaw and its controlling factors. Pages 127–131 *in* A. G. Lewkowicz and M. Allard, editors. Proceedings from Permafrost: Seventh International Conference. Collection Nordicana, no. 57.

Chambers, S. D., and F. S. Chapin III. 2002. Fire effects on surface-atmosphere energy exchange in Alaskan black spruce ecosystems: Implications for feedbacks to regional climate. Journal of Geophysical Research 108, 8145, doi:10.1029/2001JD000530 [printed in 108(D1), 2003].

Collins, C. M., R. K. Haugen, and R. A. Kreig. 1988. Natural ground temperatures in upland bedrock terrain, interior Alaska. Pages 56–60 *in* Proceedings, Fifth International Conference on Permafrost, Vol. 1. Tapir, Trondheim, Norway.

Dingman, S. L. 1966. Characteristics of summer runoff from a small watershed in central Alaska. Water Resources Research 2:751–754.

Dingman, S. L., and F. R. Koutz. 1974. Relations among vegetation, permafrost, and potential insolation in central Alaska. Arctic and Alpine Research 6:37–42.
Ecoregions Working Group. 1989. Ecoclimatic Regions of Canada, First Approximation. Ecoregions Working Group of the Canada Committee of Ecological Land Classification. Ecological Land Classification Series, No. 23, Sustainable Development Branch, Canadian Wildlife Service, Conservation and Protection, Environment Canada, Ottawa, Ontario. 119 pp. and map at 1:7 500 000.
Ferrians, O. J. 1965. Permafrost map of Alaska. U.S. Geological Survey Miscellaneous Geologic Investigations Map 1–445.
Gallant, A. L., E. F. Binnian, J. M. Omernik, and M. B. Shasby. 1995. Ecoregions of Alaska. U.S. Geological Survey Professional Paper 1567. U.S. Department of the Interior, Washington, DC.
Gieck, R. E., and D. L. Kane. 1986. Hydrology of two subarctic watersheds. Pages 283–291 *in* D. L. Kane, editor. Cold Regions Hydrology Symposium. American Water Resources Association, Fairbanks, AK.
Goodrich, L. E. 1978. Some results of a numerical study of ground thermal regimes. Pages 29–34 *in* Proceedings, Third International Conference on Permafrost, Vol. 1. National Research Council of Canada, Ottawa.
Goulden, M. L., S. C. Wofsy, J. W. Harden, S. E. Trumbore, P. M. Crill, S. T. Gower, T. Fries, B. C. Daube, S.-M. Fan, D. J. Sutton, A. Bazzaz, and J. W. Munger. 1998 Sensitivity of boreal forest carbon balance to soil thaw. Science 279:214–217.
Hamilton, T. D., J. L. Craig, and P. V. Sellmann. 1988. The Fox permafrost tunnel: A late Quaternary geologic record in central Alaska. Geological Society of America Bulletin 100:948–969.
Hammond, T., and J. Yarie. 1996. Spatial prediction of climatic state factor regions in Alaska. Ecoscience 3:490–501.
Hare, F. K., and J. C. Ritchie. 1972. The boreal bioclimates. Geographical Review 62:333–365.
Hinzman, L.D., M. Fukuda, D. V. Sandberg, F. S. Chapin III, and D. Dash. 2003. FROSTFIRE: An experimental approach to predicting the climate feedbacks from the changing boreal fire regime. Journal of Geophysical Research-Atmospheres 108: 8153, doi:10.1029/2001JD00415 [printed in 108(D1), 2003].
Hopkins, D. M. 1959. Some characteristics of the climate in forest and tundra regions in Alaska. Arctic 12:215–220.
Kane, D. L., and J. Stein. 1983. Field evidence of groundwater recharge in interior Alaska. Pages 572–577 *in* Proceedings, International Conference on Permafrost. National Academy Press, Washington, DC.
Kane, D. L., J. D. Fox, R. D. Seifert, G. S. Taylor. 1978. Snowmelt infiltration and movement in frozen soils. Pages 200–206 *in* Proceedings, Third International Conference on Permafrost. National Research Council of Canada, Ottawa.
Kudryavtsev, V. A., L. S. Garagula, K. A. Kondrat'yeva, and V. G. Melamed. 1974. Osnovy merzlotnogo prognoza. MGU [CRREL Translation: Kudryavtsev, V. A., S. Garagula, K. A. Kondrat'yeva, and V. G. Melamed.1977. Fundamentals of frost forecasting in geological engineering investigations, CRREL Draft Translation 606].
Lees, J. C. 1964. Tolerance of white spruce seedlings to flooding. Forestry Chronicle 40:221–225.
Magnuson, J., D. Robertson, B. Benson, R. Wynne, D. Livingstone, T. Arai, R. Assel, R. Barry, V. Card, E. Kuusisto, N. Granin, T. Prowse, K. Steward, V. Vuglinski. 2000. Historical trends in lake and river ice cover in the northern hemisphere. Science 289: 1743–1746.

Malmstrom, C. M., and K. F. Raffa. 2000. Biotic disturbance agents in the boreal forest: Considerations for vegetation change models. Global Change Biology 6:35–48.

Morrissey, L. A., and L. L. Strong. 1986. Mapping permafrost in the boreal forest with Thematic Mapper satellite data. Photogrammatic Engineering and Remote Sensing 52:1513–1520.

Murphy, P. J., B. J. Stocks, E. S. Kasischke, D. Barry, M. E. Alexander, N. H. F. French, and J. P. Mudd. 2000. Historical fire records in the North American boreal forest. Pages 274–288 *in* E. S. Kasischke and B. J. Stocks, editors. Fire, Climate Change, and Carbon Cycling in the North American Boreal Forest. Springer-Verlag, New York.

Oechel, W. C., G. L. Vourlitis, S. J. Hastings, R. C. Zulueta, L. Hinzman, and D. Kane. 2000. Acclimation of ecosystem CO_2 exchange in the Alaskan Arctic in response to decadal climate warming. Nature 406: 978–981.

Osterkamp, T. E., and V. E. Romanovsky. 1999. Evidence for warming and thawing of discontinuous permafrost in Alaska. Permafrost and Periglacial Processes 10:17–37.

Pewe, T. L. 1983. Alpine permafrost in the contiguous United States: A review. Arctic and Alpine Research 15:145–156.

Rieger, S., C. E. Furbush, D. B. Schoephorster, H. Summerfield Jr., and L. G. Geiger. 1972. Soils of the Caribou-Poker Creeks Research Watershed. Technical Report 236. U.S. Army Cold Regions Research and Engineering Laboratory, Hanover, NH.

Romanovsky, V. E., and T. E. Osterkamp. 1995. Interannual variations of the thermal regime of the active layer and near-surface permafrost in Northern Alaska. Permafrost and Periglacial Processes 6:313–335.

Romanovsky, V. E., and T. E. Osterkamp. 1997. Thawing of the active layer on the coastal plain of the Alaskan Arctic. Permafrost and Periglacial Processes. 8:1–22.

Romanovsky, V. E. and T. E. Osterkamp. 2000. Effects of unfrozen water on heat and mass transport processes in the active layer and permafrost. Permafrost and Periglacial Processes 11:219–239.

Romanovsky, V. E., T. E. Osterkamp, T. S. Sazonova, N. I. Shender, and V. T. Balobaev. 2001. Permafrost temperature dynamics along the East Siberian transect and an Alaskan transect. Tohoku Geophysical Journal 36:224–229.

Rouse, W. R. 1990. The regional energy balance. Pages 187–206 *in* T. D. Prowse and C. S. L. Ommanney, editors. Northern Hydrology: Canadian Perspectives. National Hydrology Research Institute Science Report No. 1. Saskatoon, Canada.

Sarvas, R. 1970. Temperature sum as a restricting factor in the development of forests in the subarctic. Pages 79–82 *in* Ecology of Subarctic Regions, Proceedings of the Helsinki Symposium, UNESCO, Paris.

Serreze, M. C., and A. J. Etringer. 2003. Precipitation characteristics of the Eurasian Arctic drainage system. International Journal of Climatology 23:1267–1291.

Serreze, M. C., J. E. Walsh, F. S. Chapin III, T. Osterkamp, M. Dyurgerov, V. Romanovsky, W. C. Oechel, J. Morison, T. Zhang, and R. G. Barry. 2000. Observational evidence of recent change in the northern high-latitude environment. Climate Change 46:159–207.

Slaughter, C.W., and L. P. Long. 1975. Upland climatic parameters on subarctic slopes, central Alaska. Pages 276–280 *in* Climate of the Arctic: Proceedings, Twenty-fourth Alaska Science Conference. University of Alaska Fairbanks.

Slaughter, C. W., and L. A. Viereck. 1986. Climatic characteristics of the taiga in interior Alaska. Pages 9–21 *in* K. Van Cleve, F. S. Chapin III, P. W. Flanagan, L. A. Viereck, and C. T. Dyrness, editors. Forest Ecosystems in the Alaskan Taiga. Springer-Verlag, New York.

Viereck, L. A. Wildfire in the taiga of Alaska. 1973. Quaternary Research. 3:465–495.

Viereck, L.A., and K. Van Cleve. 1984. Some aspects of vegetation and temperature relationships in the Alaska Taiga. Pages 129–142 *in* J. H. McBeath, G. P. Juday, G. Weller, and M. Murray, editors. Proceedings of a Symposium on the Potential Effects of CO_2 Induced Climate Change in Alaska. School of Agriculture and Land Resources, University of Alaska Fairbanks Misc. Pub. 83–1.

Viereck, L. A., K. Van Cleve, P. C. Adams, and R. E. Schlentner. 1993. Climate of the Tanana River floodplain near Fairbanks, Alaska. Canadian Journal of Forest Research 23:899–913.

Weber, M. G., and M. D. Flannigan. 1997. Canadian boreal forest ecosystem structure and function in a changing climate: Impact on fire regimes. Environmental Reviews 5:145–166.

Woo, M. K. 1986. Permafrost hydrology in North America. Atmosphere-Ocean 24:201–234.

Yoshikawa, K., L. D. Hinzman, and P. Gogineni. 2002. Ground temperature and permafrost mapping using an equivalent latitude/elevation model. Journal of Glaciology and Geocryology 24:526–531.

Yoshikawa, K., W. R. Bolton, V. E. Romanovsky, M. Fukuda, and L. D. Hinzman. 2003. Impacts of wildfire on the permafrost in the boreal forests of Interior Alaska. Journal of Geophysical Research 107, 8148, doi:10.1029/2001JD000438 [printed in 108(D1), 2003].

Zackrisson, O., M. C. Nilsson, I. Steijlen, and G. Hornberg. 1995. Regeneration pulses and climate-vegetation interactions in nonpyrogenic boreal Scots pine stands. Journal of Ecology 83:469–483.

5

Holocene Development of the Alaskan Boreal Forest

Andrea H. Lloyd   Jason A. Lynch
Mary E. Edwards   Valerie Barber
Bruce P. Finney   Nancy H. Bigelow

Introduction

Paleoecological data provide insight into patterns of change in vegetation and in the factors, such as climate and disturbance, that cause vegetation change. Disturbance by fire, insect, and mammalian herbivores and, in floodplains, flooding are the primary drivers of changes in population structure, community composition, and species distribution in the boreal forest on time scales of years to decades (Chapter 7). On longer time scales, such as centuries to millennia, the role of variation in regional climate in determining compositional changes in the boreal forest is also clearly visible. Variability in regional climate may act directly on boreal species (e.g., causing changes in species distributions) or indirectly, by altering disturbance regimes. Proxy records of environmental and ecological change (e.g., pollen and macrofossils in lake sediments, tree rings) are selective in the kind of information they record. Evidence of fires, for example, is more persistent and thus better represented in the paleoecological record than is evidence of mammalian herbivory. For this reason, our understanding of long-term patterns of compositional and structural change in the boreal forest is limited to an analysis of the effects of a few key drivers of change, primarily climate and fire.

In this chapter, we offer a long-term perspective on changes in climate and disturbance regimes and their relationship to major changes in vegetation. We first consider multimillennial time scales and discuss the role of climate and disturbance in driving the two major vegetation transitions that have occurred during the Holocene (the past 12,000 years). We then explore evidence for spatial and temporal variation in disturbance regimes during the late Holocene.

Variability in Climate, Disturbance, and Community Dynamics during the Holocene

Much of the terrain that is currently occupied by the Alaskan boreal forest remained ice-free during the glacial episodes of the Quaternary period (Pleistocene and Holocene), which spans the past 1.8 million years. Alaska forms part of the largely unglaciated Beringian region (named after the Bering Strait that lies at its heart; see Hopkins 1967) that extends from the Kolyma River in Siberia to the MacKenzie in northwest Canada and constitutes ca. 30% of the circumboreal zone. Although Beringia remained largely ice-free during the Quaternary, the boreal forest has repeatedly appeared and disappeared in the region as a result of large climatic fluctuations (Colinvaux 1964, Matthews 1974, McDowell and Edwards 2001). Although boreal forest taxa such as spruce (*Picea*) probably survived the most recent glacial period in Beringia (Brubaker et al. 2005), the landscape during that time was dominated by treeless ecosystems (e.g., Ager 1983, Ager and Brubaker 1985; Anderson et al. 2004). The boreal forest re-emerged as an important community type in Alaska beginning approximately 13,000 years ago, as the global climate warmed and continental ice sheets disappeared (Chapter 6).

Data Sources

Paleoecological studies require detective work; we attempt to reconstruct entire landscapes from the often microscopic evidence that is left in the sediments underlying lakes or forests. Because what we can learn is so completely tied to the availability of that evidence and because of the complications inherent in interpreting it, it is worth discussing briefly the sources of data on which the reconstructions discussed in this chapter are based. Alaskan vegetation history is reconstructed primarily from fossil pollen grains that have been retrieved from lake sediments. The taxonomic composition of pollen in lake sediments is interpreted in terms of catchment- to regional-scale vegetation, taking into account differences among species in the amount of pollen that is produced. Temporal resolution is limited not only by time and effort (the number of samples that can be analyzed) but also by low rates of sedimentation, which limit the minimum sampling interval: 10- to 50-year resolution is typical of high-resolution records in Alaska. The timing of events in sediment-based reconstructions is determined by radiocarbon dating. Unless otherwise stated, dates here are converted from radiocarbon years to calendar years before present (yr B.P.; Bartlein et al. 1995).

Paleoclimatic information is derived from several sources. At regional and local spatial scales, past climate is typically reconstructed from proxy data—geologic, geomorphologic, and geochemical evidence of past climates. For example, long-term histories of lake levels provide a record of effective moisture changes, and time-series of stable-isotope ratios in sediments can be used to interpret effective moisture and/or temperature, depending on the particular system being studied. Records of eolian activity (e.g., ancient sand dunes or loess deposits) and evidence of permafrost (e.g., ice wedges) also provide information on past climate and, to some extent, disturbance processes. At a hemispheric scale, several major aspects of climate change since the end of the most recent glacial period appear as robust results in climate-model simulations. These aspects of climate change are both climatologically consistent and reflected in proxy-data records

Holocene-length records of disturbance also rely largely on evidence preserved in sediments. Although there are many disturbance agents in the boreal forest, most of them leave little record. Disturbances that act on larger spatial scales—those of a catchment to a region—have the best likelihood of being represented in sediment records. Such disturbances include fires, outbreaks of pests and pathogens (such as bark beetles), and extreme weather events (such as ice storms). The temporal resolution of Holocene records precludes the detection of intra-annual weather events and their effects. Fossil beetles, including bark beetles, have been retrieved from Holocene sediments (Elias 1992, Short et al. 1992), but the records are not detailed enough to tell us about stand-scale interactions or population densities. Fire, on the other hand, is acknowledged to play a key role in boreal forest disturbance dynamics, and potentially detailed evidence of long-term fire regimes is available from microscopic (<125-micron) and macroscopic (sieved; >125-micron) charcoal fragments retrieved from lake sediments.

Vegetation Change in the Holocene

One of the important goals of paleoecological research has been to understand how changes in climate and disturbance have affected the composition of the boreal forest. Reviews of Holocene vegetation history in Alaska published during the past two decades (Ager 1983, Ager and Brubaker 1985, Lamb and Edwards 1988, Anderson and Brubaker 1993, Anderson and Brubaker 1994, Edwards and Barker 1994, Anderson et al. 2004) suggest that three forest types have dominated interior Alaska during the Holocene (Table 5.1): open woodlands or scrub dominated by broadleafed species (primarily *Populus*) during the period 13,000 to 10,000 yr B.P., white spruce (*Picea glauca*) forests during the mid-Holocene (10,000 to 5,000 yr B.P.), and black spruce (*Picea mariana*) forests during the late Holocene (5,000 yr B.P. to present). Two major transitions have thus occurred during the Holocene: (1) from broad-leafed to spruce-dominated forests, approximately 10,000 years ago, and (2) from white to black spruce dominance, approximately 5,000 years ago. These transitions provide an opportunity to examine the relationship between major vegetation change and climate and disturbance.

Holocene Changes in Climate and Disturbance

Although some aspects of climate have varied continuously during the Holocene, several distinct periods can be recognized in paleoclimatic reconstructions from interior Alaska (Table 5.1). Early Holocene conditions were warm and dry compared to present conditions. Proxy data (Kaufman et al. 2004) and general circulation model (GCM) simulations (COHMAP 1988, Kutzbach et al. 1993, Bartlein et al. 1998) both indicate that conditions significantly warmer than present conditions were attained between 13,000 and 10,000 yr B.P. This reflects maximum values

Table 5.1. Summary of Holocene environmental change and inferred climate.

Age cal yr BP	Vegetation	Lake status	Inferred climate
12,000–10,000	A mosaic of deciduous trees and shrubs and grass-herb-dominated communities	Intermediate	Warm (slightly cooler to slightly warmer than present); drier than present
10,000–8,500	Expansion of white spruce in the eastern and central interior	Rising	Warmer than present; increasing moisture but drier than today
8,500–6,000	Expansion of alder in central and eastern interior	Near present	Warm and moist; precipitation slightly less than/close to modern levels
6,000–present	Expansion of black spruce; full complement of boreal forest taxa established	Fluctuating slightly above and below present values	Slight cooling toward present; precipitation similar to today

Source: Modified from Edwards et al. (2001b).

of summer insolation at high northern latitudes, related to variations in the Earth's solar orbit. Furthermore, episodes of lowered lake levels between ca. ~12,500 and 10,000 yr B.P suggest that dry conditions prevailed, at least in the second half of this period (Abbott et al. 2000, Hu et al. 1996, Barber and Finney 2000). In contrast, the mid-Holocene (10,000–5,000 years ago) remained warm, but moisture increased. About 10,000 years ago, a prolonged rise in lake levels began, indicating increasing effective moisture until ~ 4000 yr B.P. (Finney et al. 2000). Conditions in Beringia were still warmer than those at present during the growing season at 6000 yr B.P., primarily because summer insolation was still greater than at present (Kutzbach et al. 1993, Kaplan et al. 2003). However, as insolation approached modern values after 5,000 yr B.P., the climate system relaxed toward its current configuration—generally cooler summers and warmer winters than in the earlier part of the postglacial period (Bartlein et al. 1991). Drier intervals lasting several centuries within the late Holocene are inferred from isotopic ratios in carbonate sediments in the Brooks Range and in the Yukon (e.g., Anderson et al. 2001).

Reconstructions of variation in fire regimes during the Holocene suggest that changes in climate were accompanied by (and may, in fact, have caused) changes in disturbance. Data from the interior and from the northern Alaska Range include two low-resolution records (Hu et al. 1993, Hu et al. 1996) and several high-resolution records (Lynch et al. 2002, Lynch et al.2004, Franklin-Smith et al. 2004) of macroscopic charcoal. A low-resolution record of microscopic charcoal comes from Sithylemenkat Lake in central interior Alaska (Earle et al. 1996). Theory and observations suggest that macro-charcoal records fires within a lake catchment (i.e., local fires), whereas micro-charcoal integrates a regional picture of burning (Clark and Royall 1995, Whitlock and Millspaugh 1996).

The macroscopic records tend to show a lower incidence of local fires ~10,000 to 5500 yr B.P. (the white spruce zone) and a higher frequency of fires that produced greater amounts of charcoal since 5000 yr B.P. (the black spruce zone; see, for example, Dune Lake; Lynch et al. 2002; Fig. 5.1). The nature of the fire regime

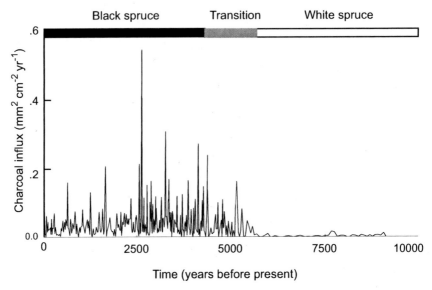

Figure 5.1. Holocene record of charcoal influx at Dune Lake, Alaska. Data from Lynch et al. (2002).

during the early Holocene deciduous period currently remains ambiguous. At Farewell Lake (Hu et al. 1996), Birch Lake, and Jan Lake (Franklin-Smith et al. 2004), macroscopic charcoal is recorded in sediments older than 10,000 yr B.P., but at other sites (e.g. Wien Lake; Hu et al. 1993) there is little evidence of charcoal in this period.

In contrast, the microscopic record shows the highest influx of charcoal to Sithylemenkat Lake during the early Holocene deciduous zone and the first part of the spruce zone (Earle et al. 1996). This implies that, although the amounts of fuel consumed within a catchment were probably small in the early Holocene (as indicated by the macroscopic charcoal records), there was nevertheless considerable burning on a regional level during the early Holocene. The fine particles represent regional atmospheric loading, and their values may also depend on how effectively they were incorporated into the regional circulation, as opposed to being rapidly "rained out" after fire; this would be a function of precipitation during the fire season. The sum of the charcoal evidence suggests that fires have occurred throughout the Holocene but that fire regimes have varied. In particular, in the earlier part of the Holocene, fires burned less severely, whereas severe—probably stand-replacing—fires typify the past 5000 years or so.

Maintenance of Broad-Leafed Dominance in the Early Holocene

Although evidence suggests that spruce was present in Alaska in the early Holocene (Brubaker et al. 2005), it became a dominant taxon only 10,000 years ago.

Why were broad-leafed species, which are now largely restricted to early successional habitats, the dominant taxa in the early Holocene boreal forest? Two hypotheses may explain the dominance of broad-leafed species during this time.

First, low-intensity but frequent fires, promoted by vegetation that produced low fuel loadings and by a dry climate, may have prevented successful spruce establishment, whereas trees such as aspen (*Populus tremuloides*) and balsam poplar (*P. balsamifera*) tolerated the fires via their ability to propagate vegetatively by sprouting. The increase in moisture at ~10,000 yr B.P. would thus have reduced fire frequency sufficiently to allow spruce to expand.

Alternatively, drought may have directly excluded spruce until moisture increased ~10,000 years ago. Evidence for climatic control comes from modern climate-vegetation relationships. White spruce growth has slowed in response to drought stress in interior Alaska (Barber et al. 2000), and there is some indication that drought stress may affect white spruce even at treeline (Jacoby and D'Arrigo 1995, Lloyd and Fastie 2002). White spruce regeneration fails due to drought beyond the present prairie-forest border in Canada (Hogg and Schwarz 1997), further suggesting the potential for dry conditions to limit spruce success.

It should be possible to distinguish between these hypotheses from sediment records of lake level, charcoal, and pollen abundances. If fire was the proximate cause of broad-leafed dominance, then a drop in fire frequency (as indicated by the influx of charcoal into lake sediments) should slightly *precede* the spruce rise. If climate was the proximate cause, then fire frequency, if it changed at all, would increase or decrease after the spruce rise. In both scenarios, however, the increase in moisture is postulated to be the ultimate cause of the change in vegetation.

At lake sites where it is possible to compare sediment and pollen changes across the spruce rise, moisture levels apparently began to rise just prior to the spruce expansion (e.g., Dune Lake; Bigelow 1997), supporting the assumption in both hypotheses that climate is the ultimate cause of the transition from hardwoods to white spruce forests. To date, parallel records of lake level, charcoal, and pollen are insufficient to define clearly the role of fire in the transition to white spruce dominance. Macro-charcoal records show no consistent pattern at the spruce rise (Hu et al. 1993, Hu et al. 1996, Franklin-Smith et al. 2004), while the micro-charcoal record shows no diminution of regional fires until about 1000 years later (Earle et al. 1996).

These interactions have a bearing on possible future dynamics at the forest-steppe ecotone of the boreal forest, which is represented locally in interior Alaska and in the Yukon on steep, south-facing slopes. Here, first white spruce and then aspen give way to azonal grassland on the most drought-affected slopes (Edwards and Armbruster 1989, Wesser and Armbruster 1991, Lloyd et al. 1994, Viereck et al. 1992; Chapter 6). Stand and landscape simulation models predict expansion of these xeric communities within the interior boreal forest region in some global warming scenarios (Bonan et al. 1990, Starfield and Chapin 1996). Indeed, the association between loss of broad-leafed dominance and increased moisture approximately 10,000 years ago suggests that the prevalence of some broad-leafed-dominated community types may be sensitive to changes in effective moisture supply.

Establishment of Black Spruce Dominance

As with the earlier transition from broad-leafed species to white spruce forests, the transition from white spruce–dominated communities in the mid-Holocene to black spruce–dominated communities in the second half of the Holocene may reflect one of two processes—or their interaction. First, changes in climate may have led to alterations in fire regime that promoted the relatively fire-tolerant black spruce over the less tolerant white spruce. A climatic basis for the shift toward more intense fires at that time is, however, difficult to argue because, from our understanding of the climatology of fire during the twentieth century (e.g., Johnson 1992), a shift toward cooler, moisture conditions would be expected to decrease fire frequency and intensity. It is possible that the explanation lies in a change in the dominant season of precipitation. In eastern Canada, Carcaillet and Richard (2000) explain an increase in fires at a time of increasing moisture by a shift to wet winters that more than compensated for drier summers. In Alaska, some change in the seasonality of precipitation is possible; winters likely became warmer in the late Holocene (more snow) and summers cooler (less convective rain), but even today, the majority of precipitation falls in the summer and early autumn. Thus, the hypothesis that drier summers—despite an overall increase in precipitation—favored a fire regime that in turn allowed black spruce to dominate is feasible, if perhaps unlikely. Second, black spruce dominance may be a direct result of the shift in climate toward cooler, moister conditions at ~5,000 yr B.P.; this climate change may also have had an indirect effect via the slow development of cold, waterlogged substrates, from which white spruce would have been excluded (Chapter 6).

The high-resolution pollen and charcoal record from Dune Lake (Bigelow 1997, Lynch et al. 2002; Figure 5.1) suggests that regional black spruce populations were increasing prior to the shift in fire frequencies. The rather rapid shift in fire frequency recorded at Dune Lake contrasts with the slower rise in black spruce values recorded by the pollen, which spans the charcoal rise. This may reflect the different spatial scales at which the pollen and charcoal are recording the environment. Whereas the vegetation record is regional and integrates changes in the boreal forest for tens to hundreds of square kilometers around Dune Lake, the charcoal largely reflects events in the Dune Lake catchment; the shift to black spruce dominance immediately around the lake, and hence the increased likelihood of fire, probably occurred over decades or a century or two at most. The data tend, therefore, to support the hypothesis that the vegetation change drove the change in fire regime. Furthermore, recently obtained records indicate that the shift in fire regime paralleled the shift to black spruce fairly consistently in different locations but that these events occurred over a spread of dates in the mid-Holocene (Lynch et al. 2004). This argues against a regional climate change that would have driven a synchronous change in fire regime and further points to the vegetation shift being a consequence of a site-dependent, climate-substrate interaction (Chapter 6).

Summary: Climate, Disturbance, and Holocene Vegetation Change

On Holocene time scales of centuries to millennia, there is a clear role for climate as a driver of major changes in Alaskan boreal forest composition. The dominance of broad-leafed species in the early Holocene, despite evidence for spruce being present in Beringia, can be attributed to the aridity of that time period: the expansion of spruce appears to have closely followed an increase in effective moisture at approximately 10,000 years ago. The subsequent expansion of black spruce seems to occur soon after a shift to cooler and moister conditions in the late Holocene. Major vegetation transitions are thus associated with large changes in both temperature and precipitation; data indicate that white spruce in particular may be highly sensitive to changes in available moisture. The degree to which fire disturbance interacted with climatic effects to drive change in the early Holocene remains unclear, as we do not yet have enough high-resolution, Holocene-length charcoal records.

When comparing events in the Alaskan forest with changes that occurred in other parts of the circumboreal zone, it should be remembered that both past climate changes and the composition of the species pool (and thus the ecology of key taxa) vary considerably among regions. Parallel shifts in dominant forest types may actually reflect responses to different climate forcing; conversely, similar climate trends may evoke a range of responses. For example, the transition from birch-dominated to pine-dominated forests in Fennoscandia has been attributed to increasingly dry conditions between 9,000 and 5,000 years BP (Seppä and Hammarlund 2000)—but *Pinus sylvestris* has a moisture response different from that of *Picea glauca*. In eastern Siberia, the spread of stone pine (*Pinus pumila*) into deciduous forest dominated by larch (*Larix*) and hardwoods appears to parallel the response to increasing moisture shown by spruce on the Alaskan side of the Bering Strait (Edwards et al. 2005). However, this species of shrub pine is sensitive to winter snow cover, which likely increased at this time in Siberia (Lozhkin et al. 1993). The apparent moisture sensitivity of boreal vegetation in Alaska is probably not unique; a shift from spruce-dominated to spruce- and pine-dominated forests in the early Holocene in eastern North America may have reflected, at least in part, drier climate conditions (Clark et al. 1996). This moisture response is worth noting, for the growth and distribution of boreal forest taxa are most often associated with temperature, rather than moisture.

The interplay of vegetation and climate with disturbance may similarly generate a range of consequences. For example, Payette and Gagnon (1985) show that the combined effect of climatic cooling and fire disturbance in the eastern Canadian forest-tundra resulted in fragmentation and a trend to deforestation. On the other hand, late-Holocene cooling in Alaska has seen the opposite trend—an expansion of the coniferous forest westward and no trend toward deforestation. Although evidence currently suggests that vegetation change drove a change in fire regime during the mid-Holocene in Alaska, climate is considered directly responsible for changes in fire frequency in the eastern Canadian boreal forest (Carcaillet and Richard 2000, Carcaillet et al. 2001). These differences suggest that the relative importance of climate and vegetation as controls over fire occurrence vary regionally and that it is probably unwise to generalize for the boreal zone as a whole.

Long-Term Perspectives on Disturbance Regimes

The Holocene history of the boreal forest shows a clear imprint of climate, but the paucity of Holocene-length records of disturbance regimes makes it difficult to assess how disturbance regimes have changed over time and the degree to which changes in vegetation cause or are caused by changes in disturbance regimes. The long-term history of disturbance, particularly by fire, is better known for the very late Holocene (the past 200–300 years) because of the availability of reconstructions based on tree rings.

Although the role of fire is most commonly recognized in interior Alaska, paleoecological studies have recorded evidence of recurrent fires in the past 1,000 years from near-maritime forests to the arctic treeline. In addition to the upland (i.e., nonfloodplain) spruce forests in interior Alaska, where the role of fire has been well studied, recurrent fires have been identified on the Kenai Peninsula in south-central Alaska (DeVolder 1999, Lynch et al. 2002), in boreal floodplain forests (Mann et al. 1995 on the Tanana River, Timoney et al. 1997 on the Peace River in northwestern Canada), and at alpine and arctic treeline (Fastie and Lloyd, unpublished data). Fire must therefore be considered a ubiquitous disturbance in the boreal forest (Chapter 17). In this section, we explore evidence for spatial and temporal variability in disturbance regimes in the boreal forest: do fire regimes vary among regions or community types? How has human activity modified fire regimes in Alaska?

Data Sources

Of all the disturbances that affect the boreal forest, fire leaves the most complete paleoecological record, so the discussion in this chapter concentrates on disturbance by fire. Fire history in Alaska may be inferred from current forest stand ages, fire-scarred or fire-killed trees (Fig. 5.2), and charcoal in soils and lake sediments. Each of these methods has limitations in terms of the fire history it reveals. Forest stand ages provide data on recent disturbance history (e.g., Yarie 1981) but can be biased by changes in disturbance regime. In Alaska, for example, estimates of fire intervals from stand ages are almost certainly biased by anthropogenic burning during the beginning of the gold-mining era (1890–1900) and thus cannot be used to infer natural (pre-European) fire regimes. Fire-scarred trees and fire-killed trees provide a temporally and spatially precise record of fire, but light surface fires may occur without killing or scarring trees, so estimates of fire frequency from fire-scarred and fire-killed trees may overestimate the length of fire intervals. Finally, burned organic matter may preserve an in situ record of fire in areas where organic matter accumulates, and airborne (or waterborne) particles of charcoal can become incorporated in lake sediments or peats. The temporal extent of sediment charcoal records is long, extending back throughout the Holocene in some cases, and may

Figure 5.2. Cross-section of a fire-scarred black spruce. The date of the fire can be obtained by determining the date in which the last complete ring was formed prior to the scar.

thus be the best source of information on pre-European fire regimes. The spatial precision of lake-sediment charcoal records, however, is somewhat limited, so resolution of intrawatershed variation in fire regime is difficult.

Spatial Variability in Historical Fire Regimes

Although fire apparently occurs throughout the boreal forest, its frequency and severity vary with topography (e.g., proximity to fire breaks such as large rivers), climate, and species composition. In particular, fires may be infrequent in floodplains, where proximity to rivers, which may act as fire breaks, and relatively nonflammable early successional vegetation may reduce fire occurrence.

Widely varying estimates have been obtained for fire intervals in upland forests; this variability probably reflects both real spatial and temporal variability in fire occurrence and methodological differences in reconstruction of fire history. Studies of current forest age structure tend to estimate higher fire frequencies than do studies that employ fire-scarred trees or charcoal in lake sediments. For example, the distribution of stand ages in the Porcupine River drainage in interior Alaska indicates that recent intervals between fires may be as low as 36 years in black spruce forests and 113 years in white spruce forests (Yarie 1981). Similarly, the age distribution of aspen forests near Denali National Park indicates that the interval between fires has been as low as 40–60 years (Mann and Plug 1999). In contrast, fire scars and stand ages estimated from large samples of tree ages suggest intervals of between 100 and >250 years between fires in upland forests in interior Alaska over the past 200–300 years (Fastie et al. 2002). Mean fire interval during the past 2,400 years was estimated from charcoal in lake sediments to be 198 ± 90 years for interior Alaska (Fig. 5.3). New paleoecological data thus suggest that natural fire intervals in upland forests may be in excess of 200 years, an interval that is substantially longer than the current distribution of forest stand ages indicates.

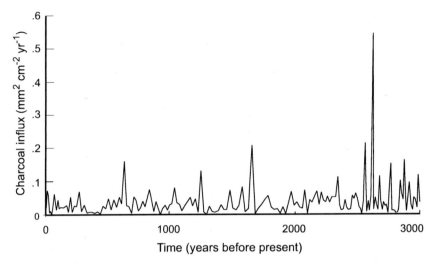

Figure 5.3. Dune Lake charcoal record for the late Holocene. Data from Lynch et al. (2002).

There is clear evidence that fires have repeatedly initiated secondary succession on active river floodplains. Mann et al. (1995) documented the occurrence of fires in forests on the Tanana River floodplain by examining fire scars and forest age structures. They found evidence for at least two widespread fires, in 1816 and in 1910, that affected both white and black spruce forests throughout the study area (including both the meander belt and the backswamp). The role of fire as a driver of forest dynamics may not, therefore, be fundamentally different on the floodplain and in upland systems. In both systems, the paleoecological records clearly indicate that stand-replacing fires repeatedly initiate secondary succession and suggest that these fires are the primary factor regulating population turnover in spruce forests. Floodplain spruce forests are probably just as flammable as upland forests and are likely ignited at a similar rate. There is at present no basis in the paleoecological data for concluding that fire is more or less common on the floodplain than in the uplands, but we emphasize that the number of published fire histories is few and that natural fire regimes thus remain very poorly known for both upland and floodplain forests.

Temporal Variability in Historical Fire Regimes

Two factors are likely to feature prominently in an analysis of changes in fire regime over time: climate and humans. The role of climate in driving changes in fire regime in Alaska remains unclear, although relationships between climate variation and fire occurrence have been identified in other boreal regions. For example, increased burning during a period of warmer temperatures between A.D. 900 and A.D. 1100 was identified from sediment charcoal in eastern Finland (Pitkänen and Huttunen 1999), and decreases in fire frequency in eastern Canada have been attributed to a reduced frequency of drought years in the twentieth century (Bergeron 1991, Bergeron and Archambault 1993, Bergeron 1998).

Human activities have clearly exerted a strong influence on fire regimes in Alaska. Historical records and paleoecological studies both indicate that Alaskan fire regimes were altered by the activity of Russian and European explorers and gold miners in Alaska and northwestern Canada. At Caribou Poker Creek Research Watershed (CPCRW), in interior Alaska, Fastie et al. (2002) identified a cluster of fires in the early 1900s, temporally associated with the arrival of gold miners in nearby valleys. The onset of gold mining in the CPCRW region was followed immediately by a swarm of severe, stand-replacing fires in 1902, 1909, 1924, and 1925 that collectively burned large areas of the studied watersheds. Although there is no direct evidence that these fires were started by people, there is clear evidence (in the form of axe-cut stumps dating to the early 1900s) that humans were present in and actively using the forests in CPCRW. The temporal relationship between the swarm of fires and the arrival of miners thus strongly suggests that human ignition of at least some of the early-twentieth-century fires is likely. Evidence that fire regimes may have been strongly influenced by European settlement has also been found in south-central Alaska, where fire frequency increased beginning in the mid-1800s (DeVolder 1999), coincident with the arrival of Russian explorers. Paleoecological studies thus seem to provide consistent support for the hypothesis that the arrival of Russian explorers and, later, American and European gold miners was accompanied by fires.

This conclusion has important implications for our understanding of fire in Alaskan boreal forests. Much of what we know about boreal fire regimes in Alaska is based on an analysis of the current age structure of forests. Our evidence is thus strongly influenced by the post-European era, in which fire regimes reflect an interaction between natural dynamics and anthropogenic influence. The development of more long-fire histories from fire scars and sediment charcoal will help us determine whether pre-European fire regimes differ from the fire regimes that have characterized the boreal forest since the arrival of Europeans and, if so, in what manner.

Conclusions

During the course of the Holocene, both climate and disturbance have been important, and probably interactive, drivers of vegetation change in the Alaskan boreal forest. The Holocene record of changes in vegetation, climate, and disturbance in Alaska yields two general insights into how the boreal forest may be affected if the recent warming trend in Alaska continues. First, although the role of disturbance in driving large-scale vegetation change in the boreal forest remains poorly understood, existing data suggest that climate, disturbance, and vegetation may interact in complex feedback loops (Fig. 5.4). There is good evidence in the Holocene history of vegetation and fire that changes in vegetation can affect disturbance regime; vegetation transitions in the early and late Holocene, for example, appear to have led to subsequent change in the fire regime. Changes in disturbance regime thus seem likely to accompany future warming-induced vegetation change in Alaska,

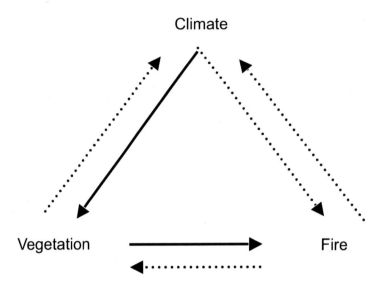

Figure 5.4. Potential linkages among climate, fire, and vegetation in the boreal forest. Pathways for which there is good paleoecological evidence are indicated by solid lines; those that are likely but for which there is *not* good evidence at present are indicated by dashed lines.

and shifts in fire intensity and/or frequency may feedback on climate by altering both carbon emissions and carbon uptake from the boreal forest.

The second insight into possible future change in Alaska's boreal forest emerges from the conclusion that, during the Holocene, different components of the boreal landscape may have differed markedly in their sensitivity to climate variation. Black spruce, for example, has remained a stable landscape dominant despite the climatic changes that have occurred during the second half of the Holocene. Black spruce may therefore be resistant to future vegetation changes. However, we do not have a record of black-spruce–dominated landscapes in the early or mid-Holocene, when temperatures were higher and conditions may be closer analogues of future warming. On the other hand, the apparent sensitivity of white spruce to water deficits, both today and in the early Holocene, suggests that upland facies of the boreal forest dominated by white spruce may be more sensitive to climatic thresholds and thus likely to undergo marked changes with future warming.

Although patterns of vegetation change during the Holocene are thus relatively well known for a large number of sites in Alaska, the mechanisms of vegetation change remain uncertain. Future research on the Holocene history of the boreal forest should focus on two primary questions. First, how do disturbance regimes vary across time and space? Modern ecological studies have revealed the importance of disturbance as a control over ecological processes in the boreal forest, but we know little about how the occurrence of disturbance changes over time or across space. Second, how do fire, climate, and vegetation covary over time, and what is the direction of causality among them? An increase in the number of paleoecological studies that utilize multiple proxy records of ecological change (e.g., isotopic records, pollen, macrofossils, and charcoal) will allow the nature of these relationships to be clarified.

References

Abbott, M. B., B. P. Finney, M. E. Edwards, and K. R. Kelts. 2000. Paleohydrology of Birch Lake, central Alaska: A multiproxy approach to lake-level records. Quaternary Research 53:154–166.

Ager, T. A. 1983. Holocene vegetation history of Alaska. Pages 128–140 *in* H. E. Wright, editor. Late Quaternary environments of the United States. Vol. 2. The Holocene. University of Minnesota Press, Minneapolis.

Ager, T. A., and L. B. Brubaker. 1985. Quaternary palynology and vegetational history of Alaska. Pages 353–384 *in* V. M. Bryant Jr. and R. G. Holloway, editors. Pollen Records of Late-Quaternary North American Sediments. American Association of Stratigraphic Palynologists Foundation, Austin, TX.

Anderson, L., M. B. Abbott, and B. P. Finney. 2001. Holocene climate inferred from oxygen isotope ratios in lake sediments, central Brooks Range, Alaska. Quaternary Research 55:313–321.

Anderson, P. M., and L. B. Brubaker. 1993. Holocene vegetation and climate histories of Alaska. Pages 386–400 *in* H. E. Wright, J. E. Kutzbach, T. Webb, W. F. Ruddiman, F. A. Street-Perrott, and P. J. Bartlein. Global Climates since the Last Glacial Maximum. University of Minnesota Press, Minneapolis.

Anderson, P. M., and L. B. Brubaker. 1994. Vegetation history of northcentral Alaska: A mapped summary of Late-Quaternary pollen data. Quaternary Science Reviews 13:71–92.

Anderson, P. M., M. E. Edwards, and L. B. Brubaker. 2004. Results and paleoclimate implications of 35 years of paleoecological research in Alaska. Pages 427–440 *in* A. E. Gillespie, S. C. Porter, and B. F. Atwater, editors. The Quaternary Period in the United States. Developments in Quaternary Science. Elsevier, New York.

Barber, V. A., and B. P. Finney. 2000. Late Quaternary paleoclimatic reconstructions for interior Alaska based on paleolake-level data and hydrologic models. Journal of Paleolimnology 24:29–41.

Barber, V., G. P. Juday, and B. Finney. 2000. Reduced growth of Alaskan white spruce in the twentieth century from temperature-induced drought stress. Nature 405:668–673.

Bartlein, J. P., P. M. Anderson, M. E. Edwards, and P. F. McDowell. 1991. A framework for interpreting paleoclimatic variations in eastern Beringia. Quaternary International 10:73–83.

Bartlein, P. J., K. H. Anderson, P. M. Anderson, M. E. Edwards, C. J. Mock, R. S. Thompson, R. S. Webb, T. Webb III, and C. Whitlock. 1998. Paleoclimate simulations for North America over the past 21,000 years: Features of the simulated climate and comparisons with paleoenvironmental data. Quaternary Science Reviews 17:549–585.

Bartlein, P. J., M. E. Edwards, S. L. Shafer, and E. D. Barker Jr. 1995. Calibration of radiocarbon ages and the interpretation of paleoenvironmental records. Quaternary Research 44:417–424.

Bergeron, Y. 1991. The influence of island and mainland lakeshore landscapes on boreal forest fire regimes. Ecology 72:1980–1992.

Bergeron, Y. 1998. Les conséquences des changements climatiques sur la fréquence des

feux et la composition forestière à sud-ouest de la forêt boréale Québécoise. Géographie physique et Quaternaire 52:167–173.
Bergeron, Y., and S. Archambault. 1993. Decreasing frequency of forest fires in the southern boreal zone of Quebec and its relation to global warming since the end of the Little Ice Age. The Holocene 3:255–259.
Bigelow, N. H. 1997. Late-Quaternary vegetation and lake-level changes in central Alaska. Ph.D. Dissertation, University of Alaska Fairbanks.
Bonan, G. B., H. H. Shugart, and D. L. Urban. 1990. The sensitivity of some high-latitude boreal forests to climatic parameters. Climatic Change 16:9–29.
Brubaker, L. B., P. M. Anderson, M. E. Edwards, and A. V. Lozhin. 2005. Beringia as a glacial refugium for boreal trees and shrubs: New perspectives from mapped pollen data. Journal of Biogeography 32:833–848.
Carcaillet, C., and P. J. H. Richard. 2000. Holocene changes in seasonal precipitation highlighted by fire incidence in eastern Canada. Climate Dynamics 16:549–559.
Carcaillet, C., Y. Bergeron, P. J. H. Richard, B. Frechette, S. Gauthier, and Y. T. Prairie. 2001. Change of fire frequency in the eastern Canadian boreal forests during the Holocene: Does vegetation composition or climate trigger the fire regime? Journal of Ecology 89:930–946.
Clark, J. S., and P. D. Royall. 1995. Particle-size evidence for source areas of charcoal accumulation in Late Holocene sediments of eastern North American lakes. Quaternary Research 4: 80–89.
Clark, J. S., P. D. Royall, and C. Chumbley. 1996. The role of fire during climate change in the eastern deciduous forest at Devil's Bathtub, New York. Ecology 77:2148–2166.
COHMAP. 1988. Climatic changes of the last 18,000 years: Observations and model simulations. Science 241:1043–1052.
Colinvaux, P. A. 1964. The environment of the Bering Land Bridge. Ecological Monographs 34:297–329.
DeVolder, A. D. 1999. Fire and climate history of lowland black spruce forests, Kenai National Wildlife Refuge, Alaska. M.S. Thesis, Northern Arizona University, Flagstaff.
Earle, C. J., L. B. Brubaker, and P. M. Anderson. 1996. Charcoal in northcentral Alaskan lake sediments: Relationships to fire and late-Quaternary vegetation history. Review of Paleobotany and Palynology 92:83–95.
Edwards, M. E., and W. S. Armbruster. 1989. A tundra-steppe transition on Kathul Mountain, Alaska, U.S.A. Arctic and Alpine Research 21:296–394.
Edwards, M. E. and E. D. Barker. 1994. Climate and vegetation in northern Alaska 18,000 yr B.P.–present. Palaeogeography, Palaeoclimatology, Palaeoecology 109:127–135.
Edwards, M.E., L. B. Brubaker, A. V. Lozhkin, and P. M. Anderson. 2005. Structurally novel biomes: A response to past warming in Beringia. Ecology 86:1696–1703.
Edwards, M. E., L. B. Brubaker, P. Anderson, B. P. Finney, and A. Lozhkin. 2001a. Mapped pollen and lake-levels indicate precipitation controlled Holocene expansion of boreal evergreen forest in Beringia. Ecological Society of America Annual Meeting, Program and Abstracts, Madison, WI.
Edwards, M. E., C. J. Mock, B. P. Finney, V. A. Barber, and P. J. Bartlein. 2001b. Potential analogues for paleoclimatic variations in eastern interior Alaska during the past 14,000 yr: Atmospheric-circulation controls of regional temperature and moisture responses. Quaternary Science Reviews 20:189–202.
Elias, S. C. 1992. Late Quaternary beetle faunas of southwest Alaska: Evidence of a refugium for mesic and hygrophilous species. Arctic and Alpine Research 24:133–144.
Fastie, C. L., A. Lloyd, and P. Doak. 2002. Fire history and postfire forest development in

an upland watershed of interior Alaska. Journal of Geophysical Research 108:8150, doi:10.1029/2001JD000570 [printed in 108(D1), 2003].
Finney, B., M. Edwards, M. Abbott, V. Barber, L. Anderson, and M. Rohr. 2000. Holocene precipitation variability in interior Alaska. Thirtieth Arctic Workshop Program and Abstracts, INSTAAR, Boulder, CO.
Franklin-Smith, L., M. E. Edwards, A. P. Krumhardt, and B. P. Finney. 2004. Interaction of boreal forest vegetation, fire and climate in the Holocene of Alaska. Ecological Society of America, Eighty-ninth Annual Meeting Abstracts, Portland, Oregon. p.166.
Hogg, E. H., and A. G. Schwarz. 1997. Regeneration of planted conifers across climatic moisture gradients on the Canadian prairies: Implications for distribution and climate change. Journal of Biogeography 24:527–534.
Hopkins, D. M. 1967. The Bering Land Bridge. Stanford University Press, Stanford.
Hu, F. S., L. B. Brubaker, and P. M. Anderson. 1993. A 12,000-year record of vegetation change and soil development from Wien Lake, central Alaska. Canadian Journal of Botany 71:1133–1142.
Hu, F. S., L. B. Brubaker, and P. M. Anderson. 1996. Boreal ecosystem development in the northwestern Alaska Range since 11,000 yr B.P. Quaternary Research 45:188–201.
Jacoby, G. C., and R. D'Arrigo. 1995. Tree-ring width and density evidence of climatic and potential forest change in Alaska. Global Biogeochemical Cycles 9:227–234.
Johnson, E. 1992. Fire and Vegetation Dynamics: Studies from the North American Boreal Forest. Cambridge University Press, Cambridge.
Kaplan, J. O., N. H. Bigelow, I. C. Prentice, S. P. Harrison, P. J. Bartlein, T. R. Christensen, W. Cramer, N. V. Matveyeva, A. D. McGuire, D. F. Murray, V. Y. Razzhivin, B. Smith, D. A. Walker, P. M. Anderson, A. A. Andreev, L. B. Brubaker, M. E. Edwards, and A. V. Lozhkin. 2003. Climate change and Arctic ecosystems. 2. Modeling, paleodata-model comparisons, and future projections. Journal of Geophysical Research-Atmospheres, 108:8171, doi:10.1029/2002JD002559 [printed in 108(D19), 2003].
Kaufman, D. S., T. A. Ager, N. J. Anderson, P. M. Anderson, J. T. Andrews, P. J. Bartlein, L. B. Brubaker, L. L. Coats, L. C. Cwynar, M. L. Duvall, A. S. Dyke, M. E. Edwards, W. R. Eisner, K. Gajewski, A. Geirsdottir, F. S. Hu, A. E. Jennings, M. R. Kaplan, M. W. Kerwin, A. V. Lozhkin, G. M. MacDonald, G. H. Miller, C. J. Mock, W. W. Oswald, B. L. Otto-Bliesner, D. F. Porinchu, K. Ruhland, J. P. Smol, E. J. Steig, and B. B. Wolfe. 2004. Holocene thermal maximum in the western Arctic (0–180°W). Quaternary Science Reviews 23:529–560.
Kutzbach, J. E., P. J. Guetter, P. J. Behling, and R. Selin. 1993. Simulated climate changes: results of the COHMAP climate model experiments. Pages 24–93 in H. E. Wright Jr., J. E. Kutzbach, T. Webb III, W. F. Ruddiman, F. A. Street-Perrott, and P. J. Bartlein. Global Climates since the Last Glacial Maximum. University of Minnesota Press, Minneapolis.
Lamb, H. F., and M. E. Edwards. 1988. The Arctic. Pages 519–555 in B. Huntley and T. Webb III, editors. Vegetation History. Kluwer Academic Publishers, Dordrecht.
Lloyd, A. H., and C. L. Fastie. 2002. Spatial and temporal variability in the growth and climate response of treeline trees in Alaska. Climatic Change 52:481–509.
Lloyd, A. H., W. S. Armbruster, and M. E. Edwards. 1994. Ecology of a steppe-tundra gradient in interior Alaska. Journal of Vegetation Science 5:897–912.
Lozhkin, A. V., P. M. Anderson, W. R. Eisner, L. G. Ravako, D. M. Hopkins, L. B. Brubaker, P. A. Colinvaux, and M. C. Miller. 1993. Late Quaternary lacustrine pollen records from southwestern Beringia. Quaternary Research 39:314–324.
Lynch, J. A., J. S. Clark, N. H. Bigelow, M. E. Edwards, and B. P. Finney. 2002. Geographical and temporal variations in fire history in boreal ecosystems of Alaska. Jour-

nal of Geophysical Research-Atmospheres 108:8152, doi:10.1029/2001JD000332 [printed in 108(D1), 2003].
Lynch, J. A., J. L. Hollis, and F. S. Hu. 2004. Climatic and landscape controls of the boreal forest fire regime: Holocene records from Alaska. Journal of Ecology 92:477–489.
Mann, D. H., and L. J. Plug. 1999. Vegetation and soil development at an upland taiga site, Alaska. Ecoscience 6:272–285.
Mann, D. H., C. L. Fastie, E. L. Rowland, and N. Bigelow. 1995. Spruce succession, disturbance, and geomorphology on the Tanana River floodplain, Alaska. Ecoscience 2:184–199.
Matthews, J. V., Jr. 1974. Quaternary environments at Cape Deceit (Seward Peninsula, Alaska): Evolution of a tundra ecosystem. Geological Society of America Bulletin 85:1353–1385.
McDowell, P. F., and M. E. Edwards. 2001. Evidence of Quaternary climatic variations in a sequence of loess and related deposits at Birch Creek, Alaska: Implications for the Stage 5 climatic chronology. Quaternary Science Reviews 20:63–76.
Payette, S., and R. Gagnon. 1985. Late Holocene deforestation and tree regeneration in the forest-tundra of Québec. Nature 313:570–572.
Pitkänen, A., and P. Huttunen. 1999. A 1300-year forest-fire history at a site in eastern Finland based on charcoal and pollen records in laminated lake sediment. The Holocene 9:311–320.
Seppä, H., and D. Hammarlund. 2000. Pollen-stratigraphical evidence of Holocene hydrological change in northern Fennoscandia supported by independent isotopic data. Journal of Paleolimnology 24:69–79.
Short, S. K., S. A. Elias, C. F. Waythomas, and N. E. Williams. 1992. Fossil pollen and insect evidence for postglacial environmental-conditions, Nushagak and Holitna lowland regions, southwest Alaska. Arctic 45:381–392.
Starfield, A. M., and F. S. Chapin III. 1996. A dynamic model of arctic and boreal vegetation change in response to global changes in climate and land-use. Ecological Applications 6:842–864.
Timoney, K. P., G. Peterson, and R. Wein. 1997. Vegetation development of boreal riparian plant communities after flooding, fire, and logging, Peace River, Canada. Forest Ecology and Management 93:101–120.
Viereck, L. A., C. T. Dyrness, A. R. Batten, and K. J. Wenzlick. 1992. The Alaska Vegetation Classification. U.S. Department of Agriculture, Forest Service, Washington, DC.
Wesser, S. D., and W. S. Armbruster. 1991. Species distribution controls across a forest-steppe transition—A causal model and experimental test. Ecological Monographs 61:323–342.
Whitlock, C., and S. H. Millspaugh. 1996. Testing the assumptions of fire-history studies: An examination of modern charcoal accumulation in Yellowstone National Park, USA. The Holocene 6:7–15.
Yarie, J. 1981. Forest fire cycles and life tables: A case study from interior Alaska. Canadian Journal of Forest Research 11:554–562.

Part II

Forest Dynamics

6

Floristic Diversity and Vegetation Distribution in the Alaskan Boreal Forest

F. Stuart Chapin III
Teresa Hollingsworth
David F. Murray
Leslie A. Viereck
Marilyn D. Walker

Introduction

Although modern forests have occupied interior Alaska for only 13,000 years, their floristic composition and patterns of distribution have remained relatively stable for the past 5,000 years (Chapter 5). Here, at the current northern limit of forests, severe environmental conditions have prevented migration of new species from the south. The Bering Sea has isolated Alaska from a taxonomically distinct flora in Eurasia. Mountains to the north (Brooks Range) and south (Alaska Range) of interior Alaska have restricted the potential for latitudinal shifts of species in response to millennial-scale variations in climate. The stability of the Alaskan vegetation mosaic for the past 5,000 years contrasts with the substantial vegetation movements that have occurred in the eastern Canadian boreal forest (Bergeron et al. 2004). Because of this long-term stability, the distribution of Alaska's boreal vegetation reflects a clear imprint of current environmental patterns on the landscape. This strong link between current environment and vegetation facilitates the use of state factors (Chapter 1) as a framework for describing floristic patterns, although barriers to past migration could lead to rapid vegetation change, if new species were to arrive in Alaska. In this chapter we describe the major patterns of diversity and distribution in interior Alaska forests and discuss their potential future changes. The common and scientific names of species mentioned in this chapter are given in Table 6.1.

Table 6.1. Scientific and common names of plants mentioned in text.

Scientific name	Common name
Current tree species	
Betula neoalaskana (B. papyrifera auct)[1]	Alaska paper birch
Larix laricina	larch, tamarack
Picea glauca	white spruce
Picea mariana	black spruce
Populus balsamifera	balsam poplar
Populus tremuloides	aspen
Miocene tree flora	
Abies	fir
Carya	hickory
Castanea	chestnut
Fagus	beech
Juglans	walnut
Metasequoia	dawn redwood
Pinus	pine
Pterocarya	wingnut
Quercus	oak
Sequoia	California redwood
Tsuga	hemlock
Exotics to interior Alaska	
Melilotus	sweetclover
Rhinanthus	yellowrattle
Taraxacum	dandelion
Shrubs, grasses, and forbs	
Alnus incana subsp. tenuifolia	thinleaf alder
Alnus viridis subsp. fruticosa (A. crispa auct)	green alder
Andromeda polifolia	bog rosemary
Arctostaphylos uva-ursi	bearberry, kinnikinnik
Arctous (Arctostaphylos) alpina	alpine bearberry
Arctous (Arctostaphylos) rubra	red-fruit bearberry
Artemisia alaskana	Alaska wormwood
Artemisia frigida	wild prairie sagewort, fringed sagebrush
Artemisia laciniatiformis	Siberian wormwood
Betula glandulosa	resin birch, bog birch
Betula nana	dwarf arctic birch
Calamagrostis canadensis	bluejoint
Calamagrostis purpurascens	purple reedgrass
Carex bigelowii	Bigelow's sedge
Carex duriuscula	needleleaf sedge
Cornus canadensis	bunchberry, dwarf dogwood
Corydalis sempervirens	rock harlequin
Cryptantha shacklettiana	Shacklett's cryptantha
Empetrum nigrum	black crowberry
Epilobium angustifolium	fireweed
Equisetum arvense	field horsetail
Equisetum fluviatale	swamp horsetail
Equisetum pratense	meadow horsetail

Table 6.1. (Continued)

Scientific name	Common name
Equisetum scirpoides	dwarf scouringrush
Equisetum sylvaticum	woodland horsetail
Erigeron ochroleucus	bluff fleabane
Eriogonum flavum	alpine golden buckwheat
Eriophorum vaginatum	tussock cottongrass
Festuca lenensis	tundra fescue
Galium boreale	northern bedstraw
Geocaulon lividum	false toadflax, northern commandra
Juniperus communis	common juniper
Ledum decumbens	marsh Labrador tea
Ledum groenlandicum	western Labrador tea
Linnaea borealis	twinflower
Lycopodium annotium	clubmoss
Menyanthes trifoliata	buckbean
Orthilia secunda	sidebells wintergreen
Oxycoccus microcarpa	bog cranberry
Pedicularis labradorica	Labrador lousewort
Phacelia sericea	silky phacelia
Plantago canescens	gray pubescent plantain
Poa glauca	glaucous bluegrass
Podistera yukonensis	Yukon podistera
Potentilla palustris	Pennsylvania cinquefoil
Pseudoroegneria (*Agropyron*) *spicata*	bluebunch wheatgrass
Pyrola asarifolia	liverleaf wintergreen
Rosa acicularis	prickly rose
Salix alaxensis	feltleaf willow
Salix bebbiana	Bebb's willow
Salix glauca	grayleaf willow
Salix lasiandra	Pacific willow
Salix pulchra	diamondleaf willow, tealeaf willow
Salix scouleriana	Scouler's willow
Shepherdia canadensis	buffaloberry
Silene repens	pink campion
Tofieldia pusilla	Scotch false asphodel
Townsendia hoookeri	Hooker's Townsend daisy
Vaccinium uliginosum	bog blueberry
Vaccinium vitis-idaea	lingonberry, low-bush cranberry
Viburnum edule	high-bush cranberry
Lichens and mosses	
Aulocomnium palustre	no common name
Cetraria cucullata	no common name
Cetraria islandica	no common name
Cetraria laevigata	no common name
Cladina arbuscula	no common name
Cladonia gracilis	no common name
Hylocomium splendens	splendid feathermoss
Pleurozium schreberi	Schreber's feathermoss
Polytrichum commune	polytrichum moss

Source: Common names are from Viereck et al. (1992) and from the PLANTS Database, Version 3.5 (http://plants.usda.gov). National Plant Data Center, U.S. Department of Agriculture, NRCS, Baton Rouge, LA 70874-4490.

[1]Previous nomenclature that is still commonly found in the literature is shown in parenthesis (auct = of authors).

Boreal Floristic Affinities

The composition of the boreal forest varies greatly throughout its circumpolar range in response to differences in both current environment and geoclimatic history. The primary species, those that give the forest is distinctive appearance, include broadleafed deciduous trees, needle-leafed evergreens, and needle-leafed deciduous trees. In Alaska, the predominant conifers are white and black spruce (*Picea glauca* and *P. mariana*, respectively); larch (*Larix laricina*) tends to be local; and pine is absent in interior Alaska but a prominent component of the forests to the east in the Yukon and the Northwest Territory, Canada. Important deciduous trees are two poplars (aspen [*Populus tremuloides*] and balsam poplar [*P. balsamifera*]) and paper birch (*Betula neoalaskana*; previously treated as *B. papyrifera*). Important shrubs include alders and willows, but the common species in Alaska differ somewhat from their ecological equivalents in eastern North America. Nomenclature for alder and birch follows Furlow (1997).

The Asian and American continents were joined for much of the Tertiary, and the initial breach did not occur until about 5 million years ago (Marincovich and Gladenkov 2001). There was an essentially continuous circumpolar Tertiary forest prior to the onset of the Ice Ages. The detailed composition of that ancient vegetation is difficult to reconstruct, but pollen and macrofossils from Tertiary deposits in arctic Canada and interior and arctic Alaska provide some clues. The rich, mixed mesophytic forest of mid-Miocene, with broadleaf trees from hickory (*Carya*), chestnut (*Castanea*), wingnut (*Pterocarya*), walnut (*Juglans*), beech (*Fagus*), and oak (*Quercus*; White and Ager 1994) and coniferous trees such as pine (*Pinus*), dawn redwood (*Metasequoia*), and California redwood (*Sequoia*; Wheeler and Arnette 1994), was gradually replaced, beginning in late Miocene, by coniferous species. The early Pliocene forest therefore consisted of spruce (*Picea*), pine (*Pinus*), fir (*Abies*), hemlock (*Tsuga*), larch (*Larix*), birch (*Betula*), and alder (*Alnus*), as well as ericaceous shrubs and several herbaceous genera (Ager et al. 1994).

During late Pliocene and early Pleistocene cooling, tundra formed, a treeline was established, and the forests of Asia and North America became isolated from one another as they were displaced southward on their respective continents, well south of today's boreal forest. Tertiary taxa of localized colder habitats, such as bogs, and aquatic taxa associated with lakes, ponds, and streams, which were all likely to be pre-adapted to the colder conditions of the Quaternary, persisted and became widespread, producing a shrub and herb element that remained as the climate cooled (Johnson and Packer 1965).

Of the 796 vascular species analyzed by Swanson (2003) that currently occupy the boreal forest of Beringia, the region extending from the Kolyma region of northeastern Russia to the Mackenzie River in northwestern Canada, about a third are shared, a third restricted to the Kolyma region, and a third restricted to Alaska and the Yukon. The elements in common are nonflowering plants known for their facile dispersal (ferns and fern-allies), bog, wetland, and aquatic plants (probable Tertiary elements), and widespread northern forest shrubs and herbs that are also common in southern tundra communities. There are currently no tree species in common to the forests of both continents.

For instance, in the Magadan region of northeastern Russia, the prominent trees are larch, poplar, and pine; the broad-leafed *Chosenia* occurs in the floodplains; and spruce is rare. In Alaska there is no *Chosenia*, and the two spruces, the larch, and the two poplars are entirely different species. Eastward in Canada, pine becomes an important element of the boreal forest.

Much of Alaska and the rest of Beringia were unglaciated at a time when much of the northern hemisphere was covered by Pleistocene ice sheets. This allowed plants to persist in the Beringian refugium throughout the glacial and interglacial intervals of the Ice Ages, whereas the flora and vegetation elsewhere in northern North America and Eurasia are entirely postglacial. Although forests were more extensive than today during the last interglacial (Sangamon; Muhs et al. 2001) and approached modern limits during the Wisconsinan interstade, the area now occupied by boreal forest was a treeless landscape throughout the Last Glacial Maximum (LGM). Presumably a portion of the herbaceous flora survived in favorable microsites and was relatively unaffected, so the primary question is when and from what sources the trees and tall shrubs of the modern forest arrived.

Some have suggested that there were forest refugia on the Bering Land Bridge. Although there is no direct evidence for such refugia, there is no reason that they could not have existed. In fact, Brubaker et al. (2004) have made a strong case for the local persistence of trees on the land bridge during the LGM. These sources best explain the early appearances of tree and shrub taxa in western Alaska. This means that trees dispersed throughout the Eastern Beringian refugium after the LGM prior to the opening of the ice-free corridor between the Laurentide and Cordilleran ice sheets east of Alaska, which ultimately permitted a major influx of trees and shrubs from south of the continental ice-sheet margins.

The general pattern for interior Alaska is first the appearance of poplar in the late-glacial—Early Holocene transition, then spruce, perhaps tree birch, and alder in Early to Mid-Holocene (Anderson et al. 2004). Alder as a genus is now widespread throughout the boreal forest of North America, but the primary taxon in Alaska suggests origins from Asia (Furlow 1997), perhaps having dispersed from the Bering Land Bridge. Similarly, the tree birch has taxonomic affinities with Asia but not central and eastern North America (Furlow 1997), but its arrival is not marked in the palynological record. Notwithstanding the many circumpolar species, Alaskan alpine tundra within the boreal zone has a distinct Asiatic element, particularly of the herbaceous component, but this influence is less important in the forest. Moreover, the common shrubs and subshrubs of Alaskan boreal forest are from the wide-ranging boreal element found throughout North America: *Arctostaphylos uva-ursi, Cornus canadensis, Linnaea borealis, Orthilia secunda, Pyrola asarifolia, Salix bebbiana, Shepherdia canadensis, Vaccinium uliginosum, V. vitis-idaea, Viburnum edule.*

Boreal Diversity

The boreal forest has fewer vascular plant species than any other forested biome (Pastor and Mladenoff 1992). Consequently, many of its species have extremely broad distributions (Chapin and Danell 2001). Interior Alaska, for example, has only six tree species—three conifers and three broad-leafed species. Typically, only one or two of these tree species dominate a given stand. By comparison, the southern transition of the boreal forest to northern hardwoods in the Great Lakes region has five times more tree species (42) and six times more shrubs (300, compared to 48 in interior Alaska; Maycock and Curtis 1960, Viereck and Little 1972). The approximately 400 herbaceous species in interior Alaska (Jorgenson et al. 1999) represent a much more depauperate flora than occurs at the southern margin with either northern hardwood forests or prairies. In contrast to vascular plants, interior Alaskan forests are more diverse in mosses and lichens than are southern forests and share most of this diversity with arctic tundra to the north (Hollingsworth 2004).

Forests contain a relatively small proportion of the total plant diversity present in interior Alaska. In the vicinity of Fairbanks, for example, forests contain all six tree species, half of the shrub species, 10% of the herbaceous species, and an unknown proportion of lichens and mosses (Jorgenson et al. 1999). Forest edge and open habitats such as alpine tundra, bogs, fens, and steppe are typically more diverse than adjacent forested regions, just as in most forests of western North America. Even in forests, most herbaceous and nonvascular taxa are more diverse in forest gaps or relatively open stands. The reduction in herbaceous and nonvascular diversity beneath the shade of a relatively homogeneous canopy explains why boreal forest diversity is less than in arctic tundra to the north (Chapin and Danell 2001).

The diversity of boreal forests is a dynamic property that responds sensitively to disturbance. The diversity of vascular plants tends to decline through postfire succession, whereas the diversity of nonvascular plants increases (Fig. 6.1; Rees and Juday 2002). There is a 40% turnover in species from the colonists present shortly after wildfire to those species present in a mature hardwood forest. After logging there is only a 30% turnover of species from early to late succession because many early successional fire specialists are excluded from recently logged sites (Rees and Juday 2002). Plant species that are either exotic to Alaska (e.g., dandelion [*Taraxacum*], yellow sweetclover [*Melilotus*]) or native only to southern Alaska (yellowrattle [*Rhinanthus*]) have appeared in some disturbed or early successional sites in the interior (Densmore et al. 2001). Many of these species are introduced along roadsides as weeds in seed mixes used to stabilize roadsides, are spread by mowing machines, then spread naturally into gravel bars and other naturally disturbed sites with low plant cover.

Vegetation Types and Their Distribution

Topography

Topography and its associated variation in microclimate (Chapter 4), soils (Chapter 3), hydrology (Chapter 16), and ecosystem processes (Chapter 15) are the predominant determinants of the distribution of mature vegetation types in interior Alaska (Fig. 6.2). Black spruce predominates on poorly drained sites, many of which

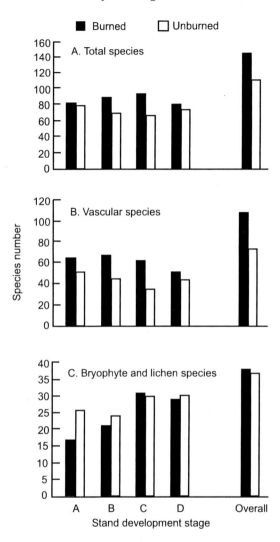

Figure 6.1. Total, vascular-plant, and nonvascular (lichen + moss) species richness (number of species encountered) in four successional stages after fire or logging. Letters represent different stages after disturbance: A = stand initiation (2–5 yr), B = stem exclusion (13–18 yr), C = understory reinitiation (30–38 yr), D = mature hardwood stage (75–95 yr). Redrawn with permission from *Forest Ecology and Management* (Rees and Juday 2002).

have permafrost. White spruce is restricted to sites with better drainage, whereas deciduous species occur in younger and more productive sites. Larch is limited in distribution and tends to occur in wet sites with flowing water. Because topography is such a key control, we describe the vegetation types within this context and then discuss how other state factors influence distribution at a range of scales. There is nearly continuous variation in community composition along topographic gradients. Rather then try to describe all these patterns in detail, we describe a few nodes in this vegetation spectrum. These types and their intermediates are described in detail elsewhere (Viereck 1975, Dyrness and Grigal 1979, Foote 1983, Viereck et al. 1983, Yarie 1983, Viereck et al. 1992, Jorgenson et al. 1999).

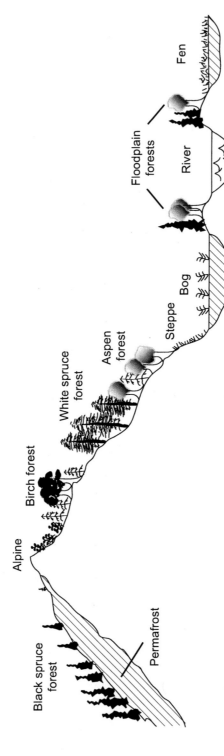

Figure 6.2. Generalized cross-section of topography, landforms, vegetation, and parent material in the Fairbanks area with locations of the major vegetation types. Modified from Viereck et al. (1983) and Jorgenson et al. (1999).

South-Facing Uplands

South-facing uplands experience the warmest, driest microclimate in interior Alaska (Chapter 4). The vegetation of these sites ranges from herb-dominated steppe-like (treeless) vegetation to the most highly productive forests in the interior. The most extreme sites, found on steep, south-facing bluffs above interior Alaskan rivers, have steppe vegetation whose genera have affinities to the Intermountain West and Asia (Murray et al. 1983). These steppe communities were first thought to be relicts of the postulated widespread tundra-steppe of the most recent glacial maximum. They are dominated by bunch grasses, sage, and a diverse assemblage of herbs (Fig. 6.3; Edwards and Armbruster 1989, Wesser and Armbruster 1991). Species growing on these south-facing bluffs are drought-tolerant, and many require high light (Wesser and Armbruster 1991). Some of the common species on these bluffs are taxa of dry forest openings like the rose *Rosa acicularis* and the juniper *Juniperus communis* but also the low sage shrubs *Artemisia frigida, A. alaskana,* the grasses *Pseudoroegneria (Agropyron) spicata, Calamagrostis purpurascens, Poa glauca,* and *Festuca lenensis,* and a variety of herbs that are, in many cases, disjunct from other steppe areas and restricted in Alaska to these sites. Whereas *Carex duriuscula* and *Artemisia laciniatiformis, Plantago canescens,* and *Silene repens* show affinities to similar sites in northeast Asia, *Eriogonum flavum* var. *aquilinum, Erigeron ochroleucus, Phacelia sericea,* and *Townsendia hookeri* are clearly North American

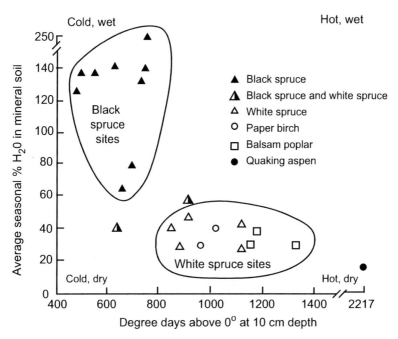

Figure 6.3. Relationship of major ecosystem types in interior Alaska to soil temperature and moisture. Redrawn with permission from the *Canadian Journal of Forest Research* (Viereck et al. 1983).

in origin. Additionally, there are the endemics *Podistera yukonensis* and *Cryptantha shacklettiana* that occur only in Alaska. These biogeographic affinities indicate the complex history of this vegetation type.

Although limited in spatial extent, these sites are hot spots of plant and insect diversity that contribute significantly to regional biodiversity. Unlike the surrounding forests and wetlands, nonvascular cover can be quite low (but species-rich) in these open sites. These steppes have, for example, a bryophyte flora of 70 mosses and five hepatics (including five genera newly reported from Alaska) from families typical of arid lands elsewhere (Murray et al. 1983, Murray 1992). Three taxa are otherwise not known in North America, and 11 more are newly reported for Alaska.

There is insufficient moisture in these sites to support the growth of trees. At transitions between steppe and forest, aspen occasionally establishes and may persist indefinitely but often dies back in dry years. Should the climate become drier, aspen might decline and steppe communities expand into areas now dominated by forest.

Aspen-dominated stands occur on more mesic sites with less extreme slopes and deeper soils. In the transition between steppe and aspen forest, some species replace one another; others, which tolerate a greater range of drought and light, span the gradient (Wesser and Armbruster 1991). With further increases in moisture availability, aspen-dominated stands include progressively more individuals of white spruce (a later successional species; Chapter 7), paper birch, and green alder (*Alnus viridis* subsp. *fruticosa*, previously treated as *A. crispa*). Black spruce-aspen stands occur on stony or sandy soils, and aspen-balsam poplar mixes occur on some floodplains. Other species commonly associated with the closed aspen stands include the shrubs *Viburnum edule, Rosa acicularis, Salix bebbiana, S. scouleriana*, and *Shepherdia canadensis.* The shrub layer may be broken to nearly continuous and is generally about 1 m tall. The herb layer is not well developed but, particularly in areas where shrubs are smaller, may include species such as *Linnaea borealis* (a prostrate semiwoody species), *Galium boreale, Geocaulon lividum, Calamagrostis canadensis,* and *Pedicularis labradorica.* As on the steeper slopes, mosses and lichens are scarce in these aspen stands. These types are generally thought to succeed to white spruce where there is sufficient moisture and seed availability for white spruce to establish (Chapter 7). Warm soils generally promote rapid nutrient cycling (Chapter 15), but moisture frequently limits production (Chapter 11). Aspen stands exhibit a wide range of productivity (Fig. 6.4), with productivity generally increasing as soil moisture increases.

Birch-dominated stands occur on east- and west-facing slopes that are typically cooler and moister than in aspen-dominated stands. These stands often have scattered individuals of white spruce or, at the cold wet end of the spectrum, black spruce, which can replace birch in late succession (Chapter 7). These stands typically have a discontinuous tall-shrub layer of alder (*Alnus viridis* subsp. *fruticosa*) about 3 m tall. Willow (*Salix bebbiana* and *S. scouleriana*), rose (*Rosa acicularis*), and highbush cranberry (*Viburnum edule*) often form a discontinuous shrub layer about 1 m tall, and there is often a ground layer with low-bush cranberry (*Vaccinium vitis-idaea*) and twinflower (*Linnaea borealis*). The grass, *Calamagrostis canadensis*,

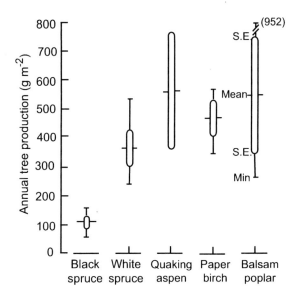

Figure 6.4. Annual aboveground tree production of major forest types in interior Alaska. Redrawn with permission from *Canadian Journal of Forest Research* (Viereck et al. 1983).

or horsetail (*Equisetum*) is often the dominant herb, and heavy leaf litter limits the cover of mosses and lichens. Although birch forests frequently have greater water availability than aspen stands, their productivity appears limited by low soil temperature, which limits decomposition rate (Chapter 15) and nutrient availability (Chapter 11).

Upland white spruce stands are the late-successional forest types on low-elevation, well-drained, south-, east-, and west-facing slopes (Chapter 7). These and their floodplain counterparts are generally less productive than the deciduous stands that precede them in succession (Fig. 6.4). This forest type is most extensive in the central portion of interior Alaska (Viereck et al. 1992). The dense canopy limits the development of a tall-shrub subcanopy, but low shrubs, such as blueberry (*Vaccinium uliginosum*), low-bush cranberry (*V. vitis-idaea*), and Labrador tea (*Ledum groenlandicum*), are common. The absence of abundant leaf litter allows proliferation of a moss layer consisting primarily of the feathermosses *Hylocomium splendens* and *Pleurozium schreberi*. The sparse herbaceous growth includes horsetails (*Equisetum sylvaticum* and *E. arvense*), *Pyrola asarifolia*, *Orthilia secunda*, *Linnaea borealis*, and *Geocaulon lividum*. The growth of white spruce is frequently limited by both water and nutrients (Chapter 11), with their relative importance varying among sites. The width of annual wood increments in white spruce correlates negatively with temperature, indicating that drought is a key factor accounting for interannual variation in the growth of white spruce (Barber et al. 2000, Juday et al. 2003). If climate continues to warm at its current rate, white spruce might be unable to sustain a positive growth rate and could disappear from many upland sites or shift its distribution to cooler sites before the end of the century (Juday et al. in press). This would radically alter patterns of community composition in interior Alaska.

North-Facing Uplands

North-facing uplands are dominated by black spruce and usually have an understory of feathermosses (*Pleurozium schreberi, Hylocomium splendens*) and abundant lichens in the genera *Cladina, Cladonia, Cetraria, Peltigera*, and *Nephroma*. *Sphagnum* mosses also occur on very wet sites with poor drainage. There are often patches of green alder (*Alnus viridis* subsp. *fruticosa*), particularly in sites of intermediate and high moisture. Other common shrubs include *Salix pulchra, S. glauca, S. scouleriana, Ledum groenlandicum, L. decumbens, Vaccinium uliginosum, V. vitis-idaea, Linnaea borealis*, and *Empetrum nigrum*. Tree birch and white spruce occasionally occur but are not dominant. Permafrost is usually present at depths ranging from 30–100 cm except on well-drained coarse alluvium or on shallow soils over bedrock. These forests occupy the coldest wettest sites (Fig. 6.3) and are the least productive of any forests in interior Alaska, with aboveground tree production only 20% of values observed in other forest types (Fig. 6.4). Most of the production in these forests occurs belowground and in the moss layer (Chapter 12).

Lowlands

The major late-successional forest types in the lowlands are white spruce forests on active floodplains with well-drained permafrost-free soils and black spruce forests on poorly drained permafrost-dominated soils.

Floodplain white spruce forests are structurally and functionally similar to those in the uplands except that balsam poplar and thinleaf alder (*Alnus incana* subsp. *tenuifolia*) are the dominant deciduous tree and tall shrub, respectively. Upland and floodplain white spruce forests account for most of the commercial timber production in interior Alaska, although they occupy only 10% of the landscape (Chapter 18).

Lowland black spruce forests are also structurally and functionally similar to their upland counterparts. Together these black spruce forests account for nearly half (44%) of the forest cover in interior Alaska. The large aerial extent and high flammability of this forest type account for the importance of fire as the dominant disturbance type in interior Alaska (Chapter 17). Larch is a sparse component of lowland black spruce forests and occurs primarily where there is flowing groundwater.

Other Types

Nonforested lands occupy a large proportion of interior Alaska. Although they are important as sources of biodiversity, nesting habitat for migratory waterfowl, carbon storage, and methane emissions, they are less well characterized ecologically than are forests and provide important opportunities for future research.

Within interior Alaska there are extensive lowland bogs where shallow permafrost impedes drainage and keeps the soil too wet to support tree growth. These acidic bogs are dominated by the sedges *Eriophorum vaginatum* or *Carex bigelowii*. The community composition of these lowland bogs is similar to that of acidic tussock tundra occurring on the North Slope of Alaska and includes low shrubs *Betula glandulosa, B. nana, Ledum decumbens, L. groenlandicum, Andromeda polifolia*, and *Oxycoccus microcarpa*. Along gradients in soil moisture, there is a gradual transition from treeless sites dominated by *Eriophorum* and *Carex* at the wet extreme to black spruce forests on drier sites. These sites are extremely nutrient-poor (Shaver and Chapin 1995). They receive most of their nutrients from precipitation and tightly recycle these nutrients between plants and soils (Chapter 14). Rivers that drain these lowlands are therefore tannin-rich, nutrient-poor blackwater streams (Chapter 10).

There are also extensive nonacidic wetlands (fens) south of the Tanana River in interior Alaska, where upwelling of groundwater from the Alaska Range delivers abundant cations (Jorgenson et al. 1999). These areas are dominated by herbaceous taxa (*Menyanthes trifoliata*, *Equisetum fluviatile*, and *Potentilla palustris*) and are critical summering areas for waterfowl.

Alpine shrublands generally occur above the elevational limits of forests (about 250 m) in interior Alaska. Common dominants of shrub communities include *Betula glandulosa*, *B. nana*, *Vaccinium uliginosum*, *Ledum decumbens*, *Arctous alpina*, *A. rubra*, and *Empetrum nigrum*. A mat of feathermosses and lichens typically provides complete cover beneath the shrubs. Herbaceous tundra occurs at even higher elevations, where shrubs decrease in abundance. As in the case of acidic bogs, these types are similar to arctic tundra that extends north of the boreal forest and to the understory of adjacent forests.

Climate

The major climate gradient in interior Alaska is from a maritime climate in the west to a more continental climate in the east (Chapter 4; Fleming et al. 2000). There is not a strong latitudinal climate gradient between the Alaska Range to the south and the Brooks Range to the north, although there are altitudinal climate gradients associated with the Alaska Range, Brooks Range, and various uplands within interior Alaska, as described earlier.

The east-west gradient in continentality alters the relative abundance of vegetation types, due primarily to changes in summer moisture and temperature. In the east, there is less precipitation, and fire is more frequent (Yarie 1981), leading to a younger average age of stands. The dry conditions also result in more extensive steppe communities (Edwards and Armbruster 1989). In contrast, the Seward Peninsula, to the west, has forest stands interspersed with tundra.

Time, Potential Biota, and Parent Material

In addition to topography and climate, time is the state factor with the strongest impact on vegetation distribution in interior Alaska and elsewhere in the boreal forest, particularly the effects of fire and postfire succession on community composition. Because of its critical importance, and because it has been the central focus of the Bonanza Creek LTER, we treat time and disturbance regime separately in Chapter 7.

Because of the relative stability of vegetation in interior Alaska for several millennia, the biota that is potentially available to colonize sites is essentially constant across interior Alaska and does not strongly influence the vegetation patterns observed. The *actual* vegetation that is distributed across the landscape is a product of vegetation response to other state factors. If new species enter Alaska in response to climate warming, variation in regional biota may become a more important determinant of community composition than it is at present.

Most LTER studies in interior Alaska have been conducted on relatively uniform parent materials—loess in the uplands and alluvium on the floodplains (Chapter 3)—making it difficult to separate the effects of parent material from those of topography. However, as described in the next section, studies that have been conducted across a range of parent materials find this state factor to be an important influence on boreal vegetation composition and diversity.

Variation within Vegetation Types

A superficial view of interior Alaskan forests from the air suggests a relatively homogeneous and monotonous landscape, with nearly half of the forested area dominated by a single vegetation type—black spruce. More careful examination, however, reveals substantial diversity and variability within that type. As in the patterning of major forest types across the landscape, the diversity within black spruce forests is controlled primarily by gradients in state factors (Hollingsworth 2004).

Two alternative approaches have been used to describe the range of variation in black spruce communities. One approach, based primarily on canopy structure, documents patterns of variation associated with topography and landscape position (Viereck et al. 1992). Short-statured stands with low density predominate on steep north-facing slopes and poorly drained lowlands. These stands experience the coldest, most waterlogged conditions of any forests in interior Alaska (Chapter 4). They have low productivity (Chapter 11), with half or more of aboveground production contributed by mosses. Unproductive black spruce forests also occur in well-drained sites near treeline, where low summer temperatures constrain growth. More productive black spruce stands intergrade with white spruce in less extreme environments. This physiognomic (structural) approach to analyzing vegetation distribution is particularly useful for satellite-based remote sensing studies that receive a strong signal from the canopy. They are also useful for understanding processes dominated by the tree canopy, such as primary production and water and energy exchange with the atmosphere, both of which have potential impacts on climate (Chapter 19).

A second approach to understanding the range of variation in black spruce communities relies on patterns of floristic composition. Although this approach has been widely used in Europe and eastern Canada, it has only recently been applied to Alaskan black spruce forests (Hollingsworth 2004). Because there is only one tree species in most black spruce forests, the floristic composition of black spruce communities is determined entirely by the floristics of the understory. Three main community types emerge from this analysis:

Floristic Diversity and Vegetation Distribution in the Alaskan Boreal Forest 95

1. An acidic black spruce/lichen forest has diagnostic species (i.e., those which distinguish it from other communities that include a lichen (*Cetraria islandica*) and a moss (*Polytrichum commune*) and a clubmoss (*Lycopodium annotium*). This community is dominated by the shrubs *Vaccinium uliginosum*, *Ledum groenlandicum*, and *Vaccinium vitis-idaea*, the gramminoid *Calamagrostis canadensis*, the mosses *Hylocomium splendens* and *Pleurozium schreberi*, and the lichen *Cladina arbuscula*.
2. A nonacidic black spruce/rose/horsetail forest has as its most diagnostic species the vascular plants rose (*Rosa acicularis*) and various species of horsetail (*Equisetum scirpoides*, *E. pratense*, and *E. sylvaticum*). This community is dominated by the shrubs *Ledum groenlandicum*, *Vaccinium vitis-idaea*, *Vaccinium uliginosum*, and *Empetrum nigrum*, the gramminoid *Calamagrostis canadensis*, the forb *Mertensia paniculata*, the mosses *Hylocomium splendens*, *Pleurozium schreberi*, and *Aulacomnium palustre*, and the lichens *Cladina arbuscula* and *Cladonia gracilis*.
3. A treeline black spruce woodland has as its most diagnostic species a shrub (*Ledum decumbens*) and a forb (*Tofieldia pusilla*) and is dominated by the shrubs *Empetrum nigrum* and *Vaccinium uliginosum*, the sedge *Carex bigelowii*, the forb *Pedicularis labradonica*, the mosses *Hylocomium splendens* and *Pleurozium scherberi*, and the lichens *Cetraria laevigata* and *Cetraria cucullata*.

As these community names imply, the environmental factors that correlate most strongly with floristic composition are soil pH and topography/soil drainage. These two factors and a third factor related to productivity account for 81% of the variability in floristic composition (Fig. 6.5). The prominent role of pH in governing floristic composition was not evident in physiognomic studies, suggesting that the canopy and understory of black spruce forests respond to quite different environmental patterns on the landscape. The sensitivity of black spruce floristics to pH is a characteristic that it shares with tundra vegetation to the north (Walker and Everett 1991), in part because of the many species in common between the two vegetation types. The causes of pH gradients that drive the floristic composition of black spruce forests are unclear but probably include variations in landscape age, parent material, fire history, and ecosystem processes related to species composition and topography (Hollingsworth 2004). Floristic approaches to vegetation analysis are particularly important in understanding processes that are sensitive to understory species composition, such as permafrost dynamics (Chapter 4), succession (Chapter 7), and nutrient cycling (Chapter 15).

Conclusions

The vegetation composition of the Alaskan boreal forest is a product of both its history and the current environment. Alaska has, in the past, alternated between being attached to Eurasia and being attached to North America and therefore shares a third of its species with each continent. Alaska has also been the bridge leading to a broad circumpolar distribution of the remaining third of the interior Alaskan flora. Many of the common understory species in Alaskan forests were part of a Tertiary

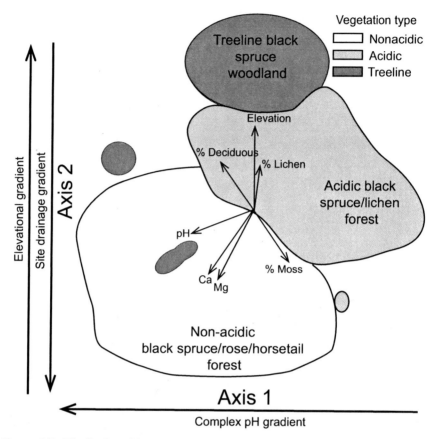

Figure 6.5. Distribution of the three black spruce community types in interior Alaska along complex ordination gradients of pH (axis 1) and soil drainage and elevation (axis 2; data from Hollingsworth et al. 2004).

bog flora that appears to have persisted in interior Alaska through many glacial and interglacial cycles. The trees are more recent arrivals. There are only six tree species, most with North American affinities. These trees define the major forest types of interior Alaska. Because of the relatively stable forest composition of interior Alaska for the past 5,000 years, these species are distributed predictably across environmental gradients associated with topography. These tree species and their underlying environmental gradients determine the floristic composition of the understory.

These highly predictable patterns related to environment raise important questions about the future of Alaska's boreal forest if recent warming trends continue. What will happen to white spruce, which is interior Alaska's major timber species and which shows reduced growth in warm years? At the same time, lodgepole pine, a drought-adapted tree that promotes fire and that was present in interior Alaska in the previous interglacial, is actively migrating north- and westward in western

Canada (Johnstone and Chapin 2003). It is currently within 100 km of Alaska's eastern border and is being grown experimentally in Alaska as a potential timber species. This species produces a thick litter layer that inhibits the growth of many of the understory species characteristic of Alaska's current forests.

The most recent major change in forest composition occurred in Alaska 5,000 years ago with the arrival of black spruce and its associated increase in fire frequency. We may be on the brink of the next major vegetation transition in Alaska's boreal forest.

References

Ager, T. A., J. Matthews, J. V., and W. Yeend. 1994. Pliocene terrace gravels of the ancestral Yukon River near Circle, Alaska: Palynology, paleobotany, paleoenvironmetal reconstruction and regional correlation. Quaternary International 22/23:185–206.

Anderson, P. M., M. E. Edwards, and L. B. Brubaker. 2004. Results and paleoclimate implications of 35 years of paleoecological research in Alaska. Pages 427–440 in A. E. Gillespie, S. C. Porter, and B. F. Atwater, editors. The Quaternary Period in the United States. Developments in Quaternary Science. Elsevier, New York.

Barber, V. A., G. P. Juday, and B. P. Finney. 2000. Reduced growth of Alaskan white spruce in the twentieth century from temperature-induced drought stress. Nature 405:668–673.

Bergeron, Y., M. Flannigan, S. Gauthier, A. Leduc, and P. Lefort. 2004. Past, current and future fire frequency in the Canadian boreal forest: Implications for sustainable forest management. Ambio 33:356–360.

Brubaker, L. B., P. A. Anderson, M. E. Edwards, and A. V. Lozhkin. 2004. Beringia as a glacial refugium for boreal trees and shrubs: New perspectives from mapped pollen data. Journal of Biogeography 31:1–16.

Chapin, F. S., III, and K. Danell. 2001. Boreal forest. Pages 101–120 in F. S. Chapin III, O. E. Sala, and E. Huber-Sannwald, editors. Global Biodiversity in a Changing Environment: Scenarios for the 21st Century. Springer-Verlag, New York.

Densmore, R. V., P. C. McKee, and C. Roland. 2001. Exotic Plants in Alaskan National Park Units. U.S. National Park Service, Anchorage.

Dyrness, C. T., and D. F. Grigal. 1979. Vegetation-soil relationships along a spruce forest transect in interior Alaska. Canadian Journal of Botany 57:2644–2656.

Edwards, M. E., and W. S. Armbruster. 1989. A tundra-steppe transition on Kathul Mountain, Alaska, USA. Arctic and Alpine Research 21:296–304.

Fleming, M. D., F. S. Chapin III, W. Cramer, G. Hufford, and M. C. Serreze. 2000. Geographic patterns and dynamics of Alaskan climate interpolated from a sparse station record. Global Change Biology 6(Suppl. 1):49–58.

Foote, M. J. 1983. Classification, description, and dynamics of plant communities after fire in the taiga of interior Alaska. Research Paper PNW–307, U.S. Department of Agriculture, Forest Service, Portland, OR.

Furlow, J. J. 1997. Betulaceae. Pages 507–538 in Flora of North America Editorial Committee, editor. Flora of North America North of Mexico. Vol. 3. Oxford University Press, New York.

Hollingsworth, T. N. 2004. Quantifying variability in the Alaskan black spruce ecosystem: Linking vegetation, carbon, and fire. Ph.D. Dissertation. University of Alaska, Fairbanks.

Johnson, A. W., and J. G. Packer. 1965. Polyploidy and environment in arctic Alaska. Science 148:237–239.

Johnstone, J. F., and F. S. Chapin III. 2003. Non-equilibrium succession dynamics indicate continued northern migration of lodgepole pine. Global Change Biology 9:1401–1409.

Jorgenson, M. T., J. E. Roth, M. K. Raynolds, M. D. Smith, W. Lentz, A. L. Zusi-Cobb, and C. H. Racine. 1999. An ecological land survey for Fort Wainwright, Alaska. CRREL Report 99–9, U.S. Army Corps of Engineers, Cold Regions Research & Engineering Laboratory, Hanover, NH.

Juday, G. P., V. Barber, T. S. Rupp, J. Zasada, and M. W. Wilmking. 2003. A 200-year perspective of climate variability and the response of white spruce in Interior Alaska. Pages 226–250 in D. Greenland, D. Goodin, and R. Smith, editors. Climate Variability and Ecosystem Response at Long-Term Ecological Research (LTER) Sites. Oxford University Press, Oxford.

Juday, G. P., V. Barber, E. Vaganov, T. S. Rupp, S. Sparrow, J. Yarie, and H. Linderholm. In press. Forests, land management, agriculture. Pages 781–862 in ACIA, editors. Arctic Climate Impact Assessment. Cambridge University Press, Cambridge, UK.

Marincovich, L., Jr., and A. Y. Gladenkov. 2001. New evidence for the age of Bering Strait. Quaternary Science Reviews 20:329–335.

Maycock, P. F., and J. T. Curtis. 1960. The phytosociology of boreal conifer-hardwood forests of the Great Lakes Region. Ecological Monographs 30:1–35.

Muhs, D. R., T. A. Ager, and J. E. Beget. 2001. Vegetation and paleoclimate of the last interglacial period, central Alaska. Quaternary Science Reviews 20:41–61.

Murray, B. M. 1992. Bryophyte flora of Alaskan steppes. Bryobrothera 1:9–33.

Murray, D. F., B. M. Murray, B. A. Yurtsev, and R. Howenstein. 1983. Biogeographic significance of steppe vegetation in subarctic Alaska. Pages 883–888 in Permafrost: Fourth International Conference, Proceedings. National Academy Press, Fairbanks, AK.

Pastor, J., and D. J. Mladenoff. 1992. The southern boreal-northern hardwood forest border. Pages 216–240 in H. H. Shugart, R. Leemans, and G. B. Bonan, editors. A systems analysis of the global boreal forest. Cambridge University Press, Cambridge.

Rees, D. C., and G. P. Juday. 2002. Plant species diversity and forest structure on logged and burned sites in central Alaska. Forest Ecology and Management 155:291–302.

Shaver, G. R., and F. S. Chapin, III. 1995. Long-term responses to factorial NPK fertilizer treatment by Alaskan wet and moist tundra sedge species. Ecography 18:259–275.

Swanson, D. K. 2003. A comparison of taiga flora in north-eastern Russian and Alaska-Yukon. Journal of Biogeography 30:1109–1121.

Viereck, L. A. 1975. Forest ecology of the Alaska taiga. Pages 1–22 in Circumpolar Conference on Northern Ecology. National Research Council of Canada, Ottawa.

Viereck, L. A., and E. L. Little Jr. 1972. Alaska Trees and Shrubs. U.S. Department of Agriculture, Forest Service, Washington, DC.

Viereck, L. A., C. T. Dyrness, A. R. Batten, and K. J. Wenzlick. 1992. The Alaska vegetation classification. General Technical Report PNW-GTR–286. U.S. Department of Agriculture, Forest Service, Pacific Northwest Research Station, Portland, OR.

Viereck, L. A., C. T. Dyrness, K. Van Cleve, and M. J. Foote. 1983. Vegetation, soils, and forest productivity in selected forest types in interior Alaska. Canadian Journal of Forest Research 13:703–720.

Walker, D. A., and K. R. Everett. 1991. Loess ecosystems of northern Alaska: Regional gradient and toposequence at Prudhoe Bay. Ecological Monographs 61:437–464.

Wesser, S. D., and W. S. Armbruster. 1991. Species distribution controls across a forest-steppe transition: A causal model and experimental test. Ecological Monographs 61:323–242.

Wheeler, E. A., and J. Arnette, C. G. 1994. Identification of Neogene woods from Alaska—Yukon. Quaternary International 22/23:91–102.

White, J. M., and T. A. Ager. 1994. Palynology, paleoclimatology and correlation of Middle Miocene beds from Porcupine River (Locality 90-1), Alaska. Quaternary International 22/23:43-77.

Yarie, J. 1981. Forest fire cycles and life tables: A case study from interior Alaska. Canadian Journal of Forest Research 11:554-562.

Yarie, J. 1983. Forest community classification of the Porcupine River drainage, interior Alaska, and its application to forest management. General Technical Report PNW-154, U.S. Department of Agriculture, Forest Service, Pacific Northwest Forest and Range Experiment Station, Portland, OR.

7

Successional Processes in the Alaskan Boreal Forest

F. Stuart Chapin III
Leslie A. Viereck
Phyllis C. Adams
Keith Van Cleve

Christopher L. Fastie
Robert A. Ott
Daniel Mann
Jill F. Johnstone

Introduction

Superimposed on the topographic and climatic gradients in vegetation described in Chapter 6 are mosaics of stands of different ages reflecting the interplay between disturbance and *succession*, that is, the ecosystem changes that follow disturbance. The nature of disturbance governs vegetation succession, and vegetation properties, in turn, influence disturbance regime. Both disturbance and succession are controlled by state factors and by stochastic variation in local conditions such as weather and the abundance of herbivores. Even in this relatively simple biome, the interactions among site, chance, and disturbance history result in a vast array of possible successional trajectories following a disturbance event, generating at least 30 forest types in interior Alaska (Viereck et al. 1992). Despite this broad range of possible dynamics, certain patterns recur more frequently than others (Drury 1956, Viereck 1970). In this chapter, we discuss selected successional pathways that commonly occur on river floodplains and on permafrost-free or permafrost-dominated upland sites in interior Alaska.

Floodplain Succession

River floodplains occupy only 17% of interior Alaska, but they account for 80% of the region's commercial forests and therefore have attracted considerable attention from forest managers (Adams 1999). These forests provide an excellent example of *primary succession*, that is, the succession that occurs on surfaces that have not been previously vegetated. Although many successional pathways are possible on interior Alaska's floodplains (Fig. 7.1; Drury 1956), the trajectory that actually occurs in a particular place is usually determined by the patterns of colonization during the first decades (Egler 1954). This, in turn, depends primarily on physical environment, flood events, and seed availability. For example, fine-textured sediments, which are common along the gradual grade of the Tanana River near Fairbanks (Chapter 3), retain more moisture than gravelly substrates and favor establishment of thinleaf alder (*Alnus incana* subsp. *tenuifolia*) following the initial colonization by willow (*Salix*). Alder is therefore a more important component of this successional sequence than along some other rivers. In this chapter, we focus on the alder-mediated pattern of floodplain succession, which has been the major focus of the LTER research program. The common and scientific names of species mentioned in this chapter are given in Table 6.1.

Geomorphic Dynamics of the Tanana River Floodplain

The successional dynamics of the Tanana River are driven by its fluvial geomorphology (Chapter 2). Glacial-fed tributaries from the Alaska Range contribute 85% of the water in the Tanana River (Yarie et al. 1998). Flooding can occur at almost any time, due to ice jams in spring, rapid melting of glaciers in midsummer, or heavy late-summer and autumn precipitation (Chapter 16). In August 1967, for example, record-high precipitation caused the river to rise 5 m above the winter level and

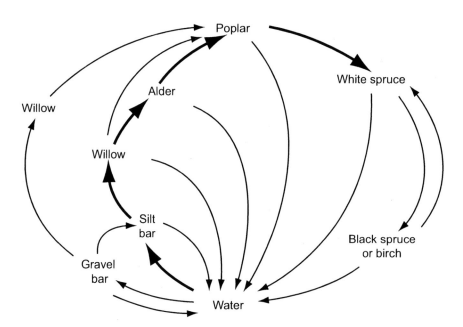

Figure 7.1. Common pathways of primary succession on the Tanana River of interior Alaska. Species names designate communities in which that species is dominant. Bold arrows indicate the pathway studied by the Bonanza Creek LTER program.

tripled the mean August discharge. Summer and autumn floods are particularly likely to erode banks, suspend sediments, and produce new alluvial deposits (Mason and Begét 1991).

Because of its glacial origin and steep gradient, the Tanana River carries a heavy sediment load, averaging 24 million tons of suspended sediment and 321,000 tons of bedload per year at Fairbanks (Burrows et al. 1981). This sediment derives originally from the Alaska Range but is repeatedly deposited and reclaimed by the river, as channels shift laterally across the floodplain. Floodplain surfaces increase in elevation as successive floods deposit new sediments (Fig. 7.2). As surface height and distance from active channels increase, the frequency of overbank deposition declines. The river has been in a phase of net aggradation of about 4.5 cm per century for at least the past 400 years (Collins 1990).

Colonization Stage

A common pattern of primary succession studied by the LTER program begins with the invasion of willows on newly formed alluvial bars, followed by communities

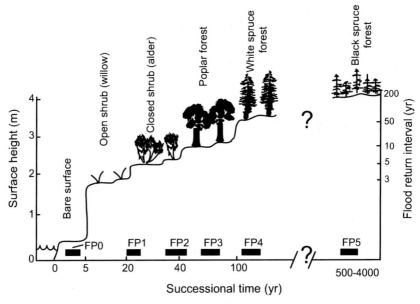

Figure 7.2. Diagram of floodplain succession (Viereck et al. 1993), showing representative surface heights above the river (Yarie et al. 1998, Adams 1999), flood return intervals (Yarie et al. 1998), and surface ages (Walker et al. 1986). The turning points in floodplain succession are the emergence of a new alluvial bar (FP0); the switch from abiotic to biotic controls over nutrient cycles at the closed shrub stage (FP1); the emergence of a poplar canopy above the alder thicket (FP2); the emergence of a spruce overstory and moss understory (FP3); the development of a mature spruce forest (FP4); and the development of a black spruce forest (FP5). The timing and mechanisms for succession from white to black spruce forests are less certain than for other transitions.

dominated by alder, balsam poplar, white spruce, and sometimes black spruce (Viereck 1970, Van Cleve et al. 1983b, Viereck et al. 1993, Yarie et al. 1998).

Interactions between river geomorphology and the semi-arid nature of the climate determine the chemical and the physical environment in which colonization occurs. Salt crusts composed of gypsum ($CaSO_4 \cdot 2H_2O$) and calcite ($CaCO_3$) form on the surface of the exposed mineral soil along the Tanana River (Slaughter and Viereck 1986, Marion et al. 1993), although they are absent from many other Alaskan rivers (Shaw et al. 2001). The salt crust results from capillary rise of calcium-rich groundwater from shallow water tables, with subsequent salt precipitation after evaporation of water (Dyrness and Van Cleve 1993). The salt crust disappears (Fig. 7.3) once a continuous plant canopy develops because the litter layer prevents surface evaporation, and the canopy vegetation moves most water from the rooting zone to the atmosphere by transpiration. As succession proceeds, calcium salts are weathered (i.e., solubilized) by organic acids and by carbonic acid derived from soil respiration and are leached from the soil. This leaching occurs at rates comparable to the upper end of weathering rates observed anywhere in the world (Marion et al. 1993).

The initial nitrogen stock on the bare alluvial bars (about 50 g m^{-2}; Van Cleve et al. 1971, Walker 1989, Van Cleve et al. 1993) is only about 10% of that which eventually accumulates in the floodplain soils (Fig. 7.3). This starting nitrogen capital is, however, five times greater than that which initiates postglacial succession at Glacier Bay (Crocker and Major 1955), due to nitrogen derived from organic matter eroded from the upstream floodplain.

Successful establishment on newly deposited sediments depends on the interaction of several factors. The river level must be high enough at the time of seed dispersal for capillary rise to provide adequate moisture for germination and establishment of the small, wind-dispersed seeds of colonizing plants. Within five years, horsetail (*Equisetum*), willow (*Salix* spp.), and balsam poplar (*Populus balsamifera*) are common (Walker et al. 1986, Viereck et al. 1993). These species develop deep roots that

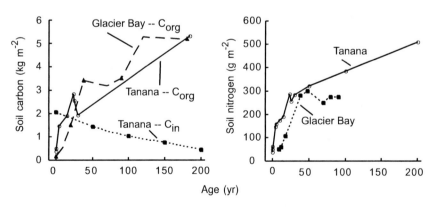

Figure 7.3. Successional changes in soil carbon (organic [C_{org}] and inorganic [C_{in}]) and total nitrogen stocks (to a depth of 60 cm) through succession on the Tanana floodplain (Van Cleve et al. 1971, Walker 1989, Marion et al. 1993, Van Cleve et al. 1993) and at Glacier Bay (Crocker and Major 1955).

stabilize soils during frequent floods and take advantage of the shallow water table during hot, dry conditions. Seedlings are vulnerable to burial by sediments during floods, but larger plants can keep up with sediment accumulation by producing adventitious roots from the stem following each deposition event (Viereck et al. 1993, Adams 1999). Although white spruce can recruit in any successional stages, the probability of establishment increases through the first 50 years of succession. White spruce seedlings that colonize young alluvial terraces grow slowly due to water and nutrient stress and have a low tolerance to the sedimentation that accompanies the frequent flooding of these sites (Zasada 1984, Walker and Chapin 1986, Krasny et al. 1988, Wurtz 1988, Viereck et al. 1993). Early-successional sites not yet dominated by trees accounted for at least 24% of the land reclaimed by the river (Fig. 7.4).

Open Willow Stage

Physical processes that drive surface evaporation and sediment deposition continue to dominate as willows increase in size and density during the open willow stage. Surface elevation increases rapidly during this stage (Fig. 7.2; Yarie et al. 1998), with as much as 10–20 cm of sediment being deposited in a single flood. Within a decade, those surfaces that survive river erosion reach an average height of 1–2 m above mean river height, and the flood return interval declines to one to three years (Walker et al. 1986, Viereck et al. 1993, Yarie et al. 1998, Shaw et al. 2001). Plant litter is sparse, and most of it blows away in strong winds or is buried by floods, leaving an exposed mineral soil surface. This generates an open nutrient cycle that

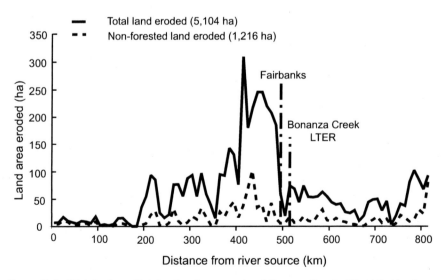

Figure 7.4. Distribution of total and early successional (i.e., nonforested) land that has been eroded between 1978–1980 and 1998–1999 along the Tanana River (adapted from Ott et al. 2001), based on interpretation of aerial photographs (1978–1980) and satellite images (1998–1999). Erosion figures include all banks and islands along a given reach of the river but do not include areas that were nonvegetated gravel bars in either time period.

is dominated by evaporation-driven capillary rise, solubility equilibria, wind erosion, and flood events that deposit sediments and leach accumulated salts (Van Cleve et al. 1991, Marion et al. 1993). Intense browsing by moose and hares recycle 30% of the aboveground production and associated nitrogen to the developing soil (Chapter 8; Kielland and Bryant 1998). Herbivory by insects, which supports a diverse community of neotropical migrant birds, is also most intense at this stage.

Plant biomass gradually increases through the open shrub stage as colonization by new seedlings and vegetative growth of established plants (Zasada 1986) gradually dominate over biomass removal by moose, hares, and other herbivores (Chapter 8). Willow species dominate the site, and poplar seedlings and suckers grow rapidly. Thinleaf alder (*Alnus incana* subsp. *tenuifolia*) establishes in the open shrub stage, if soils are fine-textured and there is a local seed source (Viereck et al. 1993). Alder colonization is more variable, both temporally and spatially, than that of willows or poplars. Its larger seeds arrive less frequently than those of willow. Its initial establishment is sensitive to both flooding and drought and may be facilitated by shading (Walker and Chapin 1986). Once alder seedlings establish, they grow quickly, because their symbiotic actinorrhizal bacteria fix nitrogen (Uliassi et al. 2000), the element that most strongly limits the productivity of all floodplain communities (Chapter 11; Walker and Chapin 1986, Yarie 1993). The establishment of a few individual alders often provides a local seed source that allows infilling of the remaining community, leading to dense alder thickets within 1–3 decades of the establishment of the first alders (Fig. 7.5). Some white spruce seedlings that establish at this stage persist through to the spruce stage, but most do not survive the deposition associated with flooding (Adams 1999).

Closed Shrub Stage

Succession in interior Alaska is not a gradual linear change in ecosystem properties but is punctuated by a series of *turning points* characterized by rapid changes in ecosystem controls (Fig. 7.2; Van Cleve et al. 1996). The shift from a physically dominated open shrub stage to a biologically regulated closed shrub stage is one of the most dramatic turning points (FP1) in floodplain succession (Van Cleve et al. 1993). Flooding occurs less frequently (every 5–10 years), and the salt crust disappears because plant transpiration dominates over soil evaporation (Dyrness and Van Cleve 1993). Alders commonly dominate the closed shrub stage in the LTER sites, with some tall willows (e.g., *Salix alaxensis* and *S. lasiandra*) persisting in the canopy. A combination of shading by alders and preferential browsing by hares gradually eliminates the willows and reduces the density of poplars that initially colonized the stand (McAvinchey 1991, Viereck et al. 1993). Those poplars that survive grow quickly due to high nutrient availability (Chapter 11), forming a canopy that gradually emerges above the alder. However, hares frequently girdle young poplars during peaks of their population cycles. Aboveground biomass and litterfall increase 10- to 50-fold from the open to the closed shrub stage (Viereck et al. 1993). The dense canopy reduces radiation at the soil surface to 20% of full sun in midsummer, causing the soil heat sum (an integrated measure of soil temperature) to decline by 40% relative to the bare alluvium (Fig. 7.5).

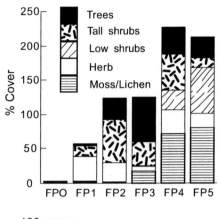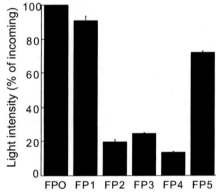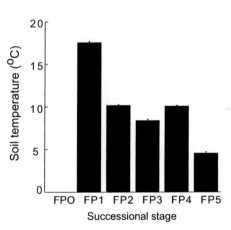

Figure 7.5. Successional changes in the cover of the major plant functional types on the Tanana River floodplain and the associated changes in understory light availability and soil temperature. Plant cover was measured separately by species in 1986–1994, then summed by plant functional type and plot (n = 3 sites). Light was measured at the soil surface and soil temperature 10 cm beneath the organic-mineral interface in 1990. Successional stages are defined in Fig. 7.2.

The nitrogen fixed by alder during the closed shrub stage (Uliassi et al. 2000, Uliassi and Ruess 2002) accounts for 60–70% of the total nitrogen that accumulates during the entire 200–year floodplain succession (Fig. 7.3; Van Cleve et al. 1971, Van Cleve et al. 1983b, Van Cleve et al. 1993). Because plants and soil microbes require a balance of nitrogen, phosphorus, potassium, and other elements, these elements are also retained and accumulate in vegetation and the developing forest floor during the closed shrub stage. The cation exchange capacity that develops in association with the organic accumulation retains cations in a highly available form (Chapter 14; Van Cleve et al. 1993, Shaw et al. 2001).

In this stage, flooding changes from being the major source of white spruce mortality to a factor that, in general, promotes spruce recruitment. Spruce seedlings that germinate in the 5–10 cm thick litter layer seldom survive. Sediment deposition from occasional floods provides a mineral soil seedbed that allows white spruce seedlings to establish (Adams 1999). Shading by the well-developed canopy reduces surface drying of soils, and rapid nutrient cycling in buried organic horizons supplies abundant nutrients (Van Cleve et al. 1996). This produces an environment in which spruce grows more rapidly than in earlier stages. Although white spruce is more shade-tolerant than floodplain deciduous species, its growth is light-limited (Walker and Chapin 1986). Snowshoe hares, which use closed shrub canopies as cover from predators, are the major cause of white spruce mortality at this stage.

Poplar Stage

The transition from alder thickets to a poplar forest (FP2) occurs gradually over 10–20 years (Fig. 7.2; Viereck et al. 1983). Poplars that extend above the alder thicket have the best of both worlds: maximal light aboveground and high availability of water and nutrients belowground. Some poplar roots extend to the water table, so they seldom experience water stress. Most fine roots of poplar are, however, concentrated near the soil surface (Viereck 1970, Tryon and Chapin 1983), particularly in buried organic horizons, where they tap the nitrogen fixed by alder and other nutrients that cycle in parallel with nitrogen in the rapidly decomposing litter (Chapter 15). Poplar tannins inhibit microbial activity in the poplar stage (Chapter 14; Schimel et al. 1996), reducing rates of nitrification and denitrification. The resulting declines in inputs (nitrogen fixation) and outputs (denitrification and nitrate leaching) reduce the openness of the nitrogen cycle, so nitrogen cycles more tightly between vegetation and soils. This is the successional stage in which soil nutrient resources, plant productivity, and hence litterfall are greatest (Chapter 11; Viereck et al. 1993).

White spruce recruitment is greatest in the poplar stage and occurs primarily in the occasional years when high seed production follows a year of alluvial deposition (10–25 year return time; Yarie et al. 1998, Adams 1999). Spruce seedlings and saplings grow relatively rapidly in the poplar stage because of the high nutrient availability in this stage and the substantial shade tolerance of spruce. Hare browse is less intense in the poplar than the alder stage because preferred food is scarce and the opening of the forest canopy reduces protection from avian predators. Feather mosses (*Hylocomium splendens* and *Pleurozium schreberi*) are absent from most of the forest floor because they are smothered by the dense deciduous litter.

However, these mosses grow rapidly on rotten logs, bases of tree trunks and other stable surfaces that rise above the litter layer (Oechel and Van Cleve 1986). These moss patches become more extensive beneath clumps of spruce saplings where deciduous litter is sparse, and the fine needles of spruce litter filter through the developing moss canopy.

White Spruce Stage

The turning point from poplar to white spruce (FP3) occurs as the shade-tolerant white spruce grows above the poplar canopy and as the short-lived poplar trees (100–150 years) begin to die (Viereck et al. 1983, Walker et al. 1986) or are felled by beaver. Removal of large (20–30 cm diameter) poplars by beaver increased substantially after 1980, in response to regional increases in beaver populations (Chapter 8). This is yet another example of the constantly changing controls over successional dynamics in Alaskan floodplains. The shift from broad-leafed to evergreen dominance initiates a chain of events that leads to a decline in rates of nutrient cycling. The turning point is facilitated by the shift from rapidly decomposing poplar litter to the more recalcitrant spruce and moss litter (Chapter 14; Van Cleve et al. 1983b, Van Cleve et al. 1993). The dense spruce canopy of the white spruce stage reduces radiation input to the forest floor, so soils thaw slowly in spring and summer. The combination of low soil temperature, high C:N ratio, and high concentrations of lignin and other polyphenolic compounds reduce rates of decomposition and nutrient cycling (Flanagan and Van Cleve 1983, Van Cleve et al. 1983b, Van Cleve et al. 1993), leading to the development of a thick organic mat between the moss layer and mineral soil. White spruce is relatively shallow-rooted, in part because low soil temperatures inhibit the growth of deep roots (Lawrence and Oechel 1983, Tryon and Chapin 1983). The growth of spruce therefore depends largely on the tight recycling of nutrients that occurs within the surface and buried organic horizons. Because of its shallow rooting depth, white spruce growing in floodplains depends on precipitation inputs to meet much of its water requirements (Chapter 11).

Most white spruce stands that the LTER program has studied on the floodplain are initially relatively even-aged. They establish over a short time period (20–40 years), primarily in the intermediate balsam poplar stages of succession, and emerge through the balsam poplar canopy to dominate at about 100 years of age. However, subsequent episodes of white spruce recruitment and survival can occur in 40- to 60-year pulses, leading to uneven-aged white spruce stands that may reach more than 300 years of age. Factors that contribute to the variability in recruitment include flood events, episodic seed production, and cyclic browsing by snowshoe hares. Winter ice storms can break off treetops, kill trees, and reduce tree cover by 30–50% (Viereck et al. 1993). The relative abundance of young, even-aged white spruce stands and older uneven-aged stands depends on the dynamics of fluvial erosion and flood deposition.

Black Spruce Stage

Black spruce stands are common on old alluvial surfaces that formed thousands of years ago (Mann et al. 1995). However, we have little information about the origin of these stands, because river erosion and fire have eliminated most transitional stands in the active floodplain (Drury 1956, Viereck et al. 1983, Viereck et al. 1993). Over millennial time scales, white spruce stands do indeed succeed to black spruce, but this transition is often associated with fire (Chapter 5), suggesting that it is more a process of secondary succession than a final stage of primary succession (Mann et al. 1995). Perhaps the most interesting result is the clear evidence that successional trajectories have changed through time, as the freely draining soils in the active floodplain change to poorly drained permafrost-dominated soils on old river terraces.

The black spruce stands on the Tanana floodplain are ecologically variable. Most have permafrost, but some do not. There is a wide range in thaw depth, productivity, and rates of nutrient cycling (Chapter 11). Biomass of these black spruce stands is only 10% of that in white spruce stands.

Upland Forest Succession in Permafrost-Free Areas

Fire is the dominant disturbance in Alaskan uplands and inactive floodplains, as in most of the boreal forest. Postfire succession can follow a maze of alternative pathways (Fig. 7.6; Dyrness and Grigal 1979, Viereck et al. 1983, 1993, Yarie 1983, Youngblood 1992, Fastie et al. 2002). The patterns range from *self-replacement* after fire, in which the prefire dominant tree species returns to dominance shortly after fire, to *relay floristics*, in which tree species show sequential patterns of dominance. In general, self-replacement predominates in extreme environments (black spruce in cold permafrost-dominated sites, aspen in extremely warm dry sites). Relay floristics is the more common successional process in intermediate sites, where the successional trajectory varies, depending on slope and aspect (Fig. 7.6). In all cases,

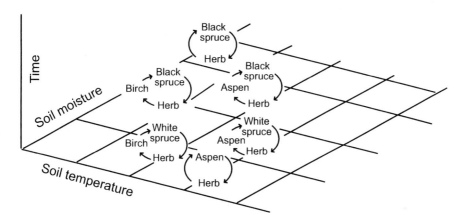

Figure 7.6. Representative successional trajectories on upland sites in interior Alaska along gradients of soil temperature and soil moisture. Upward arrows indicate successional changes in community composition; downward arrows indicate vegetation change caused by fire. Redrawn with permission from *Ambio* (Chapin et al. 2004).

the pathways become locked in place within two decades after fire, due to patterns of postfire regeneration, which reflect resprout potential of broad-leafed species, propagule availability, and species differences in physiological requirements for establishment (Johnstone et al. 2004).

The LTER program has focused most of its upland research on two successional pathways: self-replacement by black spruce, which is described in the next section, and relay floristics of stands that are dominated successively by herbs, shrubs, deciduous trees, and white spruce. Other pathways occur when at least one of these functional groups does not establish or is reduced in abundance: for example, dry bluffs where herbs dominate; deciduous forests, where spruce does not establish due to postfire seedbed conditions or timing or distance to seed source or where fire cycles are rapid enough to exclude spruce from becoming dominant (Mann and Plug 1999); and conifer-only forests, where deciduous trees do not establish. Topography, operating through its effects on interactive controls such as soil temperature and moisture, fire regime, and prefire vegetation, ultimately modulates the interplay among successional processes at landscape scales (Chapter 1).

Fire Regime

Lightning is the primary natural ignition source for interior Alaskan fires (Dissing and Verbyla 2003; Chapter 17). In most years, conifer forests burn more readily than deciduous forests because their ericaceous- and moss-dominated understories carry fire readily. Fires also spread through broad-leafed stands in spring, when dry litter and evergreen shrubs account for most of the understory fuel. More important, under conditions of severe fire weather, most stand types burn readily, and crown fires are more frequent. Most of the area burned results from fires that occur in these unusually dry years (Chapter 17).

Fire severity, that is, the proportion of the organic mat and tree canopy consumed, is patchy and depends on vegetation and topography and on local wind, weather, and moisture at the time of the fire. Some patches are untouched by fire, whereas others burn so severely that propagule banks (seeds, roots, and rhizomes) are destroyed, limiting plant establishment to species that can disperse from surrounding areas. Fire severity also influences postfire moisture and nutrient availability, with seedlings establishing most readily on severely burned mineral substrates (Zasada et al. 1983). The heterogeneous patterns of fire and postfire recruitment generate a mosaic of moderate-size forest stands (often 2–20 ha) with an intermixture of early-successional deciduous and late-successional spruce stands of variable stand composition (Dyrness et al. 1986). This fire-generated mosaic increases the exposure of deciduous stands to fire as a result of proximity to flammable conifer stands (Rupp et al. 2000) and protects some late-successional stands from fire, allowing a few stands to persist up to 300 years (Yarie 1981).

Herb-Resprout Stage

Two striking differences between floodplain and upland succession are the large initial pool of available nutrients and the rapid establishment of tree-dominated vegetation that occurs after fire in the uplands (Fig. 7.6; Viereck and Schandelmeier 1980). Some nitrogen and other nutrients are lost by volatilization (nitrogen) or carried away in smoke, but substantial quantities of nutrients return to the soil in available form. In the 1983 Rosie Creek fire that burned south-facing birch, aspen, and white spruce stands in the Bonanza Creek LTER site, for example, about half of the forest-floor carbon and nitrogen were lost during the fire, but the pool of *available* nutrients increased two- to three-fold immediately after fire (Dyrness et al. 1989). The elevated concentration of cations increased soil pH by 1.5 to 2 units, leading to a near-neutral pH (6–8) that maximizes phosphorus availability. Nutrient availability is also enhanced after fire by postfire processes. The loss of the overstory canopy and the reduction in albedo after fire increase soil temperature and moisture (Chambers and Chapin 2002), which speed decomposition and nutrient release from the remaining forest floor (Chapter 15; Van Cleve et al. 1991). The low precipitation regime of interior Alaska minimizes leaching loss to groundwater and streams (Chapter 16), so many of the nutrients released by fire may remain available to support vegetation regrowth, although we know relatively little about the retention of these postfire nutrients.

Herbs, shrubs, and trees dominate the initial postfire community in permafrost-free uplands. Many herbs, such as horsetail (*Equisetum*), resprout from rhizomes that survive fire. At the opposite extreme, *Corydalis sempervirens* germinates from a seed bank produced immediately after the previous fire, perhaps 100–200 years earlier. Other wind-dispersed species such as fireweed (*Epilobium*) and the grass bluejoint (*Calamagrostis*) colonize by seed from outside the burn and from rhizomes that survived fire. Once established, these species spread vegetatively and produce seeds that contribute to the rapidly developing cover (Viereck 1973, Viereck et al. 1983). Belowground parts of shrubs and deciduous trees that were present before fire generally survive and resprout from belowground stems and roots within days after fire. Resprouts initially grow more rapidly than seedlings and may form a large proportion of the individuals that become part of the postfire canopy. There is therefore a high probability that stands will continue to be dominated by the same deciduous tree species through multiple fire cycles (Mann and Plug 1999). The sharply delineated patches of yellow and orange aspen clones visible in autumn have probably dominated the same sites for hundreds or thousands of years, over multiple fire cycles, just as observed in the Rocky Mountains of Colorado (Mueggler 1985). White spruce, however, must recruit from seed after each fire, so its recruitment is more variable and depends on fire patterns that govern seedbed conditions and distance to seed source. Both seed rain and the number of seedlings that establish drop off exponentially with distance from seed source (Zasada et al. 1992, Rupp 1998, Cater and Chapin 2000). Just as in the floodplain, the development of a litter layer greatly reduces spruce germination and establishment, so recruitment is most successful on a mineral-soil seedbed. Unlike the floodplain, this mineral surface is only available for a brief 5- to 20-year window after fire, so the initial cohorts of both evergreen and deciduous species in upland stands are even-aged, with most trees initiating their aboveground growth within two decades after fire (Viereck 1973, Youngblood 1992, Johnstone et al. 2004). If the initial cohorts are sparse, however, additional cohorts of white or black spruce may establish in some stands, although these later recruits seldom dominate the canopy (Fastie et al. 2002). This suggests that development of even-aged stands results from both near-synchronous seedling establishment after fire and reduced survival or growth of seedlings that establish later in succession (Johnson et al. 1994).

Shrub-Sapling Stage

The turning point to the shrub-sapling stage results from changes in competitive dominance of species already present on the site. Within five years after fire, resprouting shrubs and saplings and seedlings of deciduous trees dominate the stand. The herbaceous cover declines as a result of a combination of shading and herbivory. The increasing productivity of shrubs and saplings in this stage provides forage and cover for hares and moose, which feed on willow, aspen, and birch. Green alder (*Alnus viridis* subsp. *fruticosa*, previously treated as *A. crispa*), which resprouts and establishes by seed after fire, fixes nitrogen at only 5% of the rate of the floodplain alder (Anderson et al. 2004). Nonetheless, it probably contributes to the high fertility of early postfire successional stands.

Deciduous-Forest Stage

As on the floodplain, the deciduous-forest stage is the most productive successional stage (Chapter 11). The high-quality litter produced by these trees maintains the high rates of nitrogen mineralization and nutrient supply that support this production. Deciduous litterfall prevents establishment of mosses, just as in the floodplain (Oechel and Van Cleve 1986) and minimizes the recruitment of new spruce seedlings. Spruce saplings generally grow more slowly than the deciduous trees, forming a subcanopy layer. The density of this spruce subcanopy is variable, ranging from near dominance to sites where the canopy is so sparse that deciduous trees continue to dominate in a state of gap-phase dynamics (that is, where tree replacement occurs primarily in gaps created by death of older individual trees) for two or more centuries (Fastie et al. 2002), similar to old-growth deciduous stands in western Canada (Cumming et al. 2000).

Spruce Stage

On many south-facing upland sites, a transition from deciduous forest to white spruce occurs after about 100 years in the absence of fire. This dramatic turning point is similar in timing and in ecological controls to the poplar-spruce transition on the floodplain. It is accompanied by the gradual reduction in stem density of deciduous trees and the development of a moss understory and organic mat that reduce soil temperature, litter quality, and decomposition rate. The annual nutrient uptake in the white spruce stage is only 30–40% of that in the deciduous-forest stage (Van Cleve et al. 1983b). This successional stage can persist for at least another century, during which time the stand is likely to burn. White spruce stands older than 200 years are uncommon in Alaska, so it is unclear where succession would lead if fire did not occur (Van Cleve and Viereck 1981).

Permafrost-Dominated Uplands and Lowlands

Black spruce, which dominates north-facing slopes and poorly drained forested lowlands (Chapter 6), occupies approximately 44% of interior Alaska (Chapter 6; Yarie and Billings 2002). Most of these stands are underlain by permafrost and have cold soils, slow decomposition, and low soil fertility, conditions that differ radically from those on permafrost-free floodplains and south-facing uplands. The slow growth rate, prolonged needle retention, and low nutrient requirement of black spruce enable it to exploit this environment. The cycle of fire and postfire succession in black spruce is qualitatively similar to that in south-facing uplands, but it generally lacks a deciduous forest stage and therefore returns to a flammable, spruce-dominated stage in 20–30% of the time (15–30 years) required for white-spruce succession (100 years; Fig. 7.6; Viereck 1973, Dyrness et al. 1986).

Fire Regime

Black spruce ecosystems are born to burn. The thick moss and organic mats that form beneath black spruce dry quickly during the warm, dry conditions that characterize the midsummer continental conditions of interior Alaska (Chapter 4). Oxygen diffuses readily into the low-bulk-density moss mats, promoting effective combustion. Fire moves readily through the understory of flammable evergreen shrubs, resinous dwarf birch, mosses, and the underlying organic mat (Viereck 1973, Johnson 1992). Fire also moves up a ladder of resin-rich, flammable needles and branches into the tree canopy. The arrival of black spruce in Alaska 5,000 years ago coincided with increased fire frequency, despite the cooler, moister climate that allowed black spruce to invade (Chapter 5). Today, the broad expanses of black spruce lowlands provide large homogeneous areas of flammable vegetation that are readily ignited by lightning and produce large fires (Chapter 17).

Herb-Resprout Stage

Grasses (particularly *Calamagrostis*), tall shrubs (*Salix, Alnus*), and dwarf shrubs (*Ledum, Vaccinium*) generally resprout from the partially consumed organic mat after fire, whereas fireweed and other herbaceous species germinate from wind-blown seed (e.g., *Epilobium*) or from the seed bank (e.g., *Corydalis*). The moist conditions associated with permafrost frequently result in less consumption of the organic mat in black spruce than in south-facing white-spruce uplands, although burn severity is highly variable and depends on the weather and thaw depth (time of season) at the time of the fire (Chapter 17). Black spruce seeds are released from the semiserotinous cones of the dead snags one to five years after the heat of the fire causes the cones to open (Viereck 1973, Zasada 1986). Only after the most severe fires, in which most tree crowns and cones are consumed, are there insufficient seeds to produce a fully stocked black spruce stand. This contrasts strikingly with white spruce, which disseminates seeds only from live trees and whose establishment is therefore more sensitive to distance from seed source (Zasada et al. 1992). Black spruce germinates most readily on severely burned patches with exposed mineral soil, although germination also occurs on organic substrates (Zasada et al. 1983, Johnstone et al. 2004). As in white spruce stands, there is a large return of nutrients in ash after fire in black spruce stands (Dyrness and Norum 1983). The dark color and reduction or disappearance of the moss mat results in a large ground heat flux that warms soils and increases the depth of thawed soil (Chapter 4; Chambers and Chapin 2002). The herbaceous species that recruit after fire have a much higher litter quality than the prefire moss and black spruce litter. Even within a species, such as the resprouting willows, the high nitrogen concentration and low concentrations of secondary metabolites characteristic of resprouts enhance decomposition potential (Irons et al. 1991). All these conditions favor postfire decomposition and higher soil fertility that supports the regenerating vegetation.

Willow Shrub Stage

Black spruce seedlings have a much lower growth potential than other boreal trees and shrubs in interior Alaska (Chapin et al. 1983), so black spruce remains in the herbaceous and shrub understory during early postfire regeneration. The shrub stage varies significantly in its species composition depending on burn severity. It is often dominated by willows, which resprout from the partially consumed organic mat or germinate from seed. However, ericaceous species such as Labrador tea (*Ledum groenlandicum*) and blueberry (*Vaccinium uliginosum*) resprout readily in areas of low fire severity, and aspen and birch can establish by seed in severely burned areas, where there is a bare mineral seedbed. These differences in the species composition of the shrub stage have important implications for black spruce succession. Re-establishment of ericaceous species quickly leads to a low-statured, nutrient-poor environment in which herbaceous species are replaced by black spruce within two to four decades. After more severe burns, however, this successional trajectory may diverge to include colonization by deciduous trees that can establish on exposed mineral soil (Viereck et al. 1983), converting the site to a deciduous stand or even open meadow (Foote 1983, Viereck et al. 1983). On dry glacial outwash plains in the upper Tanana River drainage, for example, aspen establishes after severe fires and retains canopy dominance for at least 50 years before black spruce and its associated evergreen understory regain dominance (Kasischke et al. 2000). The resulting doubling of the fire return interval reduces landscape-scale flammability and acts as a negative feedback to regional fire probability.

Black Spruce Stage

The return to black spruce dominance after fire is a successional turning point that is functionally similar to the shift to white spruce dominance in the floodplains and south-facing uplands (Van Cleve et al. 1991). However, the impacts on ecosystem processes are even more dramatic in black spruce succession because of the presence of permafrost (Chapters 4 and 16). The combination of cold, poorly drained soils and low litter quality reduce the abundance of deciduous species and promote growth of mosses, which further reduce soil temperature and litter quality (Van Cleve and Viereck 1981, Van Cleve et al. 1983a, Oechel and Van Cleve 1986, Bonan and

Korzuhin 1989), producing the least productive forest type in interior Alaska (Chapter 11). The declining nutritional status of black spruce through succession increases needle longevity (Hom and Oechel 1983) and the contents of lignin and resin, all of which contribute to the flammability of black spruce.

Conclusions

Successional studies in the Alaskan boreal forest shed light on many theoretical controversies that have long intrigued ecologists. Succession illustrates the importance of time and other state factors and interactive controls in shaping the properties and dynamics of ecosystems (Chapter 1; Jenny 1980, Chapin et al. 2002). Geomorphic differences associated with topography lead to radically different disturbance regimes and postdisturbance microenvironments in floodplains, permafrost-free uplands, and permafrost-dominated sites. These factors influence the composition of the initial plant community, which, to a large extent, dictates the trajectory of succession. The successional changes in community composition in turn modify the physical and chemical environment, causing changes in primary production (Chapter 11), biogeochemistry (Chapter 14), and trophic dynamics (Chapters 8 and 13). These community and ecosystem changes in turn govern disturbance regime and the species available to colonize the next disturbance. And so the cycle repeats.

The boreal forest has strikingly few tree species, each of which has quite different effects on ecosystem processes. Successional changes in species dominance therefore generate a series of turning points characterized by threshold changes in physical and chemical environment and ecosystem processes. The turning points in boreal succession provide dramatic evidence for the mechanisms by which the biota shapes its physical and chemical environment (Odum 1969). These autogenic mechanisms include facilitation, competition, and herbivory, which interact to modify the patterns of succession that might logically follow from species differences in life history traits (Clements 1916, Connell and Slatyer 1977, Walker and Chapin 1987). In the floodplain, for example, life history traits dictate the basic pattern of species replacement, in which light-seeded, fast-growing colonizers are gradually replaced by more slowly growing, longer lived, and taller late-successional species. Alder both facilitates the growth of late-successional species by adding nitrogen but also competes with them by reducing light in the understory and increases the risk of mortality by herbivores that seek cover in dense alder thickets. Succession is the net effect of these interactions (Walker and Chapin 1987, Viereck et al. 1993), and turning points describe nonlinear changes in the relative importance of these processes (Van Cleve et al. 1996).

The community dynamics observed in the Alaskan boreal forest span the range of mechanisms postulated by community ecologists to explain succession (Egler 1954, Pickett et al. 1987). On extremely cold and dry sites, black spruce and aspen, respectively, exhibit self-replacement, with no other intervening tree species. In the floodplain and in more mesic upland sites, relay floristics is the rule, with a replacement of deciduous trees by conifers. In most cases, the pattern of relay floristics is a consequence of simultaneous colonization by all species (i.e., initial floristics), followed by a shift in dominance from rapidly growing to more slowly growing species. Floodplain white spruce, which colonizes primarily in midsuccession, is the major exception to this generalization.

In summary, the successional patterns and processes observed in the Alaskan boreal forest provide clear illustrations of processes and patterns that probably occur generally but are often difficult to study in other biomes. The extreme climate and the striking topographic variation in microclimate and disturbance regime in the boreal forest accentuate the roles of state factors and interactive controls, factors that are equally important but more complex and subtly interacting in other biomes. Self-replacement, which we observe primarily in extreme environments, is also typical of dry grasslands and chaparral, whereas more mesic temperate and tropical forests exhibit relay floristics. Our observation that most species establish simultaneously after disturbance (initial floristics) is also a pattern widely observed in other ecosystems. The relatively low diversity of boreal forests clarifies the roles of individual species in ecosystem processes and community dynamics.

References

Adams, P. C. 1999. The dynamics of white spruce populations on a boreal river floodplain. Ph.D. Dissertation, Duke University, Durham, NC.

Anderson, M. D., R. W. Ruess, D. D. Uliassi, and J. S. Mitchell. 2004. Estimating N_2 fixation in two species of *Alnus* in interior Alaska using acetylene reduction and $^{15}N_2$ uptake. Ecoscience 11:102–112.

Bonan, G. B., and M. D. Korzuhin. 1989. Simulation of moss and tree dynamics in the boreal forests of interior Alaska. Vegetatio 84:31–44.

Burrows, R. L., W. W. Emmett, and B. Parks. 1981. Sediment transport in the Tanana River near Fairbanks, Alaska, 1977–1979. Water Resources Investigations Report 83–4064. U.S. Geological Survey, Menlo Park, CA.

Cater, T. C., and F. S. Chapin III. 2000. Differential species effects on boreal tree seedling establishment after fire: Resource competition or modification of microenvironment. Ecology 81:1086–1099.

Chambers, S. D., and F. S. Chapin III. 2002. Fire effects on surface-atmosphere energy exchange in Alaskan black spruce ecosystems: Implications for feedbacks to regional climate. Journal of Geophysical Research 108:8145, doi:10.1029/2001JD000530 [printed in 108(D1), 2003].

Chapin, F. S., III, P. A. Matson, and H. A. Mooney. 2002. Principles of Terrestrial Ecosystem Ecology. Springer-Verlag, New York.

Chapin, F. S., III, K. Van Cleve, and P. R. Tryon. 1983. Influence of phosphorus on the growth and biomass allocation of Alaskan taiga tree seedlings. Canadian Journal of Forest Research 13:1092–1098.

Chapin, F. S., III, T. V. Callaghan, Y. Bergeron, M. Fukuda, J. F. Johnstone, G. Juday, and S. A. Zimov. 2004. Global change and the boreal forest: Thresholds, shifting states or gradual change? Ambio 33: 361–365.

Clements, F. E. 1916. Plant Succession: An Analysis of the Development of Vegetation. Carnegie Institution of Washington Publication 242, Washington, DC.

Collins, C. M. 1990. Morphometric analyses of recent channel changes on the Tanana River

in the vicinity of Fairbanks, Alaska. Report 90–4, U.S. Army Corps of Engineers, Cold Regions Research and Engineering Laboratory, Hanover, NH.
Connell, J. H., and R. O. Slatyer. 1977. Mechanisms of succession in natural communities and their role in community stability and organization. American Naturalist 111:1119–1114.
Crocker, R. L., and J. Major. 1955. Soil development in relation to vegetation and surface age at Glacier Bay, Alaska. Journal of Ecology 43:427–448.
Cumming, S. G., F. K. A. Schmiegelow, and P. J. Burton. 2000. Gap dynamics in boreal aspen stands: Is the forest older than we think? Ecological Applications 10:744–759.
Dissing, D., and D. Verbyla. 2003. Spatial patterns of lightning strikes in interior Alaska and their relations to elevation and vegetation. Canadian Journal of Forest Research 33:770–782.
Drury, W. H. 1956. Bog flats and physiographic processes in the upper Kuskokwim River region, Alaska. Contributions from the Gray Herbarium of Harvard University 178.
Dyrness, C. T., and D. F. Grigal. 1979. Vegetation-soil relationships along a spruce forest transect in interior Alaska. Canadian Journal of Botany 57:2644–2656.
Dyrness, C. T., and R. A. Norum. 1983. The effects of experimental fires on black spruce forest floors in interior Alaska. Canadian Journal of Forest Research 13:879–893.
Dyrness, C. T., and K. Van Cleve. 1993. Control of surface soil chemistry in early-successional floodplain soils along the Tanana River. Canadian Journal of Forest Research 23:979–994.
Dyrness, C. T., K. Van Cleve, and J. Levison. 1989. The effects of wildfire on soil chemistry in four forest types in interior Alaska. Canadian Journal of Forest Research 19:1389–1396.
Dyrness, C. T., L. A. Viereck, and K. Van Cleve. 1986. Fire in taiga communities of interior Alaska. Pages 74–86 *in* C. T. Dyrness, editor. Forest Ecosystems in the Alaskan Taiga: A Synthesis of Structure and Function. Springer-Verlag, New York.
Egler, F. E. 1954. Vegetation science concepts. I. Initial floristic composition, a factor in old-field vegetation development. Vegetatio 4:414–417.
Fastie, C. L., A. H. Lloyd, and P. Doak. 2002. Fire history and post-fire forest development in an upland watershed of interior Alaska. Journal of Geophysical Research 108:8150, doi:10.1029/2001JD000570 [printed in 108(D1), 2003].
Flanagan, P. W., and K. Van Cleve. 1983. Nutrient cycling in relation to decomposition and organic matter quality in taiga ecosystems. Canadian Journal of Forest Research 13:795–817.
Foote, M. J. 1983. Classification, description, and dynamics of plant communities after fire in the taiga of interior Alaska. Research Paper PNW–307, U.S. Department of Agriculture, Forest Service, Portland, OR.
Hom, J. L., and W. C. Oechel. 1983. The photosynthetic capacity, nutrient content, and nutrient use efficiency of different needle age-classes of black spruce (*Picea mariana*) found in interior Alaska. Canadian Journal of Forest Research 13:834–839.
Irons, J. G., III, J. P. Bryant, and M. W. Oswood. 1991. Effects of moose browsing on decomposition rates of birch leaf litter in a subarctic stream. Canadian Journal of Fisheries and Aquatic Science 48:442–444.
Jenny, H. 1980. The Soil Resources: Origin and Behavior. Springer-Verlag, New York.
Johnson, E. A. 1992. Fire and Vegetation Dynamics. Studies from the North American Boreal Forest. Cambridge University Press, Cambridge.
Johnson, E. A., K. Miyanishi, and H. Kleb. 1994. The hazards of interpretation of static age structures as shown by stand reconstructions in a *Pinus contorta-Picea engelmannii* forest. Journal of Ecology 82:923–931.

Johnstone, J. F., F. S. Chapin III, J. Foote, S. Kemmett, K. Price, and L. A. Viereck. 2004. Decadal observations of tree regeneration following fire in boreal forests. Canadian Journal of Forest Research 34:267–273.

Kasischke, E. S., N. H. F. French, K. P. O'Neill, D. D. Richter, L. L. Bourgeau-Chavez, and P. A. Harrell. 2000. Influence of fire on long-term patterns of forest succession in Alaskan boreal forests. Pages 214–238 in B. J. Stocks, editor. Fire, Climate Change, and Carbon Cycling in the Boreal Forest. Springer-Verlag, New York.

Kielland, K., and J. Bryant. 1998. Moose herbivory in taiga: Effects on biogeochemistry and vegetation dynamics in primary succession. Oikos 82:377–383.

Krasny, M. E., K. A. Vogt, and J. C. Zasada. 1988. Establishment of four Salicaceae species on river bars in interior Alaska. Holarctic Ecology 11:210–219.

Lawrence, W. T., and W. C. Oechel. 1983. Effects of soil temperature on the carbon exchange of taiga seedlings. II. Photosynthesis, respiration, and conductance. Canadian Journal of Forest Research 13:850–859.

Mann, D. H., and L. J. Plug. 1999. Vegetation and soil development at an upland taiga site, Alaska. Ecoscience 6:272–285.

Mann, D. H., C. L. Fastie, E. L. Rowland, and N. H. Bigelow. 1995. Spruce succession, disturbance, and geomorphology on the Tanana River floodplain, Alaska. Ecoscience 2:184–199.

Marion, G. M., K. Van Cleve, C. T. Dyrness, and C. H. Black. 1993. The soil chemical environment along a primary successional sequence on the Tanana River floodplain, interior Alaska. Canadian Journal of Forest Research 23:923–927.

Mason, O. K., and J. E. Begét. 1991. Late Holocene flood history of the Tanana River, Alaska, USA. Arctic and Alpine Research 23:392–403.

McAvinchey, R. J. P. 1991. Winter herbivory by snowshoe hares and moose as a process affecting primary succession on an Alaskan floodplain. M.S. Thesis, University of Alaska Fairbanks.

Mueggler, W. F. 1985. Vegetation associations. Pages 45–55 in R. P. Winokur, editor. Aspen: Ecology and Management in the Western United States. U.S. Department of Agriculture, Forest Service, Washington, DC.

Odum, E. P. 1969. The strategy of ecosystem development. Science 164:262–270.

Oechel, W. C., and K. Van Cleve. 1986. The role of bryophytes in nutrient cycling in the taiga. Pages 121–137 in C. T. Dyrness, editor. Forest Ecosystems in the Alaskan Taiga. A Synthesis of Structure and Function. Springer-Verlag, New York.

Ott, R. A., M. A. Lee, W. E. Putman, O. K. Mason, G. T. Worum, and D. N. Burns. 2001. Bank erosion and large woody debris recruitment along the Tanana River, interior Alaska. Report to Alaska Department of Environmental Conservation, Division of Air and Water Quality NP–01–R9. Alaska Department of Natural Resources, Division of Forestry and Tanana Chiefs, Conference, Inc., Forestry Program, Fairbanks, Alaska.

Pickett, S. T. A., S. L. Collins, and J. J. Armesto. 1987. A hierarchical consideration of causes and mechanisms of succession. Vegetatio 69:109–114.

Rupp, T. S. 1998. Boreal forest regeneration dynamics: Modeling early forest establishment patterns in interior Alaska. Ph.D. Dissertation, University of Alaska Fairbanks.

Rupp, T. S., A. M. Starfield, and F. S. Chapin III. 2000. A frame-based spatially explicit model of subarctic vegetation response to climatic change: Comparison with a point model. Landscape Ecology 15:383–400.

Schimel, J. P., K. Van Cleve, R. G. Cates, T. P. Clausen, and P. B. Reichardt. 1996. Effects of balsam poplar *(Populus balsamifera)* tannins and low molecular weight phenolics on microbial activity in taiga floodplain soil: Implications for changes in N cycling during succession. Canadian Journal of Botany 74:84–90.

Shaw, J. D., E. C. Packee, and C. L. Ping. 2001. Growth of balsam poplar and black cottonwood in Alaska in relation to landform and soil. Canadian Journal of Forest Research 31:1793–1804.

Slaughter, C. W., and L. A. Viereck. 1986. Climatic characteristics of the taiga of interior Alaska. Pages 9–21 in C. T. Dyrness, editor. Forest Ecosystems in the Alaskan Taiga. A Synthesis of Structure and Function. Springer-Verlag, New York.

Tryon, P. R., and F. S. Chapin III. 1983. Temperature control over root growth and root biomass in taiga forest trees. Canadian Journal of Forest Research 13:827–833.

Uliassi, D. D., and R. W. Ruess. 2002. Limitations to symbiotic nitrogen fixation in primary succession on the Tanana River floodplain, Alaska. Ecology 83:88–103.

Uliassi, D. D., K. Huss-Danell, R. W. Ruess, and K. Doran. 2000. Biomass allocation and nitrogenase activity in *Alnus tenuifolia*: responses to successional soil type and phosphorus availability. Ecoscience 7:73–79.

Van Cleve, K., and L. A. Viereck. 1981. Forest succession in relation to nutrient cycling in the boreal forest of Alaska. Pages 185–211 in D. B. Botkin, editor. Forest Succession, Concepts and Application. Springer-Verlag, New York.

Van Cleve, K., L. A. Viereck, and C. T. Dyrness. 1996. State factor control of soils and forest succession along the Tanana River in interior Alaska, USA. Arctic and Alpine Research 28:388–400.

Van Cleve, K., L. A. Viereck, and R. L. Schlentner. 1971. Accumulation of nitrogen in alder (*Alnus*) ecosystems near Fairbanks, Alaska. Arctic and Alpine Research 3:101–114.

Van Cleve, K., F. S. Chapin III, C. T. Dyrness, and L. A. Viereck. 1991. Element cycling in taiga forest: State-factor control. BioScience 41:78–88.

Van Cleve, K., C. T. Dyrness, G. M. Marion, and R. Erickson. 1993. Control of soil development on the Tanana River floodplain, interior Alaska. Canadian Journal of Forest Research 23:941–955.

Van Cleve, K., C. T. Dyrness, L. A. Viereck, J. Fox, F. S. Chapin III, and W. C. Oechel. 1983a. Taiga ecosystems in interior Alaska. BioScience 33:39–44.

Van Cleve, K., L. Oliver, R. Schlentner, L. A. Viereck, and C. T. Dyrness. 1983b. Productivity and nutrient cycling in taiga forest ecosystems. Canadian Journal of Forest Research 13:747–766.

Viereck, L. A. 1970. Forest succession and soil development adjacent to the Chena River in interior Alaska. Arctic and Alpine Research 2:1–26.

Viereck, L. A. 1973. Wildfire in the taiga of Alaska. Quaternary Research 3:465–495.

Viereck, L. A., and L. H. Schandelmeier. 1980. Effects of fire in Alaska and adjacent Canada—A literature review. U.S. Department of the Interior, Bureau of Land Management, Fairbanks.

Viereck, L. A., C. T. Dyrness, and M. J. Foote. 1993. An overview of the vegetation and soils of the floodplain ecosystems of the Tanana River, interior Alaska. Canadian Journal of Forest Research 23:889–898.

Viereck, L. A., C. T. Dyrness, A. R. Batten, and K. J. Wenzlick. 1992. The Alaska vegetation classification. General Technical Report PNW-GTR-286, U.S. Department of Agriculture, Forest Service, Pacific Northwest Research Station, Portland, OR.

Viereck, L. A., C. T. Dyrness, K. Van Cleve, and M. J. Foote. 1983. Vegetation, soils, and forest productivity in selected forest types in interior Alaska. Canadian Journal of Forest Research 13:703–720.

Walker, L. R. 1989. Soil nitrogen changes during primary succession on a floodplain in Alaska, U.S.A. Arctic and Alpine Research 21:341–349.

Walker, L. R., and F. S. Chapin III. 1986. Physiological controls over seedling growth in primary succession on an Alaskan floodplain. Ecology 67:1508–1523.

Walker, L. R., and F. S. Chapin III. 1987. Interactions among processes controlling successional change. Oikos 50:131–135.

Walker, L. R., J. C. Zasada, and F. S. Chapin III. 1986. The role of life history processes in primary succession on an Alaskan floodplain. Ecology 67:1243–1253.

Wurtz, T. L. 1988. Effects of the microsite on the growth of planted white spruce seedlings. Ph.D. Dissertation, University of Oregon, Eugene.

Yarie, J. 1981. Forest fire cycles and life tables: A case study from interior Alaska. Canadian Journal of Forest Research 11:554–562.

Yarie, J. 1983. Environmental and successional relationships of the forest communities of the Porcupine River drainage, interior Alaska. Canadian Journal of Forest Research 13:703–720.

Yarie, J. 1993. Effects of selected forest management practices on environmental parameters related to successional development on the Tanana River floodplain, interior Alaska. Canadian Journal of Forest Research 23:1001–1014.

Yarie, J., and S. Billings. 2002. Carbon balance of the taiga forest within Alaska: Present and future. Canadian Journal of Forest Research 32:757–767.

Yarie, J., L. Viereck, K. Van Cleve, and P. Adams. 1998. Flooding and ecosystem dynamics along the Tanana River. BioScience 48:690–695.

Youngblood, A. 1992. Structure and dynamics in mixed forest stands of interior Alaska. Ph.D. Dissertation, University of Alaska Fairbanks.

Zasada, J. 1986. Natural regeneration of trees and tall shrubs on forest sites in interior Alaska. Pages 44–73 *in* C. T. Dyrness, editor. Forest Ecosystems in the Alaskan Taiga. A Synthesis of Structure and Function. Springer-Verlag, New York.

Zasada, J., R. A. Norum, R. M. Van Veldhuizen, and C. E. Teutsch. 1983. Artificial regeneration of trees and tall shrubs in experimentally burned upland black spruce/feather moss stands in Alaska. Canadian Journal of Forest Research 13:903–913.

Zasada, J. C. 1984. Site classification and regeneration practices on floodplain sites in interior Alaska. Forest classification at high latitudes as an aid to regeneration. Report 35–39, U.S. Department of Agriculture, Forest Service, Pacific Northwest Forest and Range Experiment Station, Portland, OR.

Zasada, J. C., T. L. Sharik, and M. Nygren. 1992. The reproductive process in boreal forest trees. Pages 85–125 *in* G. B. Bonan, editor. A Systems Analysis of the Global Boreal Forest. Cambridge University Press, Cambridge.

8

Mammalian Herbivore Population Dynamics in the Alaskan Boreal Forest

Eric Rexstad
Knut Kielland

Introduction

The population dynamics of boreal mammals differ strikingly from those of mammals in temperate and tropical ecosystems in their extraordinary fluctuations in abundance (Elton 1924). These fluctuations lead to strong top-down direct effects in which herbivores reduce the biomass of their preferred foods, such as birch and willow, and predators reduce the biomass of herbivores (Chapter 13; Sinclair et al. 2000). These effects are clearly demonstrated in experiments that exclude herbivores or their predators. Some authors have argued that bottom-up influences of food supply on herbivores are negligible because food augmentation to herbivores in the presence of predators had no detectable effect in reducing herbivore decline (Sinclair et al. 2001).

Several members of the mammalian herbivore guild are also important as a human subsistence resource. Dynamics of moose (*Alces alces*) and snowshoe hare (*Lepus americanus*) can be altered by human harvest. Overexploitation by humans may reduce moose populations to densities where they can be predator-limited—the so-called predator pit (Messier 1994).

In this chapter, we present information on dynamics of some mammalian herbivores in the Alaskan boreal forest and potential drivers that are responsible for these dynamics. We omit discussions of the dynamics of porcupines (Keith and Cary 1991), red squirrels (Boonstra et al. 2001a), and beavers (Donkor and Fryxell 1999), as studies of these species have not been conducted in Alaska's boreal forests.

Historical Population Dynamics

Moose

Moose are thought to have arrived in Alaska during the Illinoian glaciation, about 400,000 yr B.P. (Pewe and Hopkins 1967). They may have retreated to refugia in central Alaska during subsequent glacial advances (Peterson 1955) and expanded at times when climate was warmer. Moose populations in North America have more than doubled over the past 30–40 years, to approximately 890,000 animals (Kelsall 1987). The Koyukuk River drainage in the northern interior, for example, is presently known for its large moose populations. However, the oral tradition of moose hunting in the Koyukuk is relatively recent. Native elders recall that, in their youth, moose were extremely rare and that moose did not figure prominently in the local subsistence economy until the 1930s. Over the past century, Alaska moose populations expanded from their stronghold in the central and eastern interior to the southwest (Yukon-Kuskokwim Delta), west (Seward Peninsula), and north all the way to the coast of the Arctic Ocean (Coady 1980). Thus, dispersal of moose beyond their traditional boreal habitats into tundra appears to be a relatively recent event, coinciding with recent warming (Chapter 4) and changes in habitat, as well as the population dynamics of other animal species with which moose interact.

The Alaskan populations make up about 15% of the moose in North America, with about 80% in Canada and the remaining 5% in the lower 48 states. They are currently at exceptionally high densities in the Tanana Flats and northern foothills of the Alaska Range and are considered a world-class wildlife resource (Alaska Department of Fish and Game 2002). The moose population in this region grew rapidly during the 1950s and reached densities as high as 2 moose·km^{-2} in the early 1960s. Excessive hunting of female moose (34% of the total harvest), combined with hard winters and high rates of wolf predation, reduced the populations nearly 90% by the mid-1970s (Gasaway et al. 1983). However, concomitant with state-initiated predator-control programs, the moose population recovered in less than a decade and has remained fairly stable at about 1.25 moose·km^{-2} until present (Don Young, ADF&G, personal communication).

Snowshoe Hares

There is only one data set that has documented the historical dynamics of snowshoe hares in Alaska. These data were collected consecutively by the Alaska Department of Fish and Game (Ernest 1974) and the U.S. Department of Agriculture's Institute of Northern Forestry (Wolff 1980). The combined study captured six years (Fig. 8.1) of a purported 10- to 11-year snowshoe hare cycle (Sinclair et al. 1993). At the beginning of the data series, the population density was roughly 5.9 hares·ha^{-1} and subsequently dropped to fewer than 0.3 hares·ha^{-1}. In contrast, the Kluane Boreal Forest Project (Krebs et al. 2001) has monitored snowshoe hare populations in the Yukon Territory for the period 1987–1996. Densities in this ecosystem peaked in 1990 at approximately 2.0 hares·ha^{-1}, with a low of 0.1 hares·ha^{-1} in 1993. In cen-

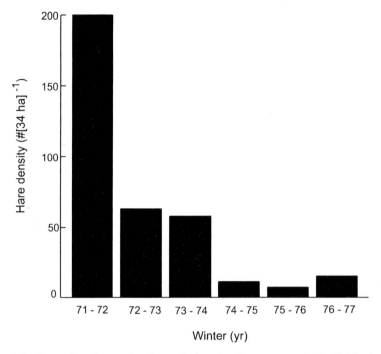

Figure 8.1. Dynamics of snowshoe hares during six winter seasons in the Fairbanks area during the 1970s. Data are compiled from different sites near Fairbanks and are tallies of animals captured. These are assumed to be a "close approximation of the total population" (Wolff 1980).

tral Alberta, in 1971, Keith and Windberg (1978) measured a snowshoe hare peak of approximately 11.8 hares·ha^{-1}.

In a study initiated at the onset of the most recent snowshoe hare peak in the Bonanza Creek LTER (Flora 2002), snowshoe hare populations peaked at 8.7 (SE = 1.9) hares·ha^{-1}. The population peak and subsequent decline roughly coincided across a 250 km "transect" from eastern to central Alaska. Whether snowshoe hare populations are geographically synchronous is a moderately contentious debate (see Hodges 2000). It has been suggested that snowshoe hares exhibit a "traveling wave" (Ranta et al. 1997) in which peaks in snowshoe hare density are initiated in the central portion of Canada and emanate north, west, and east from there.

Presently, the only geographically broad-scale snowshoe hare population index is derived from questionnaire data collected by the Alaska Department of Fish and Game (Scott and Kephardt 2002). These questionnaires are distributed annually to trappers, who respond on a qualitative scale as to whether snowshoe hares are abundant, common, or scarce in the area familiar to the trapper. These questionnaires are summarized by game management unit and depict broad-scale patterns in snowshoe hare abundance across the state of Alaska. Examination of the questionnaire data collected in the 1990s suggests a pattern of peaks first appearing in eastern Alaska and moving westward over a period of approximately six years. Snowshoe hare populations in interior Alaska have peaked in 1961, 1971, 1981, 1990, and 1999; with lows occurring in 1955, 1965, 1975, 1993–1994, and at this writing in 2002 (Flora 2002).

Microtines

Alaska's boreal forest has only a few species of small mammalian herbivores (Banfield 1974). Only four species exist in sufficient numbers to have received research attention. The northern red-backed vole (*Clethrionomys rutilus*), the tundra vole (*Microtus oeconomus*), the singing vole (*M. miurus*), and the yellow-cheeked vole (*M. xanthognathus*) have been the subject of natural history investigations (Murie 1962), but studies of the dynamics of these populations have been limited.

Whitney (1976) and West (1982) undertook investigations of dynamics of *Clethrionomys* and *Microtus oeconomus* in interior Alaska roughly 25 years ago. Their studies came to contradictory conclusions. Whitney (1976) implied that *Clethrionomys* populations did not exhibit cyclicity in their population dynamics but that *Microtus oeconomus* did, whereas West suggested that he detected a cyclic pattern in the *Clethrionomys* populations he studied. Wolff and Lidicker (1980) found that the yellow-cheeked vole, a colonial species, could grow to very high densities. However, being an obligate postfire successional species, its densities diminish as postfire vegetation stage of fireweed (*Epilobium*) and horsetail (*Equisetum*) decline.

In the 1990s, a series of graduate students investigated the dynamics of small mammal species in interior Alaska in a variety of ecosystems. Furtsch (1995) found *Clethrionomys* to exhibit amplitudes of population variation of an order of magnitude (19·ha^{-1} to 150·ha^{-1}) in Denali National Park and Preserve. McDonough (2000) found *Clethrionomys* populations to be largely insensitive to changes in forest overstory vegetation caused by infestations of spruce bark beetles but to exhibit substantial interannual variation. Lehmkuhl (2000) studied large-bodied (~150 g adults) yellow-cheeked voles and found that their populations could reach densities of 163·ha^{-1} (SE = 40) approximately 20 years after fire, with remnant stands of white spruce being the most prolific of the habitat types she studied.

Also during the 1990s, a longer (11-year) data series was collected by Rexstad and coworkers in Denali National Park and Preserve (Rexstad 1994, Oakley et al. 1999, Rexstad and Debevec 2002). This study tended to corroborate the findings of Whitney (1976) that cyclicity did not exist in populations of *Clethrionomys*. This species might more appropriately be termed "eruptive," reaching peak abundances of 80·0.8ha^{-1} but falling to lows of 6.0·0.8ha^{-1} (Fig. 8.2). At a spatial scale of 100 km separating the most distant sampling plots inside Denali National Park, there was geographic synchrony between years of high and years of low abundance, as discussed later.

In summary, although studied microtine populations in interior Alaska exhibit large population fluctuations, there is no strong evidence that these constitute popu-

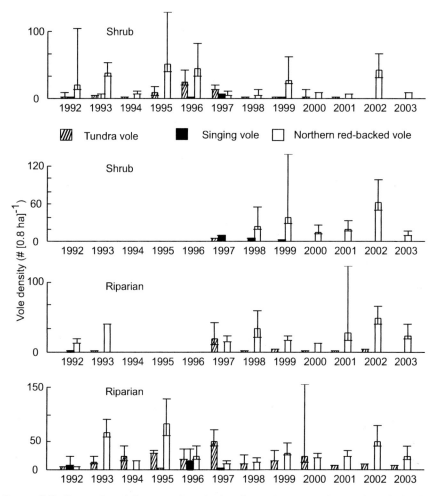

Figure 8.2. Dynamics of three species of microtine rodents—northern red-backed vole, tundra vole, and singing vole—in Denali National Park and Preserve, 1992–2002. The four plots represented are each 0.8ha in size and are located at the eastern end of the study area.

lation cycles. This contrasts with the relatively cyclic population fluctuations of snowshoe hares (Krebs et al. 2001) in western Canada and Alaska and lemmings in the tundra of Scandinavia (Boonstra et al. 1998).

Examining population dynamics of moose, snowshoe hares, and microtines in the currency of biomass rather than abundance (Table 8.1), we can see a range of 0.2 to 2.3 metric tons of herbivore biomass km^{-2} across the boreal landscape. It must be noted that peaks of biomass in microtines, hares, and moose are seldom, if ever, simultaneous, and fluctuations of microtines and hares are commonly out of phase (Boonstra et al. 2001b).

Table 8.1. Population densities and estimated biomass of moose, snowshoe hares, and microtines in interior Alaska.

Species	Population density (individuals km^{-2})	Biomass (kg km^{-2})
Moose	0.2–1.5[1]	80–600
Snowshoe hares	50–870[2]	55–957
Microtines	1000–10000[3]	50–500

Sources:
[1] Alaska Department of Fish and Game.
[2] Flora (2002).
[3] Rexstad and Debevec (2002).

Drivers of Population Change

The factors that induce population fluctuations of mammalian herbivores in boreal ecosystems fall into three categories: biotic drivers, climatic drivers, and disturbance drivers. The relative importance of these drivers depends on the body size of the herbivore.

Biotic drivers, especially predation, are most pronounced for the large-bodied mammalian herbivores, that is, moose and snowshoe hares. There is, for example, a rich and abundant literature on the influence of wolves on the dynamics of moose (Messier 1994). The coupling of lynx (*Lynx canadensis*) cycles to hare cycles (Mowat et al. 2000, O'Donoghue et al. 2001) also suggests a strong role of predation in driving hare populations. Sinclair et al. (1988) have argued that food limitation may exert an additive effect to the role of predation in diminishing snowshoe hare populations during crashes. Bryant (1981) and Fox and Bryant (1984) have also suggested that plant secondary metabolites, induced by hare browsing, contribute to hare population crashes. Palatability of forage species decreases because of induced plant defense and consequently hastens the crash. In contrast, Sinclair et al. (1988) argue that food limitation is not sufficient to induce a hare crash but rather speeds the decline in snowshoe hare abundance. Alterations of rainfall and soil moisture patterns, induced by periodicity in the solar cycle that in turn influence plant productivity, have also been implicated in influencing snowshoe hare population dynamics (Sinclair et al. 1993).

Climate has a particularly large effect on small mammalian herbivores. A model that incorporated local-scale meteorological characteristics (Rexstad and Debevec 2002) explained most of the annual variation in *Clethrionomys rutilus* population dynamics in Denali National Park and Preserve (Fig. 8.3). Overwinter survival was influenced by winter severity (a combination of air temperature and snow depth); food availability was influenced by green-up date (which depends on growing degree-day threshold); and survival of the first litter produced in the spring was influenced by early summer precipitation. Favorable survival rates among this first litter enabled them to participate in reproduction in that same season. The same suite of meteorological factors failed to adequately predict the population dynamics of either of the *Microtus* species, suggesting that these species are less sensitive to

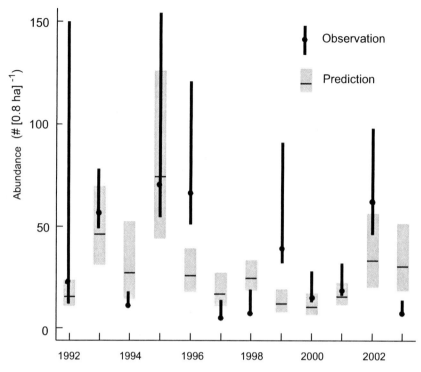

Figure 8.3. *Clethrionomys rutilus* population dynamics in eastern Denali National Park and Preserve, predicted from meteorological data. Vertical lines represent observed abundances (with 95% CI), and boxes represent predictions (with 95% prediction intervals) from the model. In only 1996 and 1999 do the confidence intervals not overlap.

interannual variation in climate. Additionally, Hansson (2001) noted that a complex interaction of snow deposition patterns and mustelid predation pressure altered the population dynamics of bank voles in Scandinavia. The quality of the forage available to microtine populations also influenced their dynamics (Oksanen et al. 1999), with populations in unproductive habitats remaining stable but populations in productive habitats fluctuating more dramatically. Together these studies indicate a strong effect of climate on microtines, with predation and food availability also being important under some circumstances.

Landscape-scale disturbances influence mammalian herbivores by determining the mosaic of habitats in which the animals find themselves. Herbivores in a tree-dominated boreal ecosystem generally seek out early successional patches of vegetation, because climax coniferous vegetation has food of extremely low palatability. As a result, riparian areas adjacent to braided rivers afford patches of high vertebrate density, partly because of the habitat mosaic created by continuous bank erosion and vegetative colonization of exposed gravel bars (Chapter 7). Fire also generates a diversity of successional stages across a landscape. Fox (1978) argued that aerial extent of area burned by fire was a good predictor of the number of lynx pelts coming to market and hence of snowshoe hare populations. Habitat alternation, via prescribed burns or "crushing," has been employed by resource management agencies for decades to enhance moose habitat.

These drivers also affect population dynamics indirectly, mediated by species interactions. For example, climatological factors operate so as to heighten either production of white spruce cone crops (in wet years) or the acreage burned by wildfire (in dry years). In years of large white spruce cone crops (Rupp 1998), red squirrel (*Tamias hudsonicus*) populations increase. Red squirrels constitute the greatest source of predation mortality for young snowshoe hare leverets (O'Donoghue 1994) and can suppress snowshoe hare populations. Another sequence of indirect effects that link the dynamics of mammalian herbivores has been put forward by Boonstra et al. (2001b). Their data from the Kluane Project suggests that *Clethrionomys* abundance peaks two to three years after a hare peak. They speculate that the pulse of nutrients moving through the snowshoe hares in the form of feces stimulates a pulse of production of forage for *Clethrionomys* while snowshoe hare populations are in their low phase.

Population Turnover

Maximum lifespan of mammalian wildlife species increases with body size. However, actual longevity is greatly affected by environment. The large ungulate species are buffered from climatic extremes but at times are susceptible to high rates of predation, especially on calves. Medium-sized animals such as the canids suffer from food shortage, trapping, and, in the case of wolves, significant intraspecific strife. The smallest bodied species are increasingly impacted by climate, especially during the shoulder seasons of freeze-up and snowmelt, in addition to predation and low food availability. The upshot of these processes is that populations turn over much more rapidly than their potential lifespan might suggest. Thus, microtines live on the average for less than six months. Snowshoe hares rarely experience more than two winters. The average wolf does not live much past three to four years (Mech et al. 1998). Last, moose populations turn over every five to six years (Fig. 8.4). These demographic phenomena demonstrate the dynamic nature of many wildlife populations.

Conclusions

The large amplitude of population fluctuations is the most striking characteristic of populations of mammalian herbivores that inhabit boreal forests. With the exception of the large-bodied, long-lived moose, the species in this guild commonly experience an order of magnitude change in density in the span of a decade. These changes in abundance precipitate radical changes in many ecosystem processes (Chapter 13).

However, the dynamics of herbivore populations are not synchronous either among or within a species. The heterogeneity of herbivore fluctuations in both time and space creates a tapestry of ecosystem effects that are important throughout boreal landscapes.

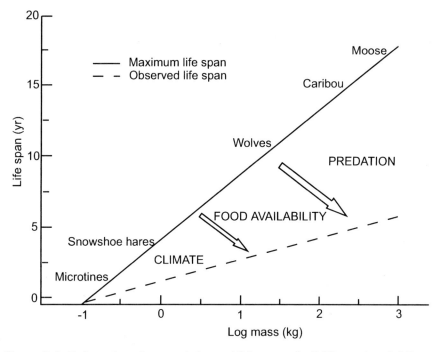

Figure 8.4. Estimated maximum and observed life spans of wildlife species of different mass in interior Alaska. The observed lifespan is estimated from field measurements of population turnover (density divided by annual recruitment). The major factors that limit lifespan are shown for animals of different sizes.

Beyond this shifting pattern of herbivores across landscapes, we have also described the differing mechanisms that induce changes in herbivore populations. Broadly classifying agents of population change into biotic, climatic, and disturbance, we find that the large-bodied herbivores are most likely to be influenced by biotic (predation) and disturbance (successional) processes, whereas the mesobodied herbivores are influenced by disturbance agents. Finally the small-bodied herbivores are most strongly influenced by climatic drivers. However, there are also subtle indirect pathways by which all three classes of drivers influence population dynamics of mammalian herbivores.

Because mammalian herbivores constitute an important food resource for people in the boreal forest, it is important to contemplate the consequences of the wide shifts in their population sizes for Indigenous People of the north. As with other predators, prey-switching is a necessary adaptation for subsistence hunters.

References

Alaska Department of Fish and Game. 2002. Moose management report of survey-inventory activities 1 July 1999–30 June 2001. C. Healy, editor. Project 1.0. Juneau, AK.

Banfield, A. W. F. 1974. The Mammals of Canada. University of Toronto Press, Toronto.

Boonstra, R, C. J. Krebs, and N. C. Stenseth. 1998. Population cycles in small mammals: the problem of explaining the low phase. Ecology 79:1479–1488.

Boonstra, R., S. Boutin, A. Byrom, T. Karels, A. Hubbs, K. Stuart-Smith, M. Blower, and S. Antpoehler. 2001a. The role of red squirrels and arctic ground squirrels. Pages 179–214 in C. J. Krebs, S. Boutin, and R. Boonstra, editors. Ecosystem Dynamics of the Boreal Forest: The Kluane Project. Oxford University Press, New York.

Boonstra, R., C. J. Krebs, S. Gilbert, and S. Schweiger. 2001b. Voles and mice. Pages 216–239 in C. J. Krebs, S. Boutin, and R. Boonstra, editors. Ecosystem Dynamics of the Boreal Forest: The Kluane Project. Oxford University Press, New York.

Bryant, J. P. 1981. The regulation of snowshoe hare feeding behaviour during winter by plant antiherbivore chemistry. Pages 720–731 in K. Myers and C. D. MacInnes, editors. Proceedings of the World Lagomorph Conference. University of Guelph, Guelph, Ontario.

Coady, J. W. 1980. History of moose in northern Alaska and adjacent regions. Canadian Field-Naturalist 94:61–68.

Donkor, N. T., and J. M. Fryxell. 1999. Impact of beaver foraging on structure of lowland boreal forests of Algonquin Provincial Park, Ontario. Forest Ecology and Management 118:83–92.

Elton, C. 1924. Periodic fluctuations in the number of animals: Their causes and effects. Journal of Experimental Biology 2:119–163.

Ernest, J. 1974. Snowshoe Hare Studies. Final Report, Alaska Department of Fish and Game, Juneau.

Flora, B. K. 2002. Spatial comparison of interior Alaska snowshoe hare populations and assessment of the hare pellet: Hare density relationship in Alaska. M.S. Thesis, University of Alaska Fairbanks.

Fox, J. F. 1978. Forest fire and the snowshoe hare–Canada lynx cycle. Oecologia 31:349–374.

Fox, J. F., and J. P. Bryant. 1984. Instability of the snowshoe hare and woody plant interaction. Oecologia 63:128–135.

Furtsch, P. R. 1995. Techniques for monitoring density and correlates of inter-annual variation for northern red-backed voles (*Clethrionomys rutilus*) in Denali National Park and Preserve, Alaska. M.S. Thesis, University of Alaska Fairbanks.

Gasaway, W. C., R. O. Stephenson, J. L. Davis, P. E. K. Shepard, and O. E. Burris. 1983. Interrelationships of wolves, prey and man in interior Alaska. Wildlife Monographs 84:1–50.

Hansson, L. 2001. Dynamics and trophic interactions of small rodents landscape or regional effects on spatial variation? Oecologia 130:259–266.

Hodges, K. E. 2000. The ecology of snowshoe hares in northern boreal forests. Pages 117–161 in L. F. Ruggiero, K. B Aubry, S. W. Buskirk, G. M. Koehler, C. J. Krebs, K. S. McKelvey, and J.R. Squires, editors. Ecology and conservation of lynx in the United States. University Press of Colorado and U.S. Department of Agriculture, Forest Service, Rocky Mountain Research Station, Fort Collins, CO.

Keith, L. B., and J. R. Cary. 1991. Mustelid, squirrel, and porcupine population trends during a snowshoe hare cycle. Journal of Mammalogy 72:373–378.

Keith, L. B., and L. A. Windberg. 1978. A demographic analysis of the snowshoe hare cycle. Wildlife Monographs 58:1–70.

Kelsall, J. P. 1987. The distribution and status of moose (*Alces alces*) in North America. Swedish Wildlife Research (Suppl.)1:1–10.

Krebs, C. J., S. Boutin, and R. Boonstra, editors. 2001. Ecosystem Dynamics of the Boreal Forest: The Kluane Project. Oxford University Press, New York.

Lehmkuhl, K. L. 2000. Population dynamics and ecology of yellow-cheeked voles (*Microtus xanthognathus*) in early post-fire series of interior Alaska. M.S. Thesis, University of Alaska Fairbanks.
McDonough, T. 2000. Response of northern red-backed vole (*Clethrionomys rutilus*) populations to a major spruce beetle infestation in the Copper River basin, Alaska. M.S. Thesis, University of Alaska Fairbanks.
Mech, L. D., L. G. Adams, T. J. Meier, J. W. Burch, and B. W. Dale. 1998. The Wolves of Denali. University of Minnesota Press, Minneapolis.
Messier, F. 1994. Ungulate population models with predation: a case study with the North American moose. Ecology 75:478–488.
Mowat, G., K. G. Poole, and M. O'Donoghue. 2000. Ecology of lynx in northern Canada and Alaska. Pages 265–306 *in* L.F. Ruggiero, K. B. Aubry, S. W. Buskirk, G. M. Koehler, C. J. Krebs, K. S. McKelvey, and J. R. Squires, editors. Ecology and Conservation of Lynx in the United States. General Technical Report RMRS-GTR–30WWW, U.S. Department of Agriculture, Forest Service, Rocky Mountain Research Station, Fort Collins, CO.
Murie. A. 1962. Mammals of Denali. Alaska Natural History Association, Anchorage.
O'Donoghue, M. 1994. Early survival of juvenile snowshoe hares. Ecology 75:1582–1592.
O'Donoghue, M., S. Boutin, D. L. Murray, C. J. Krebs, E. J. Hofer, U. Breitenmoser, C. Breitenmoser-Würsten, G. Zuleta, C. Doyle, and V. O. Nams. 2001. Coyotes and lynx. Pages 276–323 *in* C. J. Krebs, S. Boutin, and R. Boonstra, editors. Ecosystem Dynamics of the Boreal Forest: The Kluane Project. Oxford University Press, New York.
Oakley, K. L., E. M. Debevec, and E. Rexstad. 1999. Development of a Long-Term Ecological Monitoring program in Denali National Park and Preserve, Alaska (USA). Pages 307–314 *in* C. Aguirre-Bravo and C. R. Franco, editors. North American Science Symposium: Toward a Unified Framework for Inventorying and Monitoring Forest Ecosystem Resources. U.S. Department of Agriculture, Forest Service, Rocky Mountain Research Station, Fort Collins, CO.
Oksanen T., M. Schneider, U. Rammul, P. Hamback, and M. Aunapuu. 1999. Population fluctuations of voles in North Fennoscandian tundra: Contrasting dynamics in adjacent areas with different habitat composition. Oikos 86:463–478.
Peterson R. L. 1955. North American Moose. University of Toronto Press, Toronto.
Pewe, T. L., and D. M. Hopkins. 1967. Mammal remains of pre-Wisconsin age in Alaska. Pages 266–270 *in* D. M. Hopkins, editor. The Bering Land Bridge. Stanford University Press, Stanford, CA.
Ranta, E., V. Kaitala, and P. Lundberg. 1997. The spatial dimension in population fluctuations. Science 278:1621–1623.
Rexstad, E. 1994. Detecting differences in wildlife populations across time and space. Pages 219–228 *in* Fifty-ninth North American Natural Resource and Wildlife Conference, Anchorage, AK.
Rexstad, E., and E. M. Debevec. 2002. Small mammal monitoring at the landscape scale Denali National Park and Preserve. Report to the US Geological Survey, Biological Research Division (USGS/BRD) and National Park Service.
Rupp, T. S. 1998. Boreal forest regeneration dynamics: Modeling early forest establishment patterns in interior Alaska. Ph.D. Dissertation, University of Alaska Fairbanks.
Scott, R., and J. Kephardt. 2002. Trapper questionnaire. Statewide Annual Report 1 July 2001– 30 June 2002. Alaska Department of Fish and Game, Division of Wildlife Conservation.
Sinclair, A. R. E., C. J. Krebs, R. Boonstra, S. Boutin, and R. Turkington. 2001. Testing hypothesis of community organization for the Kluane ecosystem. Pages 408–436 *in* C. J. Krebs, S. Boutin, and R. Boonstra, editors. Ecosystem Dynamics of the Boreal Forest: The Kluane Project. Oxford University Press, New York.

Sinclair, A. R. E., C. J. Krebs, J. N. M. Smith, and S. Boutin. 1988. Population biology of snowshoe hares III. Nutrition, plant secondary compounds, and food limitation. Journal of Animal Ecology 57:787–806.

Sinclair, A. R. E., J.M. Gosline, G. Holdsworth, C. J. Krebs, S. Boutin, J. N. M. Smith, R. Boonstra, and M. Dale. 1993. Can the solar cycle and climate synchronize the snowshoe hare cycle in Canada? Evidence from tree rings and ice cores. American Naturalist 141:173–198.

Sinclair, A. R. E., C. J. Krebs, J. M. Fryxell, R. Turkington, S. Boutin, R. Boonstra, P. Seccombe-Hett, P. Lundberg, and L. Oksanen. 2000. Testing hypotheses of trophic level interactions: A boreal forest ecosystem. Oikos 89:313–328.

West, S. D. 1982. Dynamics of colonization and abundance in central Alaskan populations of the northern red-backed vole, *Clethrionomys rutilus*. Journal of Mammalogy 63:128–143.

Whitney, P. 1976. Population ecology of two sympatric species of subarctic microtine rodents. Ecological Monographs 46:85–104.

Wolff, J. O. 1980. The role of habitat patchiness in the population dynamics of snowshoe hares. Ecological Monographs 50:111–130.

Wolff, J. O., and W. Z. Lidicker Jr. 1980. Population ecology of the taiga vole, *Microtus xanthognathus*, in interior Alaska. Canadian Journal of Zoology 58:1800–1812.

9

Dynamics of Phytophagous Insects and Their Pathogens in Alaskan Boreal Forests

Richard A. Werner
Kenneth F. Raffa
Barbara L. Illman

Introduction

Boreal forests support an array of insects, including phytophagous (plant-eating) insects, saprophagous (detritus-eating) insects, and their associated parasites, predators, and symbionts. The phytophagous species include folivorous leaf chewers and miners, phloeophagous cambial and sapwood borers, stem gallers, and root feeders. Biological diversity and distribution of insect species exhibit predictable patterns among vegetation types (Werner 1994a). In this chapter, we discuss how phytophagous species of insects differ among plant communities and how various populations of insects react to disturbances that alter forest stand composition and density.

Phytophagous Insects

The distribution of insects differs among plant communities depending on the ecosystem type and plant height (Table 9.1; Werner 1983, 1994a). Grasses, mosses, small tree seedlings, and other herbaceous plants located on the forest floor have the highest arthropod densities. Shrubs have the lowest densities, and trees are intermediate. The herbaceous layer is inhabited primarily by scavengers, predators, and saprophages but has few defoliators (Werner 1983). Taller shrubs contain more species of phytophagous insects than do herbs, but trees have the most species of phytophagous insects, parasites, and predators (Werner 1981, 1983). Few saprophages and scavengers (carabid beetles), however, occur on shrubs and trees (Werner 1986a).

Associations of plants and phytophagous insects in boreal ecosystems are similar to temperate assemblages in that insect species differ in the range of food plants

Table 9.1. Densities (per square meter of ground surface integrated through the canopy) of arthropods collected from three different vertical levels in five major vegetation types of the boreal forest ecosystem.

Vertical level	Balsam poplar	Aspen	Paper birch	White spruce	Black spruce	Average
Herb	2,901	3,093	1,593	1,117	12,538	4,248
Shrub	727	1,059	609	420	1,334	830
Tree	582	1,172	2,088	631	4,485	1,792
Average	1,403	1,441	1,430	723	6,119	

Source: Werner (1983).

that they utilize (Bernays and Minkenberg 1997, Futuyma et al. 1993, Thorsteinson 1960). Because of low plant diversity, however, many boreal phytophagous insects feed on several species of plants (Werner 1981). For example, the spear-marked black moth, *Rheumaptera hastata* (L.), feeds primarily on paper birch, but during periods of high populations it also feeds on alder, willow, and rose species but not on poplar (Werner 1977, 1979). When population outbreaks of phytophagous insects deplete their preferred host plants, less desirable species are sometimes consumed or starvation occurs (Werner 1981, 1986a).

The biomass of phytophagous insects is greater on broad-leafed than on conifer trees (Werner 1983). Species of Coleoptera, Hemiptera, Homoptera, Hymenoptera, and Lepidoptera are common on broad-leafed trees, whereas only a few taxa of Homoptera, Hymenopera, and Lepidoptera are associated with conifers such as spruce or larch (Table 9.2; Werner 1983, 1994a). Species richness is greatest in early successional deciduous communities dominated by willow and aspen and lowest in black spruce communities (Werner 1986a, Werner and Ward 1976). Most plant species in boreal ecosystems have generalized defoliators, sap-feeders, and blotch and leaf miners, but white spruce and, to some extent, black spruce and larch are also inhabited by phloeophagous insects (Table 9.2; Werner 1986a, 1986b, Werner and Holsten 1984, Werner and Ward 1976).

Temperature is more important as a regulator of insect species composition at northern latitudes than elsewhere (Werner and Holsten 1985a). Species richness of scolytid bark beetles in Alaska is greatest where cumulative degree-day temperature sums are highest (Werner and Holsten 1984). For example, 27 species of bark beetles were found associated with white spruce in the Fairbanks area (latitude 64°37'N), where summer temperature is highest, compared to 22 species in the Brooks Range (latitude 68°15'N) in northern Alaska and 20 species in the Kenai Peninsula (latitude 60°37'N) in south-central Alaska (Table 9.3).

Folivorous Insects

Repeated defoliation by a variety of insects over several years has caused tree mortality of conifers and broad-leafed species. For example, infestations by two

Table 9.2. Biomass (g m⁻²) of the major arthropod taxa in boreal forest ecosystems of Alaska.

Taxon	Balsam poplar	Aspen	Paper birch	White spruce	Black spruce	Average
Noninsects						
Araneida	0.1186	0.1238	0.1510	0.1025	0.0846	0.1161
Acari	0.0400	0.0650	0.0742	0.0408	0.0996	0.0640
Other	0.0123	0.0126	0.0128	0.0057	0.0057	0.0098
Insects						
Coleoptera	0.0966	0.0838	0.1981	0.0820	0.1003	0.1122
Collembola	0.0153	0.0139	0.0121	0.0161	0.0378	0.0190
Diptera	0.0162	0.0704	0.0971	0.1100	0.0615	0.0710
Hemiptera	0.0397	0.0504	0.1284	0.0463	0.0423	0.0614
Homoptera	0.0351	0.0440	0.2810	0.0382	0.0307	0.0858
Hymenoptera	0.1061	0.0 836	0.1757	0.0626	0.0331	0.0922
Lepidoptera	0.0750	0.0391	0.0482	0.0564	0.1016	0.0641
Mecoptera	0.0072	0.0062	0.0038	0.0058	0.0006	0.0047
Neuroptera	0.0052	0.0316	0.0072	0.0102	0.0178	0.0144
Pscoptera	0.0035	0.0018	0.0072	0.0064	0.0064	0.0051
Thysanoptera	0.0097	0.0151	0.0151	0.0139	0.0109	0.0129
Other	0.0311	0.0208	0.0933	0.0325	0.0221	0.0400
Total	0.6805	0.7196	1.4164	0.6364	0.6583	0.8222

Source: Werner (1983).

species of budworm, *Choristoneura fumiferana* (Clem.) and *Choristoneura orae* Freeman, covered 219,125 ha of interior Alaska white spruce forests from 1990 to 1994 and 259,855 ha from 1995 to 1999 (Table 9.4; U.S. Department of Agriculture 1995, 1999). In 1993, the spruce coneworm, *Dioryctria reniculelloides* (Mutuura and Munroe), was competing for the same buds and foliage as the spruce budworm, and competition for food resources was evident (U.S. Department of Agriculture 1995, Werner 1995). White spruce trees that were defoliated for three or more years exhibited terminal shoot and upper crown mortality, loss of reproductive potential (no cones produced), and reduced vigor and growth. These trees were stressed to the point where they became susceptible to attack by the northern engraver beetle, *Ips perturbatus* (Eichhoff) (Werner 1988, 1995). All sizes of spruce from two-year-old seedlings to mature trees were defoliated. Mature trees survived the continued defoliation after the budworm populations declined, but seedlings and young saplings did not recover.

Several species of folivorous insects occur chronically in stands of larch (*Larix laricina*) in interior Alaska. From 1975 to 1979, a species of larch bud moth, *Zieraphera* sp. completely defoliated 238,773 ha of larch (Table 9.4) in interior Alaska, including stands in the Bonanza Creek Experimental Forest (Werner 1980). These trees were subsequently infested by eastern larch beetle, *Dendroctonus simplex*

Table 9.3. Diversity of scolytid bark beetles inhabiting white spruce at three sites in Alaska, 1977–1979.

	Total number of beetles caught[a]		
Species	Kenai Lake 900 degree-days[b]	Fairbanks 1225 degree-days[b]	Brooks Range 1100 degree-days[b]
Carphoborus andersoni Swaine	0	3	8
Carphoborus carri Swaine	0	21	6
Carphoborus intermedius Wood	0	5	0
Cryphalus ruficollis ruficollis Hopkins	68	4	3
Crypturgus borealis Swaine	0	163	6
Dendroctonus punctatus LeConte	0	3	3
Dendroctonus rufipennis (Kirby)	311	115	40
Dryocoetes affaber (Mannerheim)	87	655	88
Dryocoetes autographus (Ratzeburg)	36	206	15
Hylurgops rugipennis rugipennis (Mannerheim)	2	6	2
Ips borealis borealis Swaine	24	228	77
Ips perturbatus (Eichhoff)	348	3,141	43
Ips tridens tridens (Eichhoff)	954	433	6
Orthotomicus caelatus (Eichhoff)	0	274	4
Phloeosinus pini Swaine	0	2	6
Phloeotribus piceae Swaine	8	20	11
Pityophthorus bassetti Blackman	0	5	0
Pityophthorus murrayanae Blackman	14	0	0
Pityophthorus nitidulus (Mannerheim)	77	206	0
Pityophthorus nitidus Swaine	27	117	12
Pityophthorus opaculus LeConte	20	33	21
Polygraphus convexifrons Wood	3	6	15
Polygraphus rufipennis (Kirby)	128	353	634
Scierus annectans LeConte	6	103	6
Scierus pubescens Swaine	23	0	0
Scolytus piceae (Swaine)	0	752	233
Trypodendron lineatum (Olivier)	185	681	42
Trypodendron retusum (LeConte)	12	152	0
Xylechinus montanus Blackman	2	10	0
Total	2,335	7,697	1,281

[a] Beetles were caught in pheromone-baited sticky traps.
[b] Thirty-year average of cumulative degree-days above 5°C.

Source: Werner and Holsten (1984b).

LeConte, in 1977. Two successive years of defoliation may have reduced tree vigor and increased their susceptibility to beetle attack. The first reported outbreak of larch sawfly, *Pristiphora erichsonii* (Hartig), in Alaska occurred in interior Alaska from 1993 to 1999 and killed most of the larch on 651,100 ha (Table 9.4; U.S. Department of Agriculture 1999). Many of the defoliated trees were stressed and subsequently attacked by eastern larch beetle as had occurred following outbreaks of the larch bud moth (Werner 1986b). Many river bottom sites previously occupied by black spruce and larch are now devoid of larch, showing the impact of insect folivores on forest composition.

Table 9.4. Total number of hectares of Alaska boreal forest ecosystems infested by folivorous and phloeophagous insects, 1955–1999, as detected by annual aerial surveys.

Species	1955–64	1965–74	1975–79	1980–84	1985–89	1990–94	1995–99
Spruce beetle (*Dendroctonus rufipennis*)	20,235[a]	10,039	284,505	744,989	595,725	1,114,587	1,265,364
Engraver beetles (*Ips perturbatus*)	404,700	0	4,452	1,667	12,020	13,345	14,081
Larch beetle (*Dendroctonus simplex*)	0	52,611	522,063	8,134	0	1,214	1,226
Spruce budworm (*Choristoneura fumiferana*)	0	0	121	4,452	907	219,125	259,855
Larch sawfly (*Pristiphora erichsonii*)	0	0	0	0	0	45,540	651,099
Larch bud moth (*Zeiraphera* spp.)	202,350	4,047	238,773	0	36,018	4,087	651,099
Spear-marked black moth (*Rheumaptera hastata*)	2,347,260	526,110	1,092,690	159,452	32,552	4,832	0
Large aspen tortrix (*Choristoneura conflictana*)	0	2,590,080	19,426	54,877	261,367	60,379	24,036
Hemlock sawfly (*Neodiprion tsugae*)	0	16,997	0	54,027	17,807	10,045	6,609
Western black-headed budworm (*Acleris gloverana*)	2,306	39,054	2,287	10,805	66,780	209,823	18,327
Leaf beetles (*Chrysomela* spp.)	0	0	41,927	5,463	10,103	25,116	9,932
Leaf miners (*Micrurapteryx salicifolliela, Phyllocnistis* spp.)	0	0	247,271	4,856	607	5,180	146,721
Aphids (*Adelges, Cinara, Elatobium* sp.)	0	47,350	186,162	19,764	14,974	1,077	13,605
Leaf rollers (*Epinotia, Euceraphis* spp.)	0	202,350	445,777	5,969	32,990	65,892	5,017

[a] Total ha for each five-year period.

Historical records of folivorous insects indicate that populations of spear-marked black moth occurred at 10-year intervals, and high populations of the large aspen tortrix, *Choristoneura conflictana* (Wlk.), occurred at 12-year intervals (Werner unpublished data). The most recent outbreak populations of spear-marked black moth occurred in 1975 to 1979 and of large aspen tortrix from 1985 to 1989 (Table 9.4). Average annual temperatures have increased since 1980 in interior Alaska (Chapter 4; Barber et al. 2000), which may contribute to the decreased outbreak frequency. Other species of folivorous insects, such as the eastern spruce budworm, spruce coneworm, and larch sawfly, had negligible populations prior to 1990 but since then have defoliated thousands of hectares of white spruce and larch, respectively, in interior Alaska (Table 9.4). Why were some populations maintained at low levels while others exploded beyond expectations? Is atmospheric warming a factor? Barber et al. (2000) suggest that temperature-induced drought has reduced the growth of white spruce. This could predispose trees to infestation by phloeophagous species.

Phloeophagous Insects

Insects can exert markedly different impacts in different regions, even when the underlying insect and tree biologies are quite similar (Werner 1986a). The spruce beetle, *Dendroctonus rufipennis* (Kirby), is distributed throughout all regions of Alaska that contain spruce and colonizes all spruce species within its range (Werner et al. 1977). Like other bark beetles, the spruce beetle carries a variety of wood-colonizing microorganisms, such as bacteria, molds, yeast, and stain fungi. The saprophytes apparently have no effect on beetles, but sapstain fungi have the potential to play a major role in beetle-host tree interactions (Illman et al. unpublished data, Reynolds 1992, Werner and Illman 1994a, 1994b). The spruce beetle typically colonizes severely stressed or dying trees but also (depending on the region) may undergo intermittent population eruptions, which kill most healthy mature spruce trees over areas encompassing several million hectares (Holsten 1990, Werner 1994a). These natural disturbance events have pronounced landscape-level effects: accumulation of fuels, release of early succession species such as willow, poplar, birch and aspen, improved or decreased habitat for some birds, mammals, and other subcortical insects, and changes in nutrient cycling.

The same spruce beetle-host tree interaction exerts markedly different impacts at the landscape level in south-central Alaska than it does in interior Alaska. In contrast to south-central Alaska, western Canada, and the Rocky Mountains, the same insect rarely, if ever, undergoes outbreaks in interior Alaska. Spruce beetles in interior white spruce forests sometimes colonize severely stressed or dying trees on fringes of fires (Werner and Post 1985) and in areas periodically inundated with water. Thus, spruce beetles in interior Alaska might play an important role in canopy thinning and gap formation but do not exert ecosystem-level effects. The underlying reasons are poorly understood and are somewhat paradoxical. The major factors appear to be complex interactions involving abiotic and biotic influences, particularly moisture and temperature effects on translocation of defensive compounds and host suitability to the spruce beetles and to competitors such as *Ips* engraver beetles.

There are several factors that we initially hypothesized might explain the lack of outbreaks of spruce beetle in interior Alaska, including regional differences in climate, the genetic composition of beetle populations, differences in microbial associates, and different levels of natural enemies or competitors. Not only does weather fail to provide a simple explanation, but also the dichotomy between the existence of landscape-level impacts in south-central Alaska and the presence of only site-level impacts in interior Alaska is not easily explained by the relationships among phloem temperature, beetle phenology, and population dynamics. Spruce beetles exhibit facultative diapause, so a single generation can require either one or two years, depending on temperature (Werner and Holsten 1985a, 1985b). The predominant pattern in south-central and southeast Alaska is a two-year cycle. Warm temperatures, however, can favor complete brood development within one year. This has been identified as a major factor in the epidemic of the late 1980s and 1990s. In contrast, the warm summers of interior Alaska typically foster the one-year cycle (Werner and Holsten 1985a). Thus, on the basis of strictly temperature considerations, interior Alaska should be more, not less, prone to outbreaks. Likewise, lightning can play a crucial role in some bark beetle population dynamics by killing trees and making them suitable for beetle development (Coulson 1979). Lightning is more frequent in interior Alaska than in south-central Alaska, which again would generate predictions opposite to observed patterns.

One possibility for regional differences in spruce beetle population levels is that there could be genetic differences between interior and south-central Alaska populations. However, molecular (Bentz and Stock 1986, Cronin 1994) and behavioral analyses (Wallin and Raffa 2004) do not support this view. Likewise, beetles might show regional differences in their fungal complements that correspond to different population patterns. Again, we see no evidence for this possibility (Haberkern et. al. 2002, Illman et. al. unpublished data, Aukema et al. 2005). Furthermore, spruce beetles in the Great Lakes region, where forests are not susceptible to outbreaks or landscape-level infestations, carry fungal complements similar to those found in populations in south-central and interior Alaska (Haberkern et al. 2002). The increased frequency of a relatively pathogenic stain fungus has been proposed as a contributing factor to outbreaks of the European spruce beetle, *Ips typographus* L., in Norway (Krokene and Solheim 1998). The effect of fungal inoculation levels on the south-central and interior Alaska populations of spruce beetle is not known.

Predators and competitors may contribute to, but not fully explain, differences in population dynamics between populations of interior and south-central Alaska spruce beetles. For example, there is some evidence that checkered beetles (Cleridae) such as *Thanasimus undatulus* and *T. dubius* are more abundant in interior than south-central Alaskan forests (Gara et al.1995, Werner 1993, 1994b). However, there are fewer clerids and other predators collected from colonized hosts or traps baited with spruce beetle pheromones in Alaskan spruce forests than elsewhere in North America (Haberkern 2001) and fewer than are commonly seen with other bark beetles (Weslien 1994, Reeve 1997, Erbilgin and Raffa 2002). Competitors of the spruce beetle include mostly *Ips* species, *Polygraphus rufipennis, Dryocoetes affaber*, Buprestidae, and Cerambycidae (Werner and Holsten 2002). These less aggressive colonizers can cause some reductions in spruce beetle populations (Poland and Borden 1998). However, interspecific competition is reduced somewhat by partitioning of the resource, based on a combination of tree vigor, height along the bole, and colonization of upper versus lower surface of downed trees. Moreover, there is no evidence that populations of competitors are consistently higher in interior than in south-central Alaska.

Host tree species appear to have some effect on beetle population dynamics, but these do not necessarily correspond to outbreak behavior. The Kenai Peninsula in south-central Alaska, a region of historically severe outbreaks of spruce beetles, contains a combination of white, Sitka, and Lutz spruce, *Picea x lutzii* Little, the latter a hybrid between white and Sitka spruce found only on the Kenai Peninsula. Lutz spruce is the most susceptible to attack, whereas white spruce is most suitable for development of the spruce beetle (Hard et al.1983, Holsten 1984, Holsten and Werner 1990) and is typically associated with faster rates of development and more progeny. Sitka spruce is least susceptible to attack and is suitable for beetle development; it is found mostly in southeastern and coastal Alaska (Holsten and Werner 1990).

Biotic agents that predispose trees to colonization appear less important with spruce beetles in Alaska than elsewhere with other species of bark beetles. Large-scale defoliation of larch by larch bud moth in Alaska (Werner 1980), for example, is commonly followed by infestation of trees by the larch beetle (Werner 1986b). Defoliation of *Pinus banksiana* (Lamb.) by the budworm, *Choristoneura pinus* Freeman, is commonly followed by colonization of the tree by *Ips grandicollis* (Eichhoff) (Wallin and Raffa 1999, 2001), and defoliation of *Abies grandis* is usually followed by colonization of the host by *Scolytus ventralis* LeConte (Wright et al.1984). Defoliation is less common in spruce in south-central Alaska. However, repeated defoliation by eastern spruce budworm of white spruce in interior Alaska can predispose trees to *Ips* engraver beetle attack (Werner 1995). Likewise, root insects such as *Hylobius* and *Hylastes* can be important predisposing agents for *Ips pini* colonizing pine in Wisconsin (Klepzig et al.1991, 1995) and for the Douglas-fir beetle, *Dendroctonus pseudotsugae* (Witcosky et al.1986a, 1986b), colonizing Douglas-fir (*Pseudotsuga menziesii* (Mirb.) Franco) in Oregon. By contrast, populations of root colonizing beetles do not appear high in spruce forests of Alaska (Werner et al. unpublished data). The following species of weevils were caught in pit-fall traps in stands of white spruce in the Bonanza Creek Long Experimental Forest from 1996 to 2003: *Lepidophorus lineaticollis* Kirby, *Pissodes rotundatus* LeC., *Dorytomus laticollis* LeC., *Magdalis gentilis* LeC., *Orchidophilus* spp., and *Grypus equiseti* (F.) (Werner et al. unpublished data). Likewise, root pathogens are important predisposing agents to subsequent bark beetle attack in many ecosystems (Filip and Goheen 1982). However, we see no evidence for a significant role in Alaska (Illman unpublished data). One biotic agent that may be more important in disposing trees to attack than elsewhere are black bears that feed on the cambium of tree trunks, a damage that is similar to girdling (Lutz 1951). To date, however, no correlation between black bear girdling and spruce beetle attack has been found.

Interactions of abiotic and biotic environmental factors, particularly moisture with host plant physiology, could partially explain differences in eruptive spruce beetle behavior. First, the lower precipitation in interior Alaska results in phloem that has 20% less moisture than in spruce in south-central Alaska (Werner and Holsten 1985b). Spruce beetles perform poorly in dry tissue, whereas *Ips* beetles are more tolerant. The drier developmental habitat also appears to reduce the synchrony of beetle emergence, which is important for tree-killing bark beetles that overwhelm tree resistance by mass attack (Raffa and Berryman 1988, Bentz et al. 1991). Thus, interactions among precipitation, host suitability, host susceptibility, and interspecific competition may be important. There may also be interactions among temperature, host susceptibility, and beetle population dynamics. For example, in the mountainous areas of south-central Alaska, valley bottoms can remain cool well into the spring. Because the soil is cool, roots often do not begin translocation until after spruce beetle flight has begun (Hard 1987, Werner unpublished data). Thus, beetles can exploit trees that have low oleoresin flow rates, which is an important component of conifer defense against bark beetles (Hard 1985, 1987, Lorio 1993). Associated stain fungi appear to play an important role in beetle-host relationships. Stain fungi induce an energy-requiring chemical defense response in phloem tissue (Werner and Illman 1994a, 1994b, Illman et al. unpublished data) that could weaken the tree's defenses against beetles. Additionally, stain fungi could lower tree resistance by colonizing xylem tissue and blocking water transport, as occurs in Norway spruce (Krokene and Solheim 1998). Beetle success at colonization of the host tree could be highly dependent on the abundance of associated fungi attacking the tree.

Interactions with Disturbance

Disturbance factors such as wildfire, ice and wind storms, outbreaks of phytophagous folivorous insects, and timber harvest can cause fluctuations in populations of phloeophagous insects. For example, fringes of white spruce stands disturbed by wildfire or clear-cut and shelterwood harvests in the Bonanza Creek Experimental Forest had a direct impact on the diversity of species of buprestid and cerambycid wood-boring beetles and scolytid bark beetles (Table 9.5; Werner 2002). Fire and timber harvest are the two major disturbances that alter forest ecosystems of interior Alaska. Both types of disturbance provide habitats that attract wood borers and bark beetles the first year after disturbance. However, fire and clear-cutting timber significantly reduced the number of species of all three types of phloeophagous insects in stands of white spruce for 10 years following the disturbance. Shelterwood timber harvest had a less severe impact for 1 to 5 years after disturbance, but thereafter the number of species declined (Table 9.5).

Although major disturbances such as fire or insect outbreaks may appear to be independent events, they are often related (Showalter et al. 1981, Romme et al. 1986, McCullough et al.1998). Insect outbreaks that are followed by fire can also effectively disrupt or redirect plant succession in forest ecosystems (Amman 1977, Geiszler et al. 1980, Showalter et al. 1981, Romme et al. 1986, Raffa and Berryman 1987, McCullough et al.1998). For example, interactions among spruce beetle, species composition of spruce, and fire can determine the future composition of forest stands in south-central Alaska (Holsten et al.1995).

Table 9.5. Average number of wood borers and bark beetles collected from stands of disturbed white spruce in the Bonanza Creek Experimental Forest at 1-, 5-, and 10-year intervals following disturbance.

Disturbance	Buprestidae[a] years			Cerambycidae years			Scolytidae years		
	1	5	10	1	5	10	1	5	10
Burned	7	0	0	1	0	0	9	0	0
Fringe area	14	8	0	4	3	0	14	11	4
Clear-cut	17	5	0	9	5	0	15	8	2
Shelterwood	22	17	0	11	9	0	18	15	12
Control	19	18	16	6	6	6	19	19	19

[a] Beetles were collected from pheromone-baited funnel traps and window flight traps from late May to early September in the respective years following disturbance.

Conclusions

Forest insects are intrinsic components of many boreal forest ecosystems. They contribute to the biodiversity of aboveground arthropods. Episodic outbreaks of phytophagous insects, including both folivorous and phloeophagous species, can affect succession, species richness, and plant and animal diversity at the landscape level. Folivorous insects can exert landscape-level effects, or they can predispose forest stands to attack by phloeophagous insects and decay organisms. Phloeophagous insects, in turn, can predispose forest stands to risk from fire by increasing fuel density and flammable ground vegetation. Changes in ecosystem health (e.g., drought stress) are associated with greater susceptibility of stands to damaging insect pests and disease organisms and can cause increases in outbreak intensity at the landscape level. Complex interactions among abiotic and biotic factors can determine whether insects such as spruce beetle exert either stand-level or landscape-level impacts.

References

Amman, G. G. 1977. Role of mountain beetle in lodgepole pine ecosystems: Impact of succession. Pages 3–18 *in* W. J. Mattson, editor. The Role of Arthropods in Forest Ecosystems. Fifteenth International Congress of Entomology, Washington, DC. Springer-Verlag, New York.

Aukema, B. H., R. A. Werner, K. E. Haberkern, B. L. Illman, M. K. Clayton, and K. F. Raffa. 2005. Relative sources of variation in spruce beetle-fungal associations: Implications for sampling methodology and hypothesis testing in bark beetle-symbiont relationships. Forest Ecology and Management 217:187–202.

Barber, V. A., P. A. Juday, and B. Finney. 2000. Reduced growth of Alaskan white spruce in the twentieth century from temperature-induced drought stress. Nature 405:668–673.

Bentz, B. J., and M. W. Stock. 1986. Phenetic and phylogenetic relationships among ten

species of *Dendroctonus* bark beetles (Coleoptera: Scolytidae). Annals of the Entomological Society of America 79:527–534.
Bentz, B. J., J. A. Logan, G. D. Amman. 1991. Temperature-dependent development of the mountain pine beetle (Coleoptera, Scolytidae) and simulation of its phenology. Canadian Entomologist 123:1083–1094.
Bernays, E. A., and O. Minkenberg. 1997. Insect herbivores—Different reasons for being a generalist. Ecology 78:1157–1169.
Coulson, R. N. 1979. Population dynamics of bark beetles. Annual Review of Entomology 24:417–447.
Cronin, M. A. 1994. Intra- and interspecific mitochondrial DNA variation in bark beetles (*Dendroctonus*). Research Report. LGL Ecological Genetics, Inc., Anchorage, Alaska.
Erbilgin, N., and K. F. Raffa. 2002. Association of declining red pine stands with reduced populations of bark beetle predators, seasonal increases in root colonizing insects, and incidence of root pathogens. Forest Ecology and Management 164:221–236.
Filip, G. M., and D. J. Goheen. 1982. Tree mortality caused by root pathogen complex in Deschutes National Forest, Oregon. Plant Disease 66:240–243.
Futuyma, D. J., M. C. Keese, and S. J. Scheffer. 1993. Genetic constraints and the phylogeny of insect-plant associations—Responses of *Ophraella communa* (Coleoptera, Chrysomelidae) to host plants of its congeners. Evolution 47:888–905.
Gara, R. I., R. A. Werner, M.C. Whitmore, and E. H. Holsten. 1995. Arthrop associates of the spruce beetle *Dendroctonus rufipennis* (Kirby) (Col., Scolytidae) in spruce stands of south-central and interior Alaska. Journal of Applied Entomology 119:585–590.
Geiszler, D. R., R. I. Gara, C. H. Driver, V. F. Gallucci, and R. E. Martin. 1980. Fire, fungi, and beetle influences on a lodgepole pine ecosystem of south-central Oregon. Oecologia 46:239–243.
Haberkern, K. E. 2001. Subcortical insects and fungal associates colonizing white spruce in the Great Lakes Region. M.S. Thesis, University of Wisconsin, Madison.
Haberkern, K. E., B. L. Illman, and K. F. Raffa. 2002. Subcortical insects and fungal associates colonizing white spruce in the Great Lakes region. Canadian Journal of Forest Research 32:1137–1150.
Hard, J. S. 1985. Spruce beetles attack slowly growing spruce. Forest Science 31:839–850.
Hard, J. S. 1987. Vulnerability of white spruce with slowly expanding lower boles on dry, cold sites to early seasonal attack by spruce beetles in south-central Alaska. Canadian Journal of Forest Research 17:428–435.
Hard, J. S., E. H. Holsten, and R. A. Werner. 1983. Susceptibility of white spruce to attack by spruce beetles during the early years of an outbreak in Alaska. Canadian Journal of Forest Research 13:678–684.
Holsten, E. H. 1984. Factors of susceptibility in spruce beetle attack on white spruce in Alaska. J. Entomological Society of British Columbia 81:39–45.
Holsten, E. H. 1990. Spruce beetle activity in Alaska: 1920–1989. Technical Report R10–90–18. U.S. Department of Agriculture, Forest Service, Forest Pest Management, Anchorage, AK.
Holsten, E. H., and R. A. Werner. 1990. Comparison of white Sitka and Lutz spruce as hosts of the spruce beetle in Alaska. Canadian Journal of Forest Research 20:292–297.
Holsten, E. H., R. A. Werner, and R. DeVelice. 1995. Spruce bark beetle, *Dendroctonus rufipennis* (Coleoptera: Scolytidae), outbreak and prescribed fire effects on stand volume structure, and ground vegetation. Environmental Entomology 24:1539–1547.
Klepzig, K. D., E. L. Kruger, E. B. Smalley, and K. F. Raffa. 1995. Effects of biotic and abiotic stress on the induced accumulation of terpenes and phenolics in red pines inoculated with a bark beetle vectored fungus. Journal of Chemical Ecology 21:601–626.

Klepzig, K. D., K. F. Raffa, and E. B. Smalley. 1991. Association of insect-fungal complexes with red pine decline in Wisconsin. Forest Science 37:1119–1139.

Krokene, P., and H. Solheim. 1998. Pathogenicity of four blue-stain fungi associated with aggressive and nonaggressive bark beetles. Phytopathology 88:39–44.

Lorio, P. L., Jr. 1993. Environmental stress and whole-tree physiology. Pages 81–101 in T. D. Schowalter and G. M. Philip, editors. Beetle-pathogen Interactions in Conifer Forests. Academic Press, London.

Lutz, H. J. 1951. Damage to trees by black bears in Alaska. Journal of Forestry 49:522–523.

McCullough, D. G., R. A. Werner, and D. Neumann. 1998. Fire and insects in northern and boreal forest ecosystems of North America. Annual Review of Entomology 43:107–127.

Poland, T. M., and J. H. Borden. 1998. Disruption of secondary attraction of the spruce beetle, *Dendroctonus rufipennis* Kirby, by pheromones of two secondary species (Coleoptera: Scolytidae). Journal of Chemical Ecology 24:151–166.

Raffa, K. F., and A. A. Berryman. 1987. Interacting selective pressures in conifer-bark beetle systems: A basis for reciprocal adaptations? American Naturalist 129:234–262.

Reeve, J. D. 1997. Predation and bark beetle dynamics. Oecologia 112:48–54.

Reynolds, K. M. 1992. Relations between activity of *Dendroctonus rufipennis* (Kirby) on Lutz spruce and blue stain associated with *Leptographium abietinum* (Peck) Wingfield. Forest Ecology and Management 47:71–86.

Romme, W. H., D. H. Knight, and J. B. Yavitt. 1986. Mountain pine beetle in the Rocky Mountains: Regulators of primary productivity? American Naturalist 127:484–494.

Schowalter, T. D., R. N. Coulson, and D. A. Crossley Jr. 1981. Role of southern pine beetle and fire in maintenance of structure and function of the southeastern forest. Environmental Entomology 10:821–825.

Thorsteinson, A. J. 1960. Host selection in phytophagous insects. Annual Review of Entomology 5:193–218.

U.S. Department of Agriculture. 1995. Forest insect and disease conditions in Alaska—1995. U.S. Department of Agriculture, Forest Service, State and Private Forestry, Forest Health Management, Region 10, R10–TP–61. Anchorage, AK.

U.S. Department of Agriculture. 1999. Forest insect and disease conditions in Alaska—1999. U.S. Department of Agriculture, Forest Service, State and Private Forestry, Forest Health Management, Region 10, R10–TP–82. Anchorage, AK.

Wallin, K. F., and K. F. Raffa. 1999. Altered constitutive and inducible phloem monoterpenes following natural defoliation of jack pine: implications to host mediated inter-guild interactions and plant defense theories. Journal of Chemical Ecology 25:861–880.

Wallin, K. F., and K. F. Raffa. 2001. Host-mediated interactions among feeding guilds: Incorporation of temporal patterns can integrate plant defense theories to predict community level processes. Ecology 82:1387–1400.

Wallin, K. F., and K. F. Raffa. 2004. Feedback between individual host selection behavior and population dynamics in an eruptive insect herbivore. Ecological Monographs 74:101–116.

Werner, R. A. 1977. Biology and behavior of the spear-marked black moth, *Rheumaptera hastata,* in interior Alaska. Canadian Entomologist 109:1149–1152.

Werner, R. A. 1979. Influence of host foliage on development, survival, fecundity, and oviposition of the spear-marked black moth, *Rheumaptera hastata* (Lepidoptera: Geometridae). Canadian Entomologist 111:317–322.

Werner, R. A. 1980. Biology and behavior of a larch bud moth, *Zeiraphera* sp., in Alaska. Research Note PNW–356. U.S. Department of Agriculture, Forest Service, Pacific Northwest Forest and Range Experiment Station, Portland, OR.

Werner, R. A. 1981. Advantages and disadvantages of insect defoliation in the taiga ecosystem. Page 148 *in* Proceedings, Thirty-second Alaska Science Conference, Fairbanks, AK.

Werner, R. A. 1983. Biomass, density, and nutrient content of plant arthropods in the taiga of Alaska. Canadian Journal of Forest Research 13:729–739.

Werner, R. A. 1986a. Association of plants and phytophagous insects in taiga forest ecosystems. Pages 205–212 *in* K. Van Cleve, F. S. Chapin III, P. W. Flanagan, L. A. Viereck, and C. T. Dyrness, editors. Forest Ecosystems in the Alaska Taiga: A Synthesis of Structure and Function. Springer-Verlag, New York.

Werner, R. A. 1986b. The eastern larch beetle in Alaska. Research Paper PNW-RP–357. U.S. Department of Agriculture, Forest Service, Pacific Northwest Forest and Range Experiment Station, Portland, OR.

Werner, R. A. 1988. Recommendations for suppression of an *Ips perturbatus* outbreak in interior Alaska using integrated control. Pages 189–196 *in* T. L. Payne and H. Saarenmaa. Integrated Control of Scolytid Bark Beetles: Proceedings of the IUFRO Working Party and the Seventeenth International Congress of Entomology Symposium. Virginia Polytechnic Institute and State University Press, Blacksburg.

Werner, R. A. 1993. Response of the engraver beetle, *Ips perturbatus*, to semiochemicals in white spruce stands of interior Alaska. Research Paper PNW-RP–465. U.S. Department of Agriculture, Forest Service, Pacific Northwest Research Station, Portland, OR.

Werner, R. A. 1994a. Forest insect research in boreal forests of Alaska. Council of Scientific Research Integration (India). Trends in Agricultural Sciences. Entomology 2:35–46.

Werner, R. A. 1994b. Research on the use of semiochemicals to manage spruce beetles in Alaska. Pages 15–21 *in* P. J. Shea, editor. Proceedings of the Symposium on Management of Western Bark Beetles with Pheromones: Research and Development. General Technical Report PSW-GTR–150. U.S. Department of Agriculture, Forest Service, Pacific Southwest Research Station, Albany, CA.

Werner, R. A. 1995. Summary of spruce budworm activity in interior Alaska, 1990–95. Unpublished Report.

Werner, R. A. 2002. Effect of ecosystem disturbance on diversity of bark and wood-boring beetles (Coleoptera: Scolytidae, Buprestidae, Cerambycidae) in white spruce (*Picea glauca* [Moench] Voss) ecosystems of Alaska. Research Paper PNW-RP–546. U.S. Department of Agriculture, Forest Service, Pacific Northwest Research Station, Portland, OR.

Werner, R. A., and E. H. Holsten. 1984. Scolytidae associated with white spruce in Alaska. Canadian Entomologist 116:465–471.

Werner, R. A., and E. H. Holsten. 1985a. Effect of phloem temperature on development of spruce beetles in Alaska. Pages 155–163 *in* L. Safranyik, editor. The Role of the Host in the Population Dynamics of Forest Insects: Proceedings of the IUFRO Working Party Conference, 4–7 September 1983, Baniff, AB. Canadian Forestry Service, Pacific Northwest Forest Research Centre, Victoria, British Columbia.

Werner, R. A., and E. H. Holsten. 1985b. Factors influencing generation times of spruce beetles in Alaska. Canadian Journal of Forest Research 15:438–443.

Werner, R. A., and E. H. Holsten. 2002. Use of semiochemicals of secondary bark beetles to disrupt spruce beetle attraction and survival in Alaska. Research Paper PNW-RP–541. U.S. Department of Agriculture, Forest Service, Pacific Northwest Research Station, Portland, OR.

Werner, R. A., and B. L. Illman. 1994a. Response of Lutz, Sitka, and white spruce to attack by the spruce beetle, *Dendroctonus rufipennis* (Kirby) (Coleoptera: Scolytidae), and blue-stain fungi. Environmental Entomology 23:472–478.

Werner, R. A., and B. L. Illman. 1994b. The role of stilbene-like compounds in host tree resistance of Sitka spruce to the spruce beetle, *Dendroctonus rufipennis*. Pages 123–133 *in* Behavior, Population Dynamics and Control of Forest Insects. Proceedings: Joint IUFRO Working Party Conference, 6–11 February 1994, Maui, HI. OARDC/Ohio State University, Wooster, OH.

Werner, R. A., and K. E. Post. 1985. Effect of wood-boring insects and bark beetles on survival and growth of burned white spruce. Pages 14–16 *in* G. Juday and C. T. Dyrness, editors. Early Results of the Rosie Creek Fire Research Project—1984. Miscellaneous Publication 85–2. Agriculture and Forest Experiment Station, University of Alaska.

Werner, R. A., and T. Ward. 1976. Biomass and density of arthropods inhabiting the black spruce ecosystem. Page 220 *in* Proceedings, Twenty-seventh Alaska Science Conference, Science in Alaska. Fairbanks, AK.

Werner, R. A., B. H. Baker, and P. A. Rush. 1977. The spruce beetle in white spruce forests of Alaska. Technical Paper PNW–61. U.S. Department of Agriculture, Forest Service, Pacific Northwest Forest and Range Experiment Station, Portland, OR..

Weslien, J. 1994. Interactions within and between species at different densities of the bark beetle *Ips typographus* and its predator *Thanasimus formicarius*. Entomological Experimental Applications 71:133–143.

Witcosky, J. J., T. D. Schowalter, and E. M. Hansen. 1986a. *Hylastes nigrinus* (Coleoptera: Scolytidae), *Pissodes fasciatus*, and *Steremnius carinatus* (Coleoptera: Curculionidae), as vectors of black-stain root disease of Douglas-fir. Environmental Entomology 15:1090–1095.

Witcosky, J., T. D. Schowalter, and E. M. Hansen. 1986b. The influence of time of precommercial thinning on the colonization of Douglas-fir by three species of root-colonizing insects. Canadian Journal of Forest Research 16:745–749.

Wright, L. C., A. A. Berryman, and B. E. Wickman. 1984. Abundance of the fir engraver, *Scolytus ventralis*, and the Douglas-fir beetle, *Dendroctonus pseudotsugae*, following tree defoliation by the Douglas-fir tussock moth, *Orgyia pseudotsugata*. Canadian Entomologist 116:293–305.

10

Running Waters of the Alaskan Boreal Forest

Mark W. Oswood
Nicholas F. Hughes
Alexander M. Milner

Streams in the Landscape: A Boreal Perspective

Running waters reflect the character of their landscape. Landscapes influence their streams by supplying dissolved ions to the water, determining the organic matter supply to stream foodwebs, and influencing water temperature and water flows (Gregory et al. 1991, Hynes 1975). The water that feeds streams has passed over and through the vegetation, soils, and rocks of the valley. Just as urine carries the chemical imprint of metabolic activities (such as diabetes), the kinds and amounts of dissolved matter delivered to stream channels carry the signature of the valley's parent materials and biota. Riparian (streamside) vegetation similarly regulates the balance of carbon sources to stream consumers. In valleys with sparse riparian vegetation, abundant light at the streambed allows in-stream primary production by protists and plants to dominate. Where riparian vegetation forms a canopy over the stream, leaves and needles from shrubs and trees dominate carbon supplies to consumers because low light limits contributions from in-stream primary producers (Vannote et al. 1980). Water temperature and flow are complexly determined by climatic controls (e.g., air and soil temperatures, patterns of precipitation), landscape physiography (e.g., shading of streams by valley walls), and the filter of light-absorbing and water-transpiring riparian vegetation. Thus, streams in the desert biome of the American Southwest, with intermittent droughts and floods, high water temperatures, and abundant light, are very different habitats from the cool, dark waters of perennial streams in the temperate rain forest of the Pacific Northwest coast (Fisher 1995). Likewise, streams in the boreal forest of Alaska (and in the cold circumboreal forests of the world) take their cues from the landscape. Cold permeates the ecology of the boreal landscape and the running waters therein. The

148 Forest Dynamics

consequences of high-latitude climate on running waters are at least three: creation of ice in both terrestrial and running water systems; limited inputs of organic matter and nutrients to foodwebs; and thermal effects of low water temperatures on biological processes (Oswood 1997).

Making a Living: Foodwebs in Boreal Streams

For forested streams, a good case can be made for autumn as the beginning of the stream's "fiscal" year. Autumnal leaf fall from riparian vegetation provides a major proportion of the annual energy budget to stream foodwebs. The decomposition of leaf detritus and incorporation of the leaf components into stream foodwebs (Fig. 10.1) has parallels in the decay of leaf litter in forest floor soils (Anderson 1987, Wagener et al. 1998). The general sequence—leaching of soluble components, colonization by decomposer microbes, and reduction in particle size by shredding (comminuting) invertebrates—occurs in both soils and streams, with roles in soils and streams played by taxonomically distinct but functionally equivalent organisms. However, processing of leaf detritus occurs faster in streams than in soil systems and over much larger spatial scales, because the flow of water carries the products of decomposition downstream.

Leaf detritus rapidly loses soluble materials via leaching (about 10–15% of dry mass) (Cowan 1983). The dissolved organic matter (DOM) in leaf leachate is a complex mixture of organic acids, carbohydrates, tannins, and polypeptides

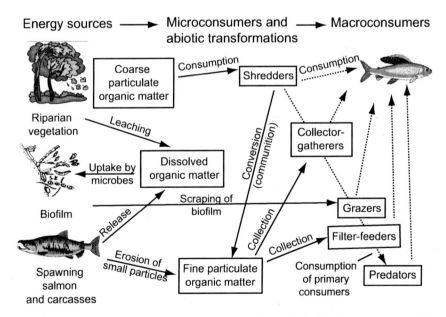

Figure 10.1. Foodwebs and carbon flow in a typical Alaskan boreal stream. Arrows show major (or especially interesting) transfers and transformations of organic carbon.

(Thurman 1985). Alder (*Alnus*), birch (*Betula*), and willow (*Salix*) dominate the riparian zone along many Alaskan boreal streams. Surprisingly, given the low water temperatures of Alaskan streams in autumn compared to streams in temperate regions, rates of decomposition of leaf litter from these and other tree species are comparable to decomposition rates in temperate streams (Irons et al. 1994). Although microbial processing of leaf detritus in streams declines with colder water temperatures at high latitudes, there is a compensating increase in the biomass of leaf-shredding macroinvertebrates (Irons et al. 1994).

Decomposer microbes, principally fungi and bacteria, rapidly colonize leaf litter and begin converting detrital tissue to carbon dioxide (Buttimore et al. 1984). Microbial colonization of leaf detritus likely facilitates shredder consumption of leaf tissue in several ways, including maceration of leaf tissue and provision of critical nutrient needs (e.g., proteins or lipids) of shredder invertebrates (Maltby 1992). In Alaskan boreal streams, the major shredder taxa are caddisfly larvae (Trichoptera) in the family Limnephilidae and stonefly nymphs (Plecoptera) in the family Nemouridae (Cowan 1983). Representatives of these two families are shown in Fig. 10.2.

Winter browsing by mammals both large (moose) and small (hares) causes shifts in community composition of trees (e.g., from deciduous to evergreen) and consequent changes in leaf litter production and soil characteristics (Pastor et al. 1988). Browsing of winter tree buds also causes physiological shifts in allocation of carbon and nutrient resources in individual trees, so that summer leaves of winter-browsed trees have higher concentrations of nitrogen and lower concentrations of tannins (Bryant et al. 1991). Consequently, autumnal leaves of winter-browsed trees have potentially higher food quality for shredder invertebrates than leaves from unbrowsed trees. Leaf packs made from leaves of moose-browsed birch trees have higher processing (decay) rates than leaves from unbrowsed trees in an Alaskan stream (Irons et al. 1991), and leachates from leaf litter of browsed trees support greater microbial respiration (Estensen 2001). Therefore, the cascading effects of vertebrate herbivory extend to stream (and soil; Chapter 13) foodwebs, suggesting that moose are keystone herbivores in the boreal landscape.

In streams where sufficient light reaches the stream bed, photosynthesis by attached algae, mosses, or vascular plants may provide an important or even dominant source of organic carbon to stream webs. Attached algae and the associated community of bacteria, fungi, and microinvertebrates are generally referred to as periphyton or biofilm (Fig. 10.1). Diatoms are a dominant element of the biofilm flora of boreal streams (Anderson 1984). Macroinvertebrates behaviorally and structurally adapted for scraping or brushing biofilm from submerged surfaces are termed *grazers* or *scrapers* (Fig. 10.2). Mayflies (Ephemeroptera) in the family Heptageniidae are common grazers in Alaskan boreal streams and demonstrate classic mouthpart adaptations for brushing biofilm from benthic surfaces. Other common grazers are found in the caddisfly (Trichoptera) families Limnephilidae (*Apatania*) and Glossosomatidae (*Glossosoma*).

Fine particulate organic matter (FPOM), derived from both living sources (e.g., algae dislodged from the stream bottom) and dead matter (products of leaf litter processing), is everywhere in streams, suspended in the water and sedimented to

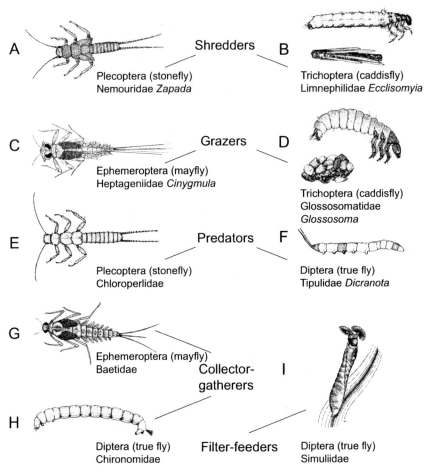

Figure 10.2. A gallery of macroinvertebrates in boreal streams. Commonly encountered representatives of each of the functional groups (Fig. 10.1) are shown (including taxa whose life histories are summarized in Fig. 10.5). Illustrations of caddisflies (Trichoptera) show both the larva and the larval case constructed by larvae. Illustrations are reproduced with permission from the following sources: A, D, E, F, H, and I from University of Alberta Press (Clifford 1991); B from G. B. Wiggins (1996); C and G from Lyons Press (Knopp and Cormier 1997).

the benthos (Fig. 10.1). Filter-feeding macroinvertebrates (Fig. 10.2) consume FPOM in suspension. Blackfly larvae are the most abundant filter feeders of boreal streams, using their cephalic fans to filter suspended FPOM (seston) from the water. A wide taxonomic range of organisms, termed *collector-gatherers*, consume FPOM from benthic surfaces or as they burrow through soft sediments (Fig. 10.2). Common collector-gatherers in Alaskan boreal streams are found in the mayfly (Ephemeroptera) family Baetidae, in the ubiquitous nonbiting midges (Diptera:

Chironomidae), and in the aquatic oligochaete worms (relatives and ecological equivalents of the terrestrial earthworms).

Finally, predacious macroinvertebrates consume other macroinvertebrates (Fig. 10.2). Common predators in Alaskan boreal streams include many of the stoneflies (e.g., Chloroperlidae), as well as some dipterans (e.g., *Dicranota*, a cranefly larvae) and caddisflies (e.g., Rhyacophilidae).

Figure 10.1 shows only the major transfers and transformations of organic matter or aspects that are especially important in Alaskan boreal streams. There are many other depictions of stream foodwebs and nutrient cycles. Winterbourn and Townsend (1991, Fig. 12.1) depict upstream-downstream fluxes, Maltby (1992, Fig. 15.2) emphasizes the central roles of particulate detritus and dissolved organic matter in stream foodwebs, and Newbold (1992) integrates stream foodwebs and transformations of inorganic nutrients.

Most taxa of macroinvertebrates are primary consumers—shredders, grazers, collector-gatherers, and filter-feeders. In contrast, only a few taxa of fish in boreal rivers and streams are primary consumers, limited to filter-feeding ammocoete larvae of arctic lampreys (*Lampetra japonica*) and longnose suckers (*Catostomus catostomus*) consuming streambed detritus and invertebrates (Fig. 10.3). All other fish species consume macroinvertebrates from the benthos or drift or are piscivores, consuming other fishes (Oswood et al. 1992). Benthic macroinvertebrates are therefore the essential trophic links between organic inputs to stream foodwebs and fishes of sport (e.g., arctic grayling; *Thymallus arcticus*), commercial (e.g., rearing salmon), or subsistence importance (Fig. 10.3).

The Gift from the Sea: Nutrients from Spawning Salmon

The migration of spawning salmon into freshwater systems transports nutrients from the productive ocean "uphill" to freshwater systems that are often oligotrophic (biotic productivity limited by low nutrient availability). Recently, research along the "low-tech–high-tech spectrum"—from basic natural history to whole-ecosystem experiments to use of stable isotopes as foodweb tracers—has permitted a better understanding of the incorporation of marine-derived nutrients (MDN) into stream and riparian foodwebs (Cederholm et al. 1999, Gende et al. 2002).

Spawning salmon contribute MDN to aquatic and terrestrial foodwebs via egg deposition during spawning and, especially, after death (Fig. 10.4). Spawning salmon dig redds in streambed gravels, depositing fertilized eggs. Some eggs are lost during spawning (or when later spawners excavate an already dug redd) and drift downstream. Such eggs are avidly consumed by drift-feeding fishes, as attested by the popularity of preserved salmon eggs as fishing bait. After spawning, salmon die, and their carcasses accumulate on streambed obstructions, in pools, and along stream margins. Comparisons of stable isotope signatures of steam reaches with and without spawning salmon show that marine-derived nitrogen is incorporated into biofilm, macroinvertebrates, and fishes (Kline et al. 1990). Experimental addition of salmon carcasses to stream reaches (or artificial streams) can increase benthic chlorophyll *a* (autotrophic component of biofilm) and abundances of macroinvertebrates (Wipfli

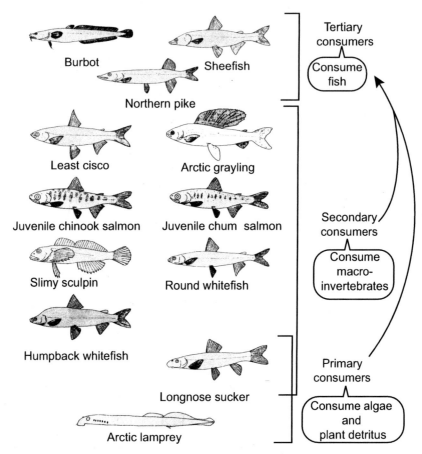

Figure 10.3. Fishes of boreal streams and their trophic roles, based on Oswood et al. (1992). Fish illustrations are reproduced with permission from Alaska Northwest Books (Morrow 1980). Spawning adults of Pacific salmon are not illustrated. Seasonal changes in habitat use of these species are shown in Table 10.1.

et al. 1998, 1999). Observations of the decomposition of salmon carcasses in streams of the Alaskan interior (Kline et al. 1997, Piorkowski 1995) show that carcasses are consumed directly by stream macroinvertebrates and by fishes when carcasses fragment in late decomposition, releasing particles of fish flesh. Carcasses exposed to air are quickly populated by dipteran maggots; these maggots may be washed into the stream to be consumed by fish (Kline et al. 1997). Bears (black bears and brown/grizzly bears) are predators, capturing migrating and spawning salmon, as well as scavengers of carcasses (Willson et al. 1998). Many other mammals and birds consume salmon carcasses (Cederholm et al. 1989, Willson et al. 1998). There is increasing evidence that the seasonal availability of salmon provides high-quality nutrients that increase survival and reproductive success of many lotic and terres-

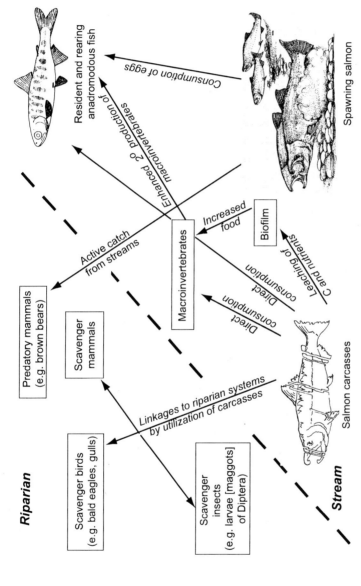

Figure 10.4. Linkages of spawning salmon to stream and riparian foodwebs. Arrows show trophic connections directly or indirectly derived from spawning or postmortem salmon. Sources of illustrations used with permission are as follows: "Spawning salmon" (Sasquatch Books, Steelquist 1992); "Resident and rearing anadromous fish" (Alaska Northwest Books, Morrow 1980); "Salmon carcasses" (Springer-Verlag, Kline et al. 1997).

trial organisms, as well as influencing seasonal migrations and timing of reproduction of consumers (Willson et al. 1998).

The high-quality organic matter (proteins and lipids) and inorganic nutrients (N, P, and others) supplied by adult salmon may increase the biotic productivity of natal spawning streams (Fig. 10.4) so that the growth and survival of young salmon are enhanced. This positive feedback loop—more spawning salmon—> greater juvenile survival—> more returning spawners—is a potential "tipping point" in the management of salmon. Decreases in the number of adults reaching natal streams can be caused by natural events (e.g., increased natural predation on adult salmon in the ocean) or human manipulations (e.g., increased harvest or barriers to migration). Such decreases may run the cycle backwards—fewer spawning salmon—> decreased juvenile survival—> fewer returning salmon (Larkin and Slaney 1997). Thus, for sustainable runs of a stock, the number of salmon spawning in natal streams must be sufficient to supply any marine-derived nutrients necessary to maintain stream productivity and growth of juvenile salmon (Levy 1997a, 1997b).

Phenology: The Seasons of Boreal Streams

Compared to lower latitudes, high latitudes receive less light annually, and the light distribution is highly seasonal, producing the cold, dark days of winter and the characteristic midnight sun of summer. In this section we briefly trace the phenology (seasonal cycles of physical and biological processes; Clifford 1978) of one of the most common types of boreal streams—clearwater systems that derive their flow from groundwater (in winter), augmented by snowmelt and rainfall during the ice-free period of the year (Fig. 10.5).

Monument Creek is a small stream (second order = fed only by headwater streams, with no tributaries of their own) northwest of Fairbanks. Water temperatures at the stream bottom ("surface" of the water/sediment interface) reach 0°C by mid-October and remain at or below freezing for about seven months each year (Fig. 10.5). Maximum water temperatures in summer in Monument Creek reach about 13°C—warmer than the temperature of streams fed by glacial meltwaters (Milner et al. 1997) or streams that drain permafrost-dominated watersheds (Irons et al. 1989, MacLean et al. 1999). The thermal regime in running waters plays a major role in determining life histories and distributions of aquatic organisms (Ward and Stanford 1982). Growth, developmental, or maturation thresholds are set by absolute water temperature or by accumulation of thermal units (degree-days; Chapter 4). Monument Creek (a clear-water stream) has approximately 950 degree-days, compared to a nearby stream that drains a permafrost-dominated valley and that has 400 degree-days (Irons and Oswood 1992). By way of comparison, Webster et al. (1995) summarized annual accumulation of degree-days for streams of the east coast of the conterminous United States—from 3000 to about 7000 between New York and Georgia.

Obviously, the boreal streams of Alaska are thermally impoverished compared to streams of temperate regions, undoubtedly a factor in the slow growth rates and the low biotic diversity of the aquatic fauna. For example, arctic grayling in the

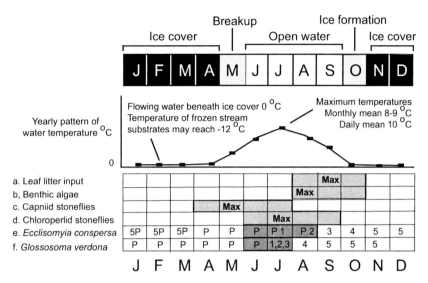

Figure 10.5. Phenology (yearly cycle of physical and biological phenomena) of a typical clear-water stream in interior Alaska (Monument Creek, near Fairbanks). Patterns are simplified and vary from stream to stream and from year to year. Rows a and b in the bottom diagram depict the timing of the major inputs of organic carbon into stream foodwebs (Fig. 10.1); gray shading shows months of major inputs. Flight periods of adult stoneflies are shown for two families in rows c and d. Capniidae (nymphs are shredders) are "winter" stoneflies; Chloroperlidae (most nymphs are predators) are summer emergers. Simplified life histories of two species of caddisflies (Trichoptera) are shown in rows e and f. *Ecclisomyia* is a shredder in the family Limnephilidae. *Glossosoma* is a grazer in the family Glossosomatidae. Caddisflies have five larval stages (instars; shown as numbers 1–5), a transitional pupal stage (P), and a winged adult stage (resembling moths). The dominant life history stage(s) for each month is shown, with the flight period of adults in gray shaded months. Sources of information are as follows: ice phenology and water temperatures (Cowan et al. 1983, Irons and Oswood 1992) and photographic record; leaf litter input (row a; Cowan and Oswood 1983); benthic algae (row b; LaPerriere et al. 1989); stonefly flight periods (rows c and d; Howe 1981); caddisfly life histories (rows e and f; Irons 1985, 1988).

Chena River near Fairbanks take five to seven years to reach the same length that grayling in Montana streams reach in two to three years, and the number of fish species in the Chena River is only 12 (Fig. 10.3; occasionally encountered lake chub (*Couesius plumbeus* not included)), compared to 90 in an Arkansas stream (Oswood et al. 1992). Life histories of insects in Alaskan streams are similarly extended (Stewart et al. 1990), and the diversity of benthic macroinvertebrates in Alaskan streams is low compared that in to temperate regions. (Oswood 1989). For both fishes and aquatic insects, the latitudinal reduction in diversity is distributed unevenly across the major taxa. Salmonids dominate boreal (Oswood et al. 1992) and Alaskan (Poff et al. 2001) running waters. The true flies (Diptera) similarly dominate the benthic insect fauna of streams in the boreal and other regions of Alaska (Oswood 1989).

The seasonal timing of organic matter inputs to foodwebs of boreal streams is truncated compared to that for streams of temperate regions. Autumn comes early; peak input of leaf detritus to streams (Fig. 10.5) occurs in mid-September (or earlier in sites with substantial permafrost) (Cowan 1983, Smidt 1997). One might expect that the abundance of benthic algae would track the seasonal cycle of solar input, peaking near summer solstice. Instead, density (and biomass, as measured by chlorophyll a) of algae in biofilm (Fig. 10.1) is greatest in late summer and early autumn (Fig. 10.5). Life cycles of macroinvertebrates show seasonal patterns, as well. The capniid stoneflies (Fig. 10.2) are members of the winter stoneflies, famous for nymphal growth at the low water temperatures of winter and emergence of adults from winter to early spring. In the more stringent climate of interior Alaska, adult emergence occurs from April to July, with a maximum in the still-cold waters of May (Fig. 10.5). In contrast, the chloroperlid stoneflies (Fig. 10.2) are summer stoneflies, emerging in Alaska as adults from June to September, with a peak in July (Fig. 10.5). Life histories of macroinvertebrate consumers show apparent parallels between availability of food resources and growth of larvae. *Ecclisomyia*, a shredder caddisfly (Fig. 10.2) grows quickly to its final larval stage from September to November (Fig. 10.5), the same period when leaf litter enters streams and is colonized by the fungi and bacteria, enhancing the food quality of the leaf detritus (Fig. 10.1). In contrast, a grazer caddisfly (*Glossosoma*; Fig. 10.2) grows rapidly to its final larval stage from July to September, approximately parallel with maximum abundance of its food resource, benthic algae (Fig. 10.5).

By midwinter, ice completely covers the surface of boreal steams and rivers (Fig. 10.5) except for small areas kept ice-free by upwelling ground water. Ice extends vertically to the benthos in the shallower portions of streams (Fig. 10.6). Temperatures at the streambed surface can be colder than -10°C, a temperature that exceeds the physiological capacity (freezing tolerance or supercooling capacity) of nearly all taxa of streambed macroinvertebrates (Irons et al. 1993, Oswood et al. 1991). Only two taxa of Diptera (Empididae "danceflies" and Chironomidae "nonbiting midges") show substantial survival when frozen blocks of streambed sediments are thawed. Laboratory experiments show that these macroinvertebrates actively move away from the freezing front as streambed sediments freeze from the surface (Irons et al. 1993). Movements of benthic invertebrates to deeper streambed sediments (which serve as a thermal refuge for invertebrates) are analogous to the long-distance migrations of grayling and other fishes along river systems to overwintering habitats (Table 10.1; discussed later).

The Diversity of Boreal Streams: A Sampler from Denali National Park and Preserve

The running waters of Alaska's boreal forest range from low-order headwater streams to the Yukon River, one of the 25 largest rivers (in drainage area) in the world (Milner et al. 1997). Water sources determine many of the physical characteristics and seasonality of streams, including the hydrothermal regime and the stability of substrates (Fig: 10.7). In the summer, some streams and rivers are fed by

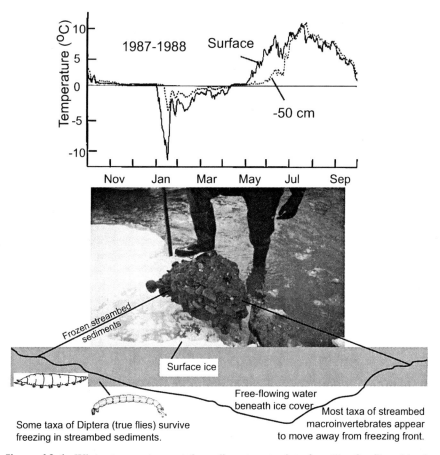

Figure 10.6. Winter temperatures at the sediment–water interface ("surface") and in the streambed (at 50 cm depth); photograph of a piece of frozen streambed exhumed from a small stream. Illustrations of Diptera larvae are used with permission from University of Alberta Press (Clifford 1991).

the meltwater of glaciers and are milky with glacial flour (the suspended sediments of ground rock), whereas others derive their flow from groundwater or precipitation and are clear. Other lowland streams drain catchments with organic-rich (peaty) soils and have brown waters, with high concentrations of dissolved organic matter. Annual thermal and hydrologic regimes of boreal streams and rivers differ greatly depending on water source (Milner et al. 1997). The stability of the substrate materials that make up the streambeds and banks is jointly determined by the kinds of bed and bank materials and the daily and seasonal fluctuations in water flow. Stream size, water clarity, and streambed stability—along with availability of inorganic nutrients supporting in-stream plant growth and riparian leaf detritus to fuel detrital food chains—constitute a checklist of the major habitat variables that should determine the kinds of organisms found in these boreal streams. Recent studies in

Table 10.1. Seasonal use of running water habitats by fishes in interior Alaska (see Fig. 10.3).

Species	Clearwater (lower)	Clearwater (middle)	Clearwater (upper)	Bog-fed	Spring-fed	Glacial	Shallow Lakes
Slimy sculpin	F S O	F S O	F S O	F S O	F S O	F S O	- - -
Arctic grayling	F S O	F S O	F S O	F S -	F - -	- - O	- - -
Round whitefish	F-O	F S O	F - -	- - -	F - -	F S O	- - -
Humpback whitefish	F - O	F S -	- - -	- - -	F - -	F S O	F - -
Least cisco	F-O	F S -	- - -	- - -	- - -	F S O	F - -
Longnose sucker	F S O	F - -	- - -	- - -	- - -	F S O	F - -
Burbot	F-O	F - O	F - O	- - -	- - -	F S O	F - -
Northern pike	F S O	F - -	- - -	- - -	- - -	F - O	F S -
Sheefish	F - -	F - -	- - -	- - -	- - -	F S O	F - -
Chinook salmon	F - -	F S O	F - -	- - -	- - -	- - -	- - -
Chum salmon	- - -	- S -	- - -	- - -	- - -	- S -	- - -
Lamprey	F - O	F S -	- - -	- - -	- - -	- - -	- - -

Key: F = feed, S = spawn, O = overwinter.

Note: For fall spawning species, the eggs overwinter in the spawning habitat, but this is not indicated in the table.

Denali National Park link these constellations of physical-chemical factors to the biotic structure of streams in Alaska's Interior.

As part of the Long-Term Ecological Monitoring Program (LTEM) in Denali National Park, 45 streams representing a wide range of sizes and physical characteristics were studied. These streams fell into seven groups, each with distinctive physicochemical features and each supporting characteristic macroinvertebrate communities (Conn 1998). Sixty-five species of Chironomidae (nonbiting midges; Fig. 10.2) have been identified, over two-thirds of the macroinvertebrate taxa found in these streams (Ray 2002) The rivers and streams fall into groups based largely on streambed stability and the influence of water source on turbidity, conductivity, and alkalinity (Conn 1998). These physical and chemical factors, in turn, influence the biological communities of the streams. We discuss four of these groups of streams and rivers.

Small (clearwater) stable streams (Fig. 10.7) are similar to spring-fed systems (discussed later) in having a very stable streambed with a close border of vegetation but are typically fed by snowmelt and rain runoff. The overhanging trees and shrubs provide inputs of leaves, which serve as food for shredder invertebrates. These streams supported the highest proportion of shredders of all the streams types sampled (Fig. 10.8c), particularly in the fall, when values reached 25%. These streams also supported the highest levels of epilithic algal growth, with chlorophyll a levels averaging the highest of the stream groups (Fig. 10.8a); scrapers averaged about 8% of the macroinvertebrate fauna (Fig. 10.8c). These productive streams support fish and a wide diversity of invertebrates, including the highest proportions of stoneflies (Plecoptera) and mayflies (Ephemeroptera; Fig. 10.2) among the stream types (Fig. 10.8b).

Spring/groundwater-fed streams (Fig. 10.7) are typically small, shallow systems. Because groundwater is the predominant water source, the water in these streams is cooler in summer and warmer in winter than in streams fed by surface runoff. In addition, their flow is nearly constant, creating a stable channel, which in some streams

Figure 10.7. Major stream types of interior Alaska. Water sources determine many of the physical characteristics and seasonality of streams, including the hydrothermal regime and the stability of substrates.

permits a close border of riparian plants, principally willow and alder, to develop. Spring/groundwater-fed streams support the highest abundance of macroinvertebrates, with densities averaging over 4,000 individuals m^{-2} (Fig. 10.8a). There is a wide diversity of taxa, with Chironomidae (Fig. 10.2) forming less than 50% of the community and Ephemeroptera approximately 25% in 1995 (Fig. 10.8b). Chlorophyll a levels (Fig. 10.8a) were sufficient to ensure that scrapers were well represented (approximately 10%; Fig. 10.8c) within the macroinvertebrate community. Spring/groundwater-fed streams typically support abundant populations of arctic grayling.

Larger river systems partially fed by glacier-melt water (Fig. 10.7) are often slightly turbid during summer because of the input of some glacial water. However, unlike the glacier-fed rivers (discussed later), they are more stable, although at times they form multiple channels across the floodplain (Fig. 10.7). Nevertheless, they are relatively productive streams, with densities of benthic macroinvertebrates averaging over 1,500 individuals m^{-2} (Fig. 10.8a) even though chlorophyll a levels average 142 mg m^{-2}.

Large glacier-fed rivers are characterized by highly turbid water in the summer, with multiple (braided) channels flowing across wide gravel floodplains (Fig. 10.7) and negligible vegetation. These rivers have large drainage basins and are fed by glaciers with peak flows in July, when long days induce maximum ice melt (Chapter 16). Large glacier-fed rivers carry high loads of suspended sediments in the summer (up to 1,800 mg L^{-1} have been reported; Edwards and Tranel 1998), causing high turbidity and low water clarity. However, in early spring and late fall, when glacial discharge is low, these rivers become clear, and mats of filamentous green algae may appear. The beds of large glacier-fed rivers are very unstable and continually

160 Forest Dynamics

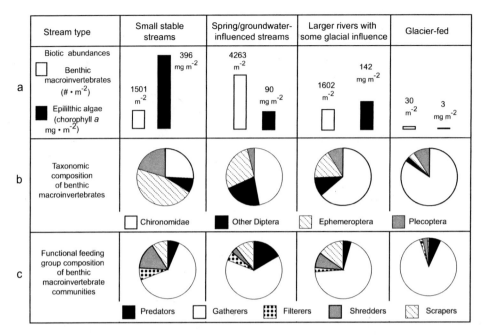

Figure 10.8. Synoptic ecology of stream types (Fig. 10.7) in Denali National Park. Top panel: mean abundance of benthic macroinvertebrates across three seasons and abundance of streambed (epilithic) primary producers (mean chlorophyll *a* values). Middle panel: composition of the major taxonomic groups of benthic macroinvertebrates. Lower panel: relative abundance of the functional feeding groups of benthic macroinvertebrates across three seasons.

shift across the floodplain during the summer, making it difficult for macroinvertebrates and fish to establish communities. Consequently, these rivers support a low diversity of invertebrates (typically less than 10 species), most of which belong to the family Chironomidae (Fig. 10.8b). Because of the dominance by the Chironomidae, functional feeding group composition is dominated by collector-gatherers (Fig. 10.8c). Scrapers are poorly represented as a result of the low levels of epilithic algal growth on the stones (chlorophyll *a* levels < 3 mg m^{-2} for this group; Fig. 10.8a). Overall, densities of macroinvertebrates are low, averaging fewer than 200 individuals m^{-2} (Fig. 10.8a). However, there are "hot spots" of activity across these glacial floodplains where upwelling water creates small channels with patches of riparian vegetation; these channels support a diverse and abundant macroinvertebrate community (as yet unsampled).

Fish Migrations: Life Histories and the Habitat Template

Like macroinvertebrates, riverine fishes have life histories shaped by the yearly cycle of temperature and water flow, food availability, and the bottleneck of winter, but with seasonal migrations that may encompass thousands of kilometers along river corridors. To understand the seasonality of fish in boreal running waters, we must consider how the range of fish habitats, from headwater streams to large rivers to the ocean, changes over a yearly cycle. Southwood (1977, 1988) argues persuasively that the spatial and temporal distribution of opportunities provided by the habitat creates the template that shapes the evolution of life histories and migration.

Anadromous fish illustrate this role of the habitat as a template on a grand scale. These fishes spawn (and often live their juvenile lives) in freshwater, migrate to the ocean to feed and to grow to adult size, and then return to freshwater to spawn. The global distribution of anadromy is strongly related to the biotic productivity of the oceans relative to the biotic productivity of freshwaters at any given latitude (Gross et al. 1988). At cool temperate and higher latitudes, ocean productivity is greater than freshwater productivity. It is likely that this greater food availability in the ocean has favored the evolution of anadromy at northern latitudes and partly explains the abundance of salmon in Alaskan streams.

On a smaller scale, the match between the timing of salmon runs and the water temperatures their eggs experience while incubating in the gravel also illustrates Southwood's argument. Summer-run chum salmon (*Oncorhynchus keta*) in the Tanana drainage (a major tributary of the Yukon River) spawn in July and August, in clearwater rivers like the Chena, where winter water temperatures in the gravel are near freezing. The later fall-run fish select upwelling areas in the glacial-silt-laden Tanana River, where winter water temperatures in the gravel are warmer. The difference in the timing of spawning between the two runs is well matched to the subgravel temperature regime in their spawning habitat so that, in both cases, eggs accumulate enough degree-days to hatch before spring break-up.

Migration in subarctic stream fish is not limited to the anadromous species; it is also almost ubiquitous among stream resident species. The reason is probably that dramatic seasonal shifts in habitat characteristics favor the evolution of life histories that involve seasonal movements between habitats particularly well suited for spawning, feeding, or overwintering (Reynolds 1997). Of the 12 species of fish common in the Chena River system (a tributary of the Tanana River) and in boreal rivers of Alaska (Fig. 10.3), perhaps only the slimy sculpin (*Cottus cognatus*) could be considered essentially sedentary. All other species conduct substantial migrations. In general, fish overwinter in deep pools in the Tanana River or the lower Chena, where they find protection from the thick ice that forms in winter, although arctic grayling may also use upwelling areas in headwaters (see later discussion). In summer, many individuals migrate to feeding grounds in small and middle-size rivers or in shallow lakes connected to the river system (Table 10.1). Spring-spawning species (e.g., arctic grayling, longnose suckers, and northern pike) make a special-purpose spawning migration before traveling to their summer feeding areas. Fall-spawning species (e.g., round whitefish, humpback whitefish, least cisco, and sheefish) make their spawning migration before returning to overwintering areas. Burbot are unusual in that they spawn in February, under the ice, in the glacial Tanana River. Piscivores (burbot, northern pike, and sheefish) often feed actively in the winter, perhaps finding it easier to capture prey when they are concentrated in overwintering areas. Feeding opportunities in winter are scant for the other species.

Arctic grayling are, after slimy sculpin, the most widespread and abundant fish in the boreal streams of Alaska. Thanks to the research of Tack (1980), we know a great deal about how migration maps the life history of arctic grayling onto the habitats available in the Tanana River. We also know enough about the biology of arctic grayling (Hughes 1992a, 1992b, 1998, 1999, Hughes and Dill 1990, Hughes and Reynolds 1994, Northcote 1995, Vascotto and Morrow 1973) to speculate on how these migrations adapt arctic grayling to their demanding environment.

Migration maps the life histories of arctic grayling onto the four stream types described in Fig. 10.7 in at least three different patterns. Some populations complete their whole life cycle in clearwater streams. In these populations, spawning activity is concentrated in the lower river, and fish segregate by size during the summer feeding period, with larger fish further upstream. Fish generally move downstream to overwinter, as discussed earlier, but juvenile fish may overwinter in headwater reaches, where there is just enough water between the bottom of the ice and the stream bed to permit survival. In other populations, spawning and juvenile feeding occur in the lower reaches of a clearwater river, but larger fish switch to feeding in a spring-fed stream. In these populations, fish may overwinter in the clearwater stream or in the glacially influenced Tanana River. The third common life history pattern is spawning and juvenile rearing in bog-fed streams, with large fish feeding in spring-fed streams. In these populations, all fish overwinter in the Tanana River.

This variety of migratory patterns may seem bewildering, but there is a logical pattern to it that demonstrates how well these remarkable fish are adapted to their environment. Adults spawn in the lower reaches of clearwater streams and in bog-fed streams, because these are the warmest areas available. Natural selection favors adults that select warm areas for spawning because this maximizes the growth rate of their offspring during the first year of life, when growth is temperature-limited. Spawning fish select against spring-fed streams and the headwaters of clearwater streams because they are too cold. Glacial streams are entirely unsuitable as spawning habitat. Habitat selection by feeding fish produces a larger-fish/cooler-water pattern, with small fish in the lower reaches of clearwater streams or bog-fed streams and larger fish in the upper reaches of clearwater streams or spring-fed streams. This pattern of habitat selection may be adaptive because, in food-poor boreal streams, fish find it increasingly hard to obtain maximum rations as they grow. Since the optimum temperature for growth decreases as ration size falls, this results in a bioenergetic incentive for fish to move into colder water as they grow (Hughes 1998). Finally, the selection of overwintering habitat is geared to maximize survival. This involves tradeoffs that take into account the risk of freezing, starving, and being eaten. Small fish may overwinter in headwater habitats, where there is considerable risk of freezing, because piscivores are absent, and their small size renders them vulnerable to piscivory in deeper downstream reaches. Larger fish may prefer to take their chances with the piscivores downstream because their larger size reduces the risk they will be eaten but increases their risk of freezing in the headwaters. Fish may avoid overwintering in spring-fed streams because these are relatively warm in the winter. This higher temperature increases metabolic rates and the risk of starvation. Fish avoid small bog-fed streams because they freeze solid.

Conclusions and Implications for Management

The running waters of the Alaskan boreal forest reflect the landscapes they drain, from crystalline meltwater streams high in the Alaska Range to the brown waters of muskeg streams to the turbid waters of rivers under the influence of glaciers. Each of these kinds of running waters has a characteristic yearly regime of flow, temperature, and sediment transport and so has a characteristic biota. Movement between habitats suitable for reproduction and feeding and habitats with sufficient free-flowing water for overwintering is a life-history theme for many benthic invertebrates and fishes. Production at higher trophic levels (fish and their consumers) in boreal streams is likely constrained jointly by cold water temperatures (limiting growth rates and slowing maturation) and low inputs of carbon to running water foodwebs (Oswood 1997). In consequence, the management of resident fishes for sport, commercial, or subsistence fisheries must consider the age structure of fish populations (big fish are often old fish) and the slow rate of biomass replacement. The exceptions are, of course, the spectacular runs of anadromous fishes, especially Pacific salmon. The productive ocean ecosystems of the North Pacific provide the bioenergetic means to grow large salmon stocks. The lack of dams on mainland Alaskan rivers allows free passage upstream for spawning adults and seaward for smolts. High-latitude limitations on human ecosystems in interior Alaska have so far limited the degradations to spawning habitats (from urbanization, forestry, agriculture, and water extraction) that have reduced freshwater survival of salmon stocks in the Pacific Northwest. It is hoped that the lessons learned from the political, economic, social, and biological mélange that has decimated most populations of Pacific salmon (Cone 1995, Lichatowich 1999) will allow populations of wild salmon in Alaska to connect the forest and the sea for millennia to come.

References

Anderson, J. M. 1987. Forest soils as short dry streams: Effects of invertebrates on transport processes. Verhandlungen Gesellshaft für Okologie 17:33–46.

Anderson, P. R. 1984. Seasonal changes of attached algae in two Alaskan subarctic streams. M.S. Thesis, University of Alaska Fairbanks.

Bryant, J. P., K. Dannell, F. Provenza, P. Reichardt, T. Clausen, and R. Werner. 1991. Effects of mammal browsing on the chemistry of deciduous woody plants. Pages 135–154 in D. W. Tallamy and M. J. Raupp, editors. Phytochemical Induction by Plants. Wiley, New York.

Buttimore, C.A., P.W. Flanagan, C.A. Cowan, and M.W. Oswood. 1984. Microbial activity during leaf decomposition in an Alaskan subarctic stream. Holarctic Ecology 7:104–110.

Cederholm, C. J., D. B. Houston, D. L. Cole, and W. J. Scarlett. 1989. Fate of coho salmon (*Oncorhynchus kisutch*) carcasses in spawning streams. Canadian Journal of Aquatic Sciences 46:1347–1355.

Cederholm, C. J., M. D. Kunze, T. Murota, and A. Sibatani. 1999. Pacific salmon carcasses: Essential contributions of nutrients and energy for aquatic and terrestrial ecosystems. Fisheries 24:6–15.

Clifford, H. F. 1978. Descriptive phenology and seasonality of a Canadian brown-water stream. Hydrobiologia 58:213–231.

Clifford, H. F. 1991. Aquatic invertebrates of Alberta. University of Alberta Press, Edmonton, Alberta.

Cone, J. 1995. A common fate: Endanagered salmon and the people of the Pacific Northwest. Henry Holt, New York.

Conn, S. C. 1998. Benthic macroinvertebrates in the rivers of Denali National Park and Preserve, Alaska: An approach for watershed classification and ecological monitoring. Ph.D. Dissertation, University of Birmingham, UK.

Cowan, C. A. 1983. Phenology of benthic detritus input, storage, and processing in an Alaskan subarctic stream. M.S. Thesis, University Alaska Fairbanks.

Cowan, C. A., and M. W. Oswood. 1983. Input and storage of benthic detritus in an Alaskan subarctic stream. Polar Biology 2:35–40.

Cowan, C. A., M. W. Oswood, C. A. Buttimore, and P. W. Flanagan. 1983. Processing and macroinvertebrate colonization of detritus in an Alaskan subarctic stream. Holarctic Ecology 6:340–348.

Edwards, P. J., and M. J. Tranel. 1998. Physical and chemical characteristics of streams and rivers within Denali National Park and Preserve. Final report submitted to Denali National Park and Preserve. U.S. Forest Service, Northeastern Forest Experiment Station, Parsons, WV.

Estensen, J. L. 2001. Winter vertebrate browsing of birch: Effects on the use of leaf litter leachates by stream microorganisms. M.S. Thesis, University of Alaska Fairbanks.

Fisher, S. G. 1995. Stream ecosystems of the western United States. Pages 117–187 in C. E. Cushing, K. W. Cummins, and G. W. Minshall, editors. River and Stream Ecosystems. Elsevier, Amsterdam.

Gende, S. M., R. T. Edwards, M. F. Willson, and M. S. Wipfli. 2002. Pacific salmon in aquatic and terrestrial ecosystems. BioScience 52:917–928.

Gregory, S. V., F. J. Swanson, W. A. McKee, and K. W. Cummins. 1991. An ecosystem perspective of riparian zones. BioScience 41:540–551.

Gross, M. R., R. M. Coleman, and R. M. McDowell. 1988. Aquatic productivity and the evolution of diadromous fish migration. Science 239:1291–1293.

Howe, A. L. 1981. Life histories and community structure of Ephemeroptera and Plecoptera in two Alaskan subarctic streams. M.S. Thesis, University of Alaska Fairbanks.

Hughes, N. F. 1992a. Ranking of feeding positions by drift-feeding arctic grayling (*Thymallus arcticus*) in dominance hierarchies. Canadian Journal of Fisheries and Aquatic Sciences 49:1994–1998.

Hughes, N. F. 1992b. Selection of positions by drift-feeding salmonids in dominance hierarchies: Model and test for arctic grayling (*Thymallus arcticus*) in subarctic mountain streams, interior Alaska. Canadian Journal of Fisheries and Aquatic Sciences 49:1999–2008.

Hughes, N. F. 1998. A model of habitat selection by drift-feeding stream salmonids at different scales. Ecology 79:281–294.

Hughes, N. F. 1999. Population processes responsible for larger-fish-upstream distribution patterns of arctic grayling (*Thymallus arcticus*) in interior Alaska runoff rivers. Canadian Journal of Fisheries and Aquatic Sciences 56:2292–2299.

Hughes, N. F., and L. M. Dill. 1990. Position choice by drift-feeding salmonids: Model and test for arctic grayling (*Thymallus arcticus*) in subarctic mountain streams, interior Alaska. Canadian Journal of Fisheries and Aquatic Sciences 47:2039–2048.

Hughes, N. F., and J. B. Reynolds. 1994. Why do arctic grayling (*Thymallus arcticus*) get bigger as you go upstream? Canadian Journal of Fisheries and Aquatic Sciences 51: 2154–2163.

Hynes, H. B. N. 1975. Edgardo Baldi memorial lecture: The stream and its valley. Verhandlungen der Internationalen Vereinigung für theoretische und angewandte Limnologie 19:1–15.
Irons, J. G., III. 1985. Life histories and community structure of the caddisflies (Trichoptera) of two Alaskan subarctic streams. M.S. Thesis, University of Alaska Fairbanks.
Irons, J. G., III. 1988. Life history patterns and trophic ecology of Trichoptera in two Alaskan (USA) subarctic streams. Canadian Journal of Zoology 66:1258–1265.
Irons, J. G., III, and M. W. Oswood. 1992. Seasonal temperature patterns in an arctic and two subarctic Alaskan (USA) headwater streams. Hydrobiologia 237:147–157.
Irons, J. G., III, J. P. Bryant, and M. W. Oswood. 1991. Effects of moose browsing on decomposition rates of birch leaf litter in a subarctic stream. Canadian Journal of Fisheries and Aquatic Sciences 48:442–444.
Irons, J. G., III, L. K. Miller, and M. W. Oswood. 1993. Ecological adaptations of aquatic macroinvertebrates to overwintering in interior Alaska (USA) subarctic streams. Canadian Journal of Zoology 71:98–108.
Irons, JG, III, M. W. Oswood, R. J. Stout, and C. M. Pringle. 1994. Latitudinal patterns in leaf litter breakdown: Is temperature really important? Freshwater Biology 32:401–411.
Irons, J. G., III, S. R. Ray, L. K. Miller, and M. W. Oswood. 1989. Spatial and seasonal patterns of streambed water temperatures in an Alaskan subarctic stream. Pages 381–390 in W. W. Woessner and D. F. Potts, editors. Symposium on Headwaters Hydrology. American Water Resources Association, Missoula, MT.
Kline, T. C. J., J. J. Goering, and R. J. Piorkowski. 1997. The effect of salmon carcasses on Alaskan freshwaters. Pages 179–204 in A. M. Milner and M. W. Oswood, editors. Freshwaters of Alaska: Ecological Syntheses. Springer-Verlag, New York.
Kline, T. C. J., J. J. Goering, O. A. Mathisen, P. H. Poe, and P. L. Parker. 1990. Recycling of elements transported upstream by runs of Pacific Salmon: I. ^{15}N and ^{13}C evidence in Sashin Creek, southeastern Alaska. Canadian Journal of Fisheries and Aquatic Sciences 47:136–144.
Knopp, M., and R. Cormier. 1997. Mayflies: An angler's study of trout water Ephemeroptera. Greycliff, Helena, MT.
LaPerriere, J. D., E. E. Van Nieuenhuyse, and P. R. Anderson. 1989. Benthic algal biomass and productivity in high subarctic streams, Alaska. Hydrobiologia 172:63–75.
Larkin, G. A., and P. A. Slaney. 1997. Implications of trends in marine-derived nutrient influx to south coastal British Columbia salmonid production. Fisheries 22:16–24.
Levy, S. 1997a. Pacific salmon bring it all back home. BioScience 47:657–660.
Levy, S. 1997b. Ultimate sacrifice. New Scientist 155:38–41.
Lichatowich, J. 1999. Salmon without rivers: A history of the Pacific Salmon crisis. Island Press, Washington, DC.
MacLean, R., M. W. Oswood, J. G. Irons III, and W. H. McDowell. 1999. The effect of permafrost on stream biogeochemistry: A case study of two streams in the Alaskan (USA) taiga. Biogeochemistry 47:239–267.
Maltby, L. 1992. Detritus processing. Pages 331–353 in P. Calow and G. E. Petts, editors. The Rivers Handbook: Hydrological and Ecological Principles. Blackwell Scientific Publications, London.
Milner, A. M., J. G. Irons III, and M. W. Oswood. 1997. The Alaskan landscape: An introduction for limnologists. Pages 1–44 in A. M. Milner and M. W. Oswood, editors. Fresh Waters of Alaska: Ecological Syntheses. Springer-Verlag, New York.
Morrow, J. E. 1980. The Freshwater Fishes of Alaska. Alaska Northwest Publishing Company, Anchorage.

Newbold, J. D. 1992. Cycles and spirals of nutrients. Pages 379–408 *in* P. Calow and G. E. Petts, editors. The Rivers Handbook: Hydrological and Ecological Principles. Blackwell Scientific Publications, London.

Northcote, T. G. 1995. Comparative biology and management of Arctic and European grayling (Salmonidae, *Thymallus*). Reviews in Fish Biology and Fisheries 5:141–194.

Oswood, M. W. 1989. Community structure of benthic invertebrates in interior Alaskan (USA) streams and rivers. Hydrobiologia 172:97–110.

Oswood, M. W. 1997. Streams and rivers of Alaska: A high latitude perspective on running waters. Pages 331–356 *in* A. M. Milner and M. W. Oswood, editors. Fresh Waters of Alaska: Ecological Syntheses. Springer-Verlag, New York.

Oswood, M. W., L. K. Miller, and J. G. Irons III. 1991. Overwintering of freshwater benthic invertebrates. Pages 360–375 *in* R. E. J. Lee and D. L. Denlinger, editors. Insects at Low Temperatures. Chapman and Hall, New York.

Oswood, M. W., J. B. Reynolds, J. D. LaPerriere, R. Holmes, J. Hallberg, and J. H. Triplehorn. 1992. Water quality and ecology of the Chena River, Alaska. Pages 5–27 *in* C. D. Becker and D. A. Neitzel, editors. Water Quality in North American River Systems. Battelle Press, Columbus, OH.

Pastor, J., R. J. Naiman, B. Dewey, and P. McInnes. 1988. Moose, microbes, and the boreal forest. BioScience 38:770–777.

Piorkowski, R. J. 1995. Ecological effects of spawning salmon on several southcentral Alaska streams. Ph.D. Dissertation, University of Alaska Fairbanks.

Poff, N. L., P. L. Angermeier, S. D. Cooper, P. S. Lake, K. D. Fausch, K. O. Winemiller, L. A. K. Mertes, M. W. Oswood, J. Reynolds, and F. J. Rahel. 2001. Fish diversity in streams and rivers. Pages 315–349 *in* F. S. Chapin III, O. E. Sala, and E. Huber-Sannwald, editors. Global Biodiversity in a Changing Environment. Springer-Verlag, New York.

Ray, J. 2002. Chironomidae (Diptera) communities of the rivers in Denali National Park and Preserve, Alaska. University of Birmingham, Birmingham, UK.

Reynolds, J. B. 1997. Ecology of overwintering fishes in Alaskan freshwaters. Pages 281–302 *in* A. M. Milner and M. W. Oswood, editors. Fresh Waters of Alaska: Ecological Syntheses. Springer-Verlag, New York.

Smidt, S. 1997. Spatial variation in the community structure of stream macroinvertebrates within a subarctic Alaskan watershed. M.S. Thesis, University Alaska Fairbanks.

Southwood, T. R. E. 1977. Habitat, the templet for ecological strategies? Journal of Animal Ecology 46:337–365.

Southwood, T. R. E. 1988. Tactics, strategies, and templets. Oikos 52:3–18.

Steelquist, R. 1992. Field Guide to the Pacific Salmon. Sasquatch Books, Seattle, WA.

Stewart, K. W., R. L. Hassage, S. J. Holder, and M. W. Oswood. 1990. Life cycles of six stonefly species (Plecoptera) in subarctic and arctic Alaska (USA) streams. Annals of the Entomological Society of America 83:207–214.

Tack, S. L. 1980. Migrations and distributions of arctic grayling in interior Alaska. Annual Report of Progress. 1979–1980. Project F–9–12, 21(R-I), Alaska Department of Fish and Game, Fairbanks, AK.

Thurman, E. M. 1985. Organic Geochemistry of Natural Waters. Dr. W. Junk, Dordrecht, Netherlands.

Vannote, R. L., G. W. Minshall, K. W. Cummins, J. R. Sedell, and C. E. Cushing. 1980. The river continuum concept. Canadian Journal of Fisheries and Aquatic Sciences 37:130–137.

Vascotto, G. L., and J. E. Morrow. 1973. Behavior of the arctic grayling, *Thymallus arcticus* in McManus Creek, Alaska. Biological Papers of the University of Alaska 14:29–38.

Wagener, S. M., M. W. Oswood, and J. P. Schimel. 1998. Rivers and soils: Parallels in carbon and nutrient processing. BioScience 48:104–108.

Ward, J. V., and J. A. Stanford. 1982. Thermal responses in the evolutionary ecology of aquatic insects. Annual Review of Entomology 27:97–117.

Webster, J. R., J. B. Wallace, and E. F. Benfield. 1995. Organic processes in streams of the eastern United States. Pages 117–187 *in* C. E. Cushing, K. W. Cummins, and G. W. Minshall, editors. River and Stream Ecosystems. Elsevier, Amsterdam.

Wiggins, G. B. 1996. Larvae of the North American Caddisfly Genera (Trichoptera). 2nd ed. University of Toronto Press, Toronto.

Willson, M. F., S. M. Gende, and B. H. Marston. 1998. Fishes and the forest: Expanding perspectives on fish-wildlife interactions. BioScience 48:455–462.

Winterbourn, M. J., and C. R. Townsend. 1991. Streams and Rivers: One-way flow systems. Pages 230–242 *in* R. S. K. Barnes and K. H. Mann, editors. Fundamentals of Aquatic Ecology. 2nd ed. Blackwell Scientific Publications, Oxford.

Wipfli, M. S., J. Hudson, and J. Caouette. 1998. Influence of salmon carcasses on stream productivity: Response of biofilm and benthic macroinvertebrates in southeastern Alaska, USA. Canadian Journal of Fisheries and Aquatic Sciences 55:1503–1511.

Wipfli, M. S., J. P. Hudson, D. T. Chaloner, and J. P. Caouette. 1999. Influence of salmon spawner densities on stream productivity in southeast Alaska. Canadian Journal of Fisheries and Aquatic Sciences 56:1600–1611.

Part III

Ecosystem Dynamics

11

Controls over Forest Production in Interior Alaska

John Yarie
Keith Van Cleve

Introduction

State factors provide a powerful conceptual basis for understanding current patterns and potential changes in forest productivity (Chapter 1). The BNZ-LTER program has focused on investigations of ecosystem structure and function related to state factors (time, topography, and climate) that account for dramatic spatial variation in productivity and provide a basis for predicting future temporal variation (e.g., climate change). Other state factors are either relatively uniform across the region (e.g., potential biota) or co-vary with topography (i.e., parent material) and are difficult to study as clearly independent factors.

State factors have many direct and indirect effects on productivity, and these controls may co-vary in a complex fashion across the landscape. The nitrogen productivity concept provides a mechanistic framework for understanding the effects of environmental variation on forest productivity (Ågren 1985). The nitrogen productivity of a tree or forest stand is defined as the amount of production per unit of foliar nitrogen (gram biomass production per gram foliar nitrogen) in the canopy of the tree or stand. At steady-state nutrition, the growth rate is proportional to the amount of foliar nitrogen and the N-productivity. Biological and chemical processes that occur in soils are an excellent example of the way in which multiple interacting factors influence productivity through their effects on N supply.

In interior Alaska, several state factors have a hierarchical influence on forest production. These factors are time (Chapter 7), parent material (Chapter 3), topography (Chapter 2), and macroclimate (Chapter 4). These factors have both direct and indirect effects, many of which vary over time and space. In this chapter we emphasize the influence of the four relatively direct state factors: parent material, topography, time, and climate and a critical "resource" (Chapter 1), soil, which represents the indirect interaction of multiple state factors.

State Factors: Parent Material, Topography, Time, and Climate

Parent Material

The parent material in lowland locations is primarily alluvium or loess over alluvium; thick silt, glacial deposits, or eolian sands are present in some areas (Chapter 3). Organic soils also occur on level surfaces that rarely or never flood. Both alluvial and organic soils usually contain a fine-grained mineral substratum. In addition, limited lowland areas in interior Alaska contain very thick loess deposits. In the uplands, the parent material usually consists of crystalline igneous and metamorphic rocks and noncalcareous sedimentary rocks overlain by loess that can be many meters thick. The parent material itself has no direct effect on forest productivity but indirectly affects productivity through interactions with topography and soils.

Topography

The three major topographic units in interior Alaska are the uplands, the floodplains, and the lowlands (Chapter 2). On the basis of an analysis of digital terrain data, the topography of interior Alaska is fairly evenly divided into the four cardinal directions. Approximately 30% of the slopes are north-facing (315° to 45°), 24% are south-facing (135° to 225°), 21% are east-facing (45° to 135°), and 25% are west-facing (225° to 315°). In general, the topography is relatively flat, with 63.6% of land area less than 5% slope; 28.8% are 5–15%, and 7.6% are greater than 15% slope. Again the effects of topography on forest productivity are closely tied to the climate on the site. North-facing slopes are colder than flat areas, which are colder than south-facing slopes (Chapter 4). The relationship of climate to productivity is addressed in the next section.

Time

Succession as a Factor of Time

Succession, or the change in vegetation structure and productivity over time, is a key component in the study of vegetation dynamics. Because time is independent of other causes of ecosystem variation, its effects can be viewed as sequential observations of changes in ecosystem characteristics and processes. In this regard, we examine phenomena related to aboveground productivity that change through secondary succession in the uplands and primary succession on the floodplains. Mortality and ingrowth (new stems that grow to be 2.5 cm diameter at breast height [dbh]) are the vegetation processes that directly account for successional changes in productivity in interior Alaska.

Mortality and Ingrowth

On floodplain sites, mortality is the primary factor causing successional changes in tree density (Table 11.1). Tree numbers declined through mortality at each location, suggesting that most establishment occurred initially and that mortality accounted for subsequent changes in species composition, a pattern consistent with the initial floristics model of succession (Egler 1954).

Only alder showed significant ingrowth during the period of study (Table 11.1). Ingrowth of alder occurred in the youngest stage (FP1 sites) as a result of seed input and root sprouting. Alder also increased in density in the older sites (FP3 and FP4) where age-related overstory mortality created canopy gaps that allowed alder to root sprout (Table 11.1).

In the early successional upland (UP1) sites, vegetation structure developed through ingrowth, following the 1983 Rosie Creek fire (Table 11.1), causing an increase in stand density. White spruce seedlings grow slowly after fire, so very few spruce are recorded in the counts of tree density (>2.5 cm at 1.3 m height) in the UP1 stands. At the mid- and late-successional UP2 and UP3 sites, the change in vegetation structure is largely the result of the mortality of the shorter-lived species. In the UP2 sites, hardwoods decreased, causing a change from hardwood to conifer dominance. In the older white spruce sites (UP3), there was also a slow decline in numbers of white spruce trees, accompanied by small increases in birch and alder (Table 11.1).

Periodic early-winter heavy snowfall events can cause both tree death and substantial top breakage. Two such events have occurred in the past 40 years (Van Cleve

Table 11.1. Changes in tree numbers (# ha^{-1}) due to mortality (–) and ingrowth (+) by species for the LTER floodplain (FP1 = open shrub stage, FP2 = alder-poplar stage, FP3 = balsam poplar stage, FP4 = mature white spruce stage) and upland (UP1 = early successional stage, UP2 = birch-aspen stage, UP3 = mature white spruce stage) research sites.

Species	Year	UP1	UP2	UP3	FP1	FP2	FP3	FP4
Alder	1989–92			–89		–2733	–800	–89
	1992–93			–178		–1467	–467	–44
	1993–98	267	333	1289	4467	–1800	1667	800
Aspen	1989–92		–200					
	1992–93	–67	–67	–44				
	1993–98	6600	–67					
Birch	1989–92		–4400					
	1992–93		–267	–44				–44
	1993–98	2733	–1067	44				
Balsam Poplar	1989–92		–67			–267	–467	
	1992–93		–67			–400	–133	
	1993–98		–133			–1400	-400	
White Spruce	1989–92		–67				-400	–133
	1992–93		–67	–44			–200	–178
	1993–98	0	200	–133			–466	–178

and Zasada 1970, Yarie personal communication). The most recent event occurred in 1990 and produced 0.666 g m^{-2} and 1130 g m^{-2} litterfall in upland and floodplain old-growth white spruce sites, respectively. These inputs are an order of magnitude larger than average yearly litterfall (Table 11.2). In additional old-growth white spruce stands, breakage was estimated to be 90 g m^{-2} (Sampson and Wurtz 1994). In a previous event, during the winter of 1967–1968, breakage was estimated to be 1120 g m^{-2} in a black spruce stand and 1330 g m^{-2} in a white spruce stand (Van Cleve and Zasada 1970). The variability among stands is high (Table 11.2). In several thinning and fertilization plots located in a younger (100-year-old) white spruce stand, the estimated breakage ranged from none to 328 g m^{-2}. Thinning and fertilization treatments carried out in this site had no effect on tree death and damage. The variability in breakage reflected variation in stand structure at the time of the storm and the variability of the storm across the landscape. Snow breakage events occur at long intervals in interior Alaska. Such rare events have a significant effect on stand structure and, therefore, productivity. The change in white spruce biomass of the upland and floodplain white spruce stands over a three-year period (1989–1992) in which the breakage event occurred was 1170 g m^{-2} and -2600 g m^{-2}, respectively, indicating that growth was not sufficiently high in the floodplain sites to offset the tree damage and mortality due to snow breakage. The damage seen in the upland sites was about half that present in the floodplain and was coupled with less total tree mortality in the uplands.

Floodplain Succession

The most productive forests in interior Alaska occur on the floodplains. Availability of water in the floodplain depends on terrace elevation. Forest productivity declines through successional development of ecosystems in floodplain sites (Fig. 11.1), a successional pattern observed in many forests. Three exceptionally productive sites on the Yukon River floodplain were not included in Figure 11.1.

Table 11.2. Woody litterfall occurring in the FP4, UP3, and UP2 LTER sites as a result of a heavy early snowfall event in September 1990 and the average quantity of litterfall for the years 1990–1998, exclusive of large woody material.

Site	Woody Litter (1990) (g m^{-2})	Foliar Litter (1990–1998 average) (g m^{-2})
UP2A	20	48
UP2B	120	39
UP2C	—	49
UP3A	520	33
UP3B	260	30
UP3C	1200	30
FP4A	1470	40
FP4B	980	32
FP4C	960	26

These three sites occurred on relatively low river terraces where moisture limitation was potentially absent. The average standardized production (measured production divided by the stand density index equals a potential production value for a fully stocked stand) on these three sites was 2856 g m^{-2} yr^{-1} at an average age of 183 years, a value that is significantly higher than the standardized values found for most of the floodplain stands in interior Alaska. The productivity of the three stands is apparently limited only by nitrogen. Nitrogen availability may be relatively high in floodplains because the alder-dominated successional stage adds large amounts of nitrogen to the soil prior to the development of white spruce stands (Chapter 15).

Upland Succession

Productivity also declines through succession in upland stands (Fig. 11.2, upper curve). The maximum value observed in the uplands (1421 g m^{-2} yr^{-1}) is only 62 percent of the maximum value observed on the floodplain (Fig. 11.1). This 40% reduction in production relative to floodplain forests presumably reflects moisture stress. The black spruce ecosystems found on flat surfaces and on north-facing slopes in the uplands represent the end of the successional sequence. We define successional age as the total time since the most recent major disturbance at the site. The age of the trees present

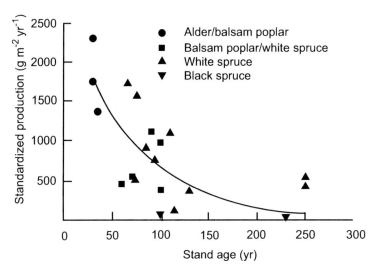

Figure 11.1. Relationship between standardized forest productivity and ecosystem age for floodplain ecosystems (Standardized Production = 2769.8e$^{0.0144}$ r^2 = 0.53). The stand age is a relative measure of the total age of the site throughout the successional process. The standardized forest productivity is equal to the measured aboveground production divided by the stand density index as a fraction of full stocking. The following sites are included in this figure: FP2, FP3, and FP4 LTER sites, and additional sites in the Fairbanks area and in the Porcupine River Drainage. The vegetation type represents the typical stages of floodplain succession.

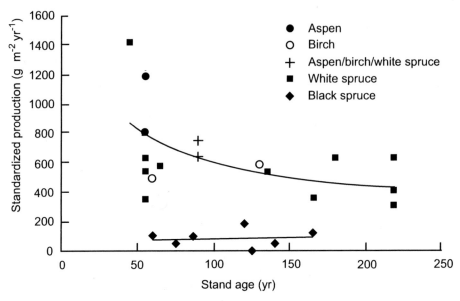

Figure 11.2. Relationship between standardized forest productivity and stand age for upland sites. The stand age is a relative measure of the total age of the site throughout the successional process through the white spruce sites. In black spruce sites, stand age does not represent the successional age of the sites. The standardized forest productivity is defined in Fig. 11.1. The data set, which includes aspen, birch, white spruce, and mixtures of these species, represents the development of vegetation from a fire to mature white spruce stands (Standardized Production = $4761.327x^{-0.4478}$ $r^2 = 0.2947$). The diamond data set represents an age sequence of black spruce stands. The vegetation type represents the typical stages of upland succession if the black spruce sites are viewed at ages in the order of 500 years.

represents the time since that disturbance only if the trees regenerated immediately after the disturbance. Otherwise, the successional age of the site would be greater than the maximum tree ages. If black spruce did regenerate immediately after disturbance, the production values presented in Figure 11.2 would have been substantially lower than those for the south-facing upland successional sequence. Aboveground production of the black spruce sites was constant over the range of stand ages (Fig. 11.2, lower curve), and it was lower than production in the successional sequence of hardwood and white spruce stands (Fig. 11.2, upper curve).

Climate

Temperature

At the global scale, mean annual air temperature is highly correlated with aboveground production, with forest production decreasing from 15 Mg ha^{-1} to less than 5 Mg ha^{-1} as mean annual air temperature drops from 15°C to -3°C (Van Cleve and

Powers 1995). In interior Alaska, however, correlations between temperature and tree growth (current and next year's) in both upland and floodplain ecosystems are negative, indicating that warmer climates would reduce growth, especially in the upland sites (Barber et al. 2000). Correlations between aboveground production and rainfall are less clear. There are both positive and negative relationships with growth in the upland stands. Fewer correlations were found for the floodplain sites, and they were all negative (Yarie et al. unpublished), for reasons that are not clear. The negative relationship for floodplain stands may be the result of cloudy days that are associated with rainy weather. In addition, moisture from rainfall on floodplain sites can be augmented by groundwater, which is related to river levels and therefore is affected by rainfall events and glacial dynamics in the headwaters.

Evaporation

Evaporation is another climatic factor that is closely tied to temperature and rainfall. A high evaporation potential can decrease tree growth if sufficient rainfall is not available to replace the evapotranspiration losses. In interior Alaska, potential evaporation exceeds summer precipitation (Yarie et al. 1990). In general, there is a negative correlation between open pan evaporation and tree growth for both the current year and the following year's basal area growth for birch and aspen (Table 11.3).

Precipitation

Positive correlations were found (Table 11.3) for next year's basal area growth and for August precipitation for birch. Evaporation and the ratio between evaporation and rainfall showed significant correlations with basal area growth measurements for all three species measured (Table 11.3). Perhaps August precipitation replenishes soil moisture and has a positive effect on early-season growth for young birch the following year. The positive relationship with evaporation might indicate that the summer temperature was higher in conjunction with higher growth rates and that the soil moisture status was sufficient to supply water to sustain the tree growth. However, in aspen sites, the water-holding capacity may have been so low that high rates of evapotranspiration resulted in high moisture stress, resulting in slower growth (as indicated by the negative correlations). The soil moisture dynamics on black spruce sites that are underlain by permafrost are quite different. Because of the slow, continuous thawing of soil during the growing season, these black spruce sites do not experience a growth reduction due to soil moisture (Van Cleve et al. 1990). In summary, there is not a simple relationship between tree growth and above- and belowground temperature, precipitation, or evaporation.

Weather-to-Be: Experimental Manipulations of Environment

Two major experimental manipulations have investigated the potential effects of future climate on the function of selected taiga forest ecosystems. The first was a study in which heat was applied to the surface soil of a black spruce ecosystem. The second

Table 11.3. Significant correlations (p = 0.05) between tree basal area growth and monthly precipitation and evaporation parameters for a 30-year study of the major forest types in interior Alaska.

Species	Site	Variable correlated with current or next year's BA growth[1]	Correlation
Birch	3BC	Next–Aug. precipitation	0.500
		Next–Aug. evaporation	−0.647
	5BC	Next–Aug. evaporation	−0.631
		Next–Summer evaporation	−0.698
Aspen	1AC	Next–June evaporation	−0.511
		Next–Aug. evaporation	−0.763
		Next–J+J evaporation	−0.554
		Next–Summer evaporation	−0.628
	6AC	Next–J+J+A evaporation	−0.541
	45AC	Current–Aug. evaporation	−0.562
	Total aspen	Current–June evaporation	−0.205
	data set	Current–June evaporation rain ratio	−0.444
White spruce	8S6	Current–June evaporation	0.593
		Current–July evaporation	0.535
		Current–J+J evaporation	0.621
		Current–J+J+A evaporation	0.643
		Next–June evaporation	0.733
		Next–J+J evaporation	0.649
		Next–J+J+A evaporation	0.589
White spruce	8SC	Current–Aug. evaporation	−0.694
		Current–J+J+A evaporation	−0.556
		Next–Aug. evaporation	−0.562
	Total white	Current–June evaporation	0.608
	spruce data	Current–July evaporation	0.555
	set	Current–J+J evaporation	0.637
		Current–J+J+A evaporation	0.658
		Next–June evaporation	0.733
		Next–J+J evaporation	0.649
		Next–J+J+A evaporation	0.589
		Current–Total evaporation rain ratio	0.720
		Next–Total evaporation rain ratio	0.763

[1] Summer = June, July, August, and part of September; J+J = June and July; J+J+A = June and July and August

Source: Yarie (unpublished).

study examined the effect of eliminating summer rainfall at upland and floodplain midsuccessional sites on the growth and nutrient uptake by the trees on the site.

Soil Temperature Manipulations

A warming experiment in which the soil of a black spruce site was heated (Hom 1986) allowed a direct test of the effect of soil temperature on ecosystem function.

Black spruce was chosen because it is the least productive interior Alaskan forest ecosystem developed on permafrost. Heat tapes were installed in the forest floor, and the summer soil temperature was increased by 9°C for three years. The heating resulted in an increase in the soil degree-day total from 563 to 1589 degree-days during the summer. This was an increase in temperature from that typical of a north-facing permafrost black spruce site to that typical of a south-slope white spruce site. This higher temperature caused increased decomposition of the forest floor and consequently higher concentrations of extractable N and P in the forest floor and elevated N concentrations in the soil solution (Van Cleve et al. 1990). This, in turn, caused a significant increase in foliar nutrient concentrations for N, P, and K (Van Cleve et al. 1990) and a significant increase in tree growth (Hom 1986). The ratio of post- to pretreatment growth was 0.79 for the control trees (they grew slower in the years following the start of treatment) and 1.03 for the trees in the treated area. Average radial increase in the control trees was 0.21 mm year^{-1} for the last year of treatment and for two subsequent years and 0.26 mm year^{-1} for the treated trees. For the six years prior to start of treatment, growth averaged 0.30 mm for the control trees and 0.26 mm for the treated trees (Hom 1986). This growth increase was the result of interaction between warmer soil and improved soil nutrient status. At the opposite end of the soil-temperature spectrum in interior Alaska (south-facing, postharvested white spruce sites), soil temperature does not show enough variability to be an important factor in the control of nutrient availability and plant growth unless the soil is severely disturbed (Pare and Van Cleve 1993). Soil temperature on these sites is high enough that moderate changes in stand density have little or no effect on temperature-mediated soil processes.

Soil Moisture Manipulations

Summer precipitation covers were constructed in three replicate stands of an upland hardwood–white spruce site type and a floodplain balsam poplar–white spruce site type. The exclusion of rainfall in the uplands was expected to reduce tree growth. On the floodplain, growth was not expected to be affected because of the shallow depth to groundwater. The average height of the floodplain sites above the river level ranged from 0.5 m to 2.0 m during the summer. Current results from this study (Yarie in review) indicate that in the floodplain sites the moisture covers significantly reduced the growth of white spruce. Basal area growth over the period 1991–1997 was 9.93 cm^2 in control plots compared to 3.85 cm^2 in rain-exclusion plots. Balsam poplar growth was also consistently reduced, although the reductions were not statistically significant. The average yearly growth for the years 1991–1997 decreased from 3.07 cm^2 on the control plot to 1.13 cm^2 on the drought plots. The water-holding capacity of the floodplain soils is sufficiently low, because of coarser soil texture, that when the river level falls, the amount of water available in the soil is reduced due to both rapid soil drainage and vegetation utilization. This rapid reduction in soil moisture availability generates plant moisture stress. Soil moisture recharge due to capillary rise is restricted because of the forest and forest floor cover of the soil surface (Dyrness and Van Cleve 1993). Periods of higher river levels are required for moisture to move toward the rooting zone either by capillarity or along water potential gradients driven by transpiration. In addition, the root system of floodplain trees may be sufficiently shallow to restrict direct access to deeper soil water levels.

In the upland sites, there was no effect of the moisture covers on the growth of trees, with the exception of birch. Exclusion of rain decreased the yearly average basal area growth of birch from 1.38 to 0.72 over the period 1991–1997. However, significant growth reductions were not consistent from year to year. These results were exactly the opposite of our predications. In the uplands, moisture stress develops after the soil moisture recharge resulting from snowmelt is utilized (Yarie in review). After snowmelt, soil moisture recharge by rainfall is insufficient to allow maximum growth during the primary months (June and July) when tree diameter growth occurs.

A more detailed examination of the depth distribution of moisture shows that recharge of soil moisture occurred only to about 10 cm depth during the late summer and fall (Fig. 11.3). At a depth of 20 cm and below, there was relatively little recharge during the frost-free season. This indicates that the primary source of soil moisture recharge to support tree growth occurs during spring snowmelt (Fig. 11.3).

Soil Nutrient Dynamics

We suggest that the interactions between soil chemistry and plant production are the primary dynamics that limit intraseason forest growth. The amount of nitrogen present in the tree controls the total response potential that would result from alleviation of additional growth-limiting factors such as temperature and moisture.

The rate of N mineralization is controlled by a small pool of rapidly cycling N that is poorly correlated to forest floor total N concentrations (Pare and Van Cleve 1993). In floodplain ecosystems, substrate chemistry exerts the primary control of net N mineralization in both the forest floor and the surface mineral soil, although the potential influence of moisture and temperature controls has not been fully investigated in each successional stage (Van Cleve et al. 1993).

In a young (15-year-old) aspen stand, diameter increment was significantly increased by the addition of N, P, K, NP, and NPK fertilizers after the first year of treatment (Van Cleve 1973). Continuation of the experiment for an additional five years yielded similar results in that leaf area index, specific leaf area, and total tree biomass were significantly greater as a result of N fertilization (Van Cleve and Oliver 1982). After five years, the P and K treatments were equal to the control and the NP and NPK treatments were substantially higher but because of the variance around the mean were not significantly higher (Van Cleve and Oliver 1982).

In a 70-year-old white spruce stand, fertilizer treatment with N, P, and K did not increase growth, even after six years of fertilization (Van Cleve and Zasada 1976). However, when combined with thinning (a reduction in the number of stems per unit land area), fertilization did produce an increase in growth both during the time period of fertilization and up to four years after the fertilization was completed (Yarie et al. 1990). In this site, thinning resulted in a decreased demand for soil moisture at the site level, an increase in water available to individual stems, and a decrease

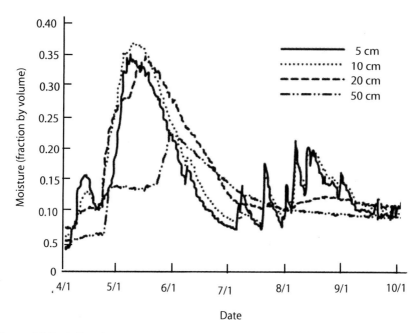

Figure 11.3. Soil moisture status by depth for an upland deciduous stand (UP2A) in 1997.

in the influence of moisture availability on tree growth (Yarie et al. 1990); added nutrients had no effect on tree growth until moisture supply was increased.

A different scheme was used to increase nutrient availability in later experiments at our LTER sites. Plots at the LTER sites received a low-level fertilization in which available N was applied at a rate equal to the estimated N-mineralization rate on a yearly basis. This treatment has been repeated in the spring of each year since 1990. The intent was to slowly build up the total N pool over a long time period and, as a result, to increase the available pool of N for tree growth. It was hoped that this slow increase in available N would raise the aboveground productivity at the site to a higher level on a permanent basis.

Low-level fertilization of the LTER upland sites resulted in a significant increase in growth of white spruce in the UP2 sites for two years (Fig. 11.4). There was no increase in growth for birch, aspen, and balsam poplar on the UP2 sites. On the UP3 sites, no significant differences were found in growth of white spruce and birch. The lack of response in the old-growth sites could be the result of tree age. These trees are currently more than 200 years old. In addition, the moss layer present in these stands may be the primary uptake site of the fertilizer. The fertilizer may not reach the tree root system until the earlier cohorts of moss production start to decay. This may require up to six years, depending on the moss species present.

Fertilizers have also been applied to forests at a higher rate (111 kg ha^{-1} N, 55 kg ha^{-1} P and 111 kg ha^{-1} of K were applied annually from 1968 through 1973). Results of the fertilization have not been consistent between species or between the age classes studied within each species (Table 11.4). In the aspen site, the number

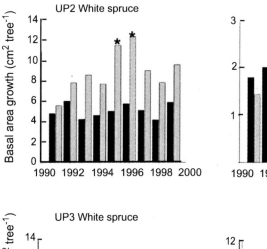
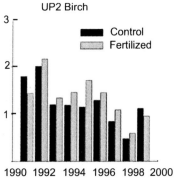
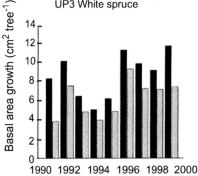
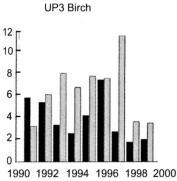

Figure 11.4. Yearly basal area growth for white spruce and birch trees in upland hardwood (UP2) and in spruce (UP3) sites. Significant increases in growth are indicated by an asterisk above the bars for individual years. Fertilization levels were equal to the annual N mineralization per year. Fertilization started in 1990 and continued throughout the study.

of significant increases in tree growth in response to fertilization addition decreased with the age of the stand. In birch, nutrient addition reduced growth or had no effect. However, in a nitrogen rate study, diameter growth did increase after a single addition of fertilizer for the two lowest levels of the fertilization treatments. Nine years of significant increases in growth between 1980 and 2002 were found for the single addition that occurred in 1979 (Table 11.4). In the 70-year-old white spruce stand, fertilization had a positive effect on growth. However, the increased growth was closely tied to stand density. Thinning to 1483 stems ha^{-1} did not result in a significant increase in growth until an NP fertilizer was applied. When the stand was thinned to the level of 741 stems ha^{-1}, fertilization did not show a consistent increase in growth above that found in the thinning treatment.

In the floodplain sites, the low-level fertilization resulted in an increase in growth for balsam poplar in the FP2 sites. This is surprising because the FP2 site type is the one that is thought to have the highest levels of naturally available nitrogen. The increase in growth due to fertilization at this site could indicate that the nitro-

Table 11.4. General results (either a statistically significant increase or decrease) of the long-term fertilizer studies. Total measurement period was 31 years as of 2000.

	Fertilizer treatment[1]	No. of years increased growth	No. of years decreased growth	Notes
Aspen	NPK	28	0	
	NPK	12	6	Growth declines started 13 years after the start of the study.
	NPK	0	2	Years 1980 and 1981.
	N, P, K factorial	27 for N 0 for P, K		
Birch	NPK	5	5	Positive and negative growth response intermixed throughout measurement period.
	NPK	0	7	
	NPK	0	10	
	N, P, K factorial	0	11 of 15 years	K only consistent nutrient but only in combination with N or P.
	N rate study	12	0	Increases with 110 and 220 Kg ha^{-1}. No increases in the higher fertilization rates.
White spruce	NPK	1	0	Difference was found in the second year of the study.
	Thinning	23	0	
	Thinning plus fertilization	10	0	
	Thinned	0	0	No significant increase was found.
	Thinning plus NP	11	0	
White spruce	Thinned	26	0	
	Thinning plus fertilization	24 to 27 years for various treatments	0	
	Thinning plus $NH_4H_2PO_4 + K_2SO_4$	12	0	

[1] The fertilization was carried out over a number of age classes of birch and aspen stands and one age class of white spruce, all in upland locations (Van Cleve 1973, Van Cleve and Zasada 1972, Van Cleve and Oliver 1982, Yarie unpublished). The initial ages of the stands were 15, 45, and 120 years for aspen; 30, 70, and 120 years for birch; and 70 years for white spruce. In addition an N, P and K factorial study was performed in the aspen and birch stands, and a nitrogen rate study was performed in the birch stand. Diameter growth has been monitored for the past 32 years.

gen fixed by alder is not readily available to other plants or that poplar has a higher nutrient requirement than do other species (Chapin and Tryon 1983). White spruce showed a decrease in growth in the FP3 sites (but only in the first two years) and a decrease in growth in the FP4 site. These results suggest that nitrogen limitation of growth is much less severe in floodplain than in upland sites.

A Generalized Forest Growth Model

Finally, on the basis of work by Yarie (1997, 2000), we find that nitrogen availability mediates the relationship between state factors and aboveground boreal forest production in interior Alaska. Therefore, forest production is a function of the nitrogen productivity of the stand, which is a function of the ecosystem's primary state factors:

$$Fp = v(N(Pm, T, C, s), t)$$

where Fp is the forest production; v(N) is the vegetation nitrogen content; N(Pm, T, C, s) is the relationship of the nitrogen cycling to the parent material, topography, climate and soils; and t is time. This function represents a general description of the relationship of the state factors in controlling vegetation nitrogen dynamics, which control vegetation production.

Selected Hierarchy of State Factor Controls of Ecosystem Function

The preceding equation presents a view of state factor structure that controls forest production in interior Alaska. This structure represents a hierarchy of environmental controls. This formulation shows plant production to be controlled primarily by the amount of nitrogen present in the plant. This, in turn, is controlled by the availability of nitrogen in the soil and the plant's ability to draw on that nutrient reserve. If all other growth-controlling factors are at their optimum values, then the growth of the plant will be totally controlled by the amount of available nitrogen. This relationship is displayed in Figure 11.5. The box in the upper right of curve 1 (Fig. 11.5) represents the potential growth of the plant if no environmental factor limits plant growth and the plant is growing at its genetic maximum rate. This curve represents the growth rate across a range of nitrogen availabilities, where growth is controlled only by the availability of nitrogen. Curve 2 shows the effect of an additional controlling factor that progresses from 0% to 100% (the x-axis scale depicts the percentage of the growth-limiting factor present compared to the optimum availability of that factor). Curve 3 (Fig. 11.5) shows the potential result if more than two factors in addition to nitrogen limit growth. In this case, one of the growth-limiting factors could be alleviated through changes in climate, stand density, insect herbivory,and so on in a particular year, but the other factor would still control growth. For example, a limitation due to low environmental temperatures may be alleviated by warmer-than-normal growing conditions, but total rainfall quantities may not be altered. As a result, a new primary controlling variable (rainfall, in this

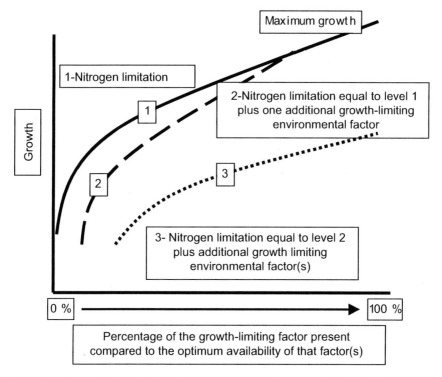

Figure 11.5. Relationship of plant nitrogen content to plant growth and the effect of varying environmental factors.

case) that may not be represented in the original depiction of the environmental controls over production could become the major control of the functional structure of the ecosystem. This interpretation of plant growth limitations suggests that the LTER FP2 (alder-balsam poplar) sites may represent the forest sites with the least potential growth limitations because of the favorable occurrence of nitrogen fixation and groundwater present at these sites.

State Factors as a Basis for Predicting Ecosystem Response to Climate Change

There are two potential effects caused by climate change in interior Alaska. It is expected that the mean annual temperature will increase. This will result primarily from increasing winter temperatures (IPCC 1995). This type of warming trend has been observed at both sea and land surfaces from 1961 to 1990 (Chapman and Walsh 1993, Serreze et al. 2000). However, summers are only marginally warmer. Winter warming should extend the growing season by lengthening the time between major frosts during the fall and spring. The net effect of this on vegetative growth and reproduction is not fully understood at this time. Climate warming may increase tree growth in interior Alaska (Keyser et al. 2000). However, any increased summer temperatures may increase moisture limitations to growth (Yarie et al. 1990, Barber et al. 2000). In addition, global climate change models that are used on large scales do not take into account specific physiological differences that exist between different hardwood (e.g., balsam poplar and aspen) and coniferous species (e.g., white and black spruce), even though these species generally occupy totally different sites.

Changes in precipitation, from both a seasonal and an annual standpoint, have been much more difficult to forecast. It is currently predicted that mean annual precipitation will increase (Serreze et al. 2000; Chapter 4). This prediction is based on the study of the recent history of precipitation in northern high-latitude systems. This change appears to be the result of increases in winter snowfall during the months of December, January, and February and a smaller increase during the spring (Serreze et al. 2000). The summer and fall months (June through November) have not shown any increase in precipitation over the past 20 years. The indication is that the increased snowfall will have to recharge soil moisture during the spring melt period to higher levels than are currently observed to have a significant effect on forest growth during the summer growing season. Recharge during the fall will then depend on the pre-freeze-up precipitation events.

Conclusions

The state factors—climate, biotic components, topography, parent material, and the age of the system—represent key ecosystem components that control aboveground forest productivity in interior Alaska. We suggest that the primary control on forest productivity is related to the nitrogen content of the tree foliage. Further reductions in tree growth result from limitations caused by the additional state factors.

Soil parent material and topography represent two relatively stable controls on ecosystem production. The parent materials represent three major groups that are correlated to three major physiographic positions on the landscape. Upland locations are generally underlain by loess material laid down by wind, floodplains are underlain by a water-deposited silty or sandy material, and the broad flats north of the Alaska Range and in the Yukon Flats have developed primarily from water-deposited silt and sand and from eolian-deposited silt. Topography affects radiation received at specific sites (controlled by slope and aspect) and the soil depth to groundwater (controlled by the site elevation above the major river levels and the location of the site in relationship to the major mountain ranges).

The key components of climate related to productivity are the cold environmental temperatures, both air and soil, found in the subarctic and the relatively dry summer growing season, especially the early part (May and June) of the season. On the basis of moisture limitation experiments and growth-climate correlations, it now appears that current summer rainfall is insufficient for maximum growth of trees in upland locations. In addition, the rooting structure of white spruce on the floodplains may be such that the trees have limited access to groundwater and rely primarily on summer precipitation for moisture supply.

Soil environments (both physical and chemical) also control forest productivity. Soil organic matter turnover rates are limited by soil temperature, which limits both the uptake potential of nitrogen by the trees and the decomposition dynamics of the organic matter. Thinning and fertilization experiments have indicated that both nitrogen availability and tree density control growth rates of individual trees. Short-term, high-dosage fertilization studies combined with thinning have shown that the thinning results in long-term increased tree growth. High-dosage fertilizer addition results in a short-term increase in tree growth. A low-dosage fertilization treatment continued for many years is starting to show consistent increases in growth rates in unthinned forest stands. These relationships between fertilization experiments and the growth responses indicate that nutrient availability is the primary growth controlling factor on nonpermafrost-dominated sites but that climate factors are intermingled with the nutrient dynamics as the major controls of forest growth.

The biotic components, which are related to the successional stage (time) on a patch of the landscape, co-determine the potential productivity of a patch. Balsam poplar, found on the floodplain, has the highest growth rates found in interior Alaska. White spruce, also found on the floodplain, can also have high growth rates, especially if the river terrace is low enough to allow the trees a constant supply of water.

Finally, there are a number of short-term events that can have long-term effects on forest productivity. These events include fire (chapter 17), insect outbreaks (chapter 9), flooding, and permafrost dynamics (chapter 4). The information presented in this chapter should give the reader a better understanding of the complexity of the state factors in the relatively simple environmental setting of the taiga forest in interior Alaska.

References

Ågren, G. I. 1985. Theory for growth of plants derived from the nitrogen productivity concept. Physiologia Plantarum 64:17–28.

Barber, V. A., G. P. Juday, and B. P. Finney. 2000. Reduced growth of Alaskan white spruce in the twentieth century from temperature-induced drought stress. Nature 405:668–673.

Chapin, F. S., III, and P. R. Tryon. 1983. Habitat and leaf habit as determinants of growth, nutrient absorption, and nutrient use by Alaska taiga forest species. Canadian Journal of Forest Research 13:818–826.

Chapman, W. L., and J. E. Walsh. 1993. Recent variations of sea ice and air temperature in high latitudes. Bulletin of American Meteorological Society 74:33–47.

Dyrness, C. T., and K. Van Cleve. 1993. Control of surface soil chemistry in early-successional floodplain soils along the Tanana River, interior Alaska. Canadian Journal of Forest Research 23:979–994.

Egler, F. E. 1954. Vegetation science concepts. I. Initial floristic composition—A factor in old-fields vegetation development. Vegetatio 4:412–418.

Hom, J. L. 1986. Investigations into some of the major controls on the productivity of a black spruce (*Picea mariana* [Mill.] B.S.P.) forest ecosystem in the interior of Alaska. Ph.D. Dissertation, University of Alaska Fairbanks.

IPCC. 1995. Climate Change: The Science of Climate Change. Cambridge University Press, Cambridge.

Keyser, A. R., J. S. Kimball, R. R. Nemani, and S. W. Running. 2000. Simulating the effects of climate-change on the carbon balance of North American high-latitude forests. Global Change Biology 6:185–195.

Pare, D., and K. Van Cleve. 1993. Soil nutrient availability and relationships with aboveground biomass production on post-harvested upland white spruce sites in interior Alaska. Canadian Journal of Forest Research 23:1223–1232.

Sampson, G. R., and T. L. Wurtz. 1994. Record interior Alaska snowfall effect on tree breakage. Northern Journal of Applied Forestry 11:138–140.

Serreze, M. C., J. E. Walsh, F. S. Chapin III, T. Osterkamp, M. Dyurgerov, V. Romanovsky, W. C. Oechel, J. Morison, T. Zhang, and R. G. Barry. 2000. Observational evidence of recent change in the northern high-latitude environment. Climatic Change 46:159–207.

Van Cleve, K. 1973. Short-term growth response to fertilization in young quaking aspen. Journal of Forestry 71:758–759.

Van Cleve, K., and L. K. Oliver. 1982. Growth response of postfire quaking aspen (*Populus tremuloides* Michx.) to N, P, and K fertilization. Canadian Journal of Forest Research 12:160–165.

Van Cleve, K., and R. Powers. 1995. Soil carbon, soil formation, and ecosystem development. Pages 155–200 *in* W.W. McFee and J. M. Kelly, editors. Carbon Forms and Functions in Forest Soils. Soil Science Society of America, Madison, WI.

Van Cleve, K., and J. C. Zasada. 1970. Snow breakage in black and white spruce stands in interior Alaska. Journal of Forestry 68:82–83.

Van Cleve, K., and J. C. Zasada. 1976. Response of 70-year-old white spruce to thinning and fertilization in interior Alaska. Canadian Journal of Forest Research 6:145–152.

Van Cleve, K., W. C. Oechel, and J. L. Hom. 1990. Response of black spruce (*Picea mariana*) ecosystems to soil temperature modification in interior Alaska. Canadian Journal of Forest Research 20:1530–1535.

Van Cleve, K., J. Yarie, and R. Erickson. 1993. Nitrogen mineralization and nitrification in successional ecosystems on the Tanana River Floodplain, interior Alaska. Canadian Journal of Forest Research 23:970–978.

Yarie, J. 1997. Nitrogen productivity of Alaskan tree species at an individual tree and landscape level. Ecology 78:2351–2358.

Yarie, J. 2000. Boreal forest ecosystem dynamics. I. A new spatial model. Canadian Journal of Forest Research 30:998–1009.

Yarie, J. In review. Effects of moisture limitation on tree growth upland and floodplain forest ecosystems in interior Alaska. Ecology.

Yarie, J., K. Van Cleve, and R. Schlentner. 1990. Interaction between moisture, nutrients and growth of white spruce in interior Alaska. Forest Ecology and Management 30:73–89.

12

The Role of Fine Roots in the Functioning of Alaskan Boreal Forests

Roger W. Ruess
Ronald L. Hendrick
Jason G. Vogel
Bjartmar Sveinbjörnsson

Introduction

The patterns of production described in Chapter 11 tell only half of the story about boreal forest production because a large proportion of the carbon (C) acquired by plants is allocated belowground in ways that have traditionally been extremely difficult to quantify. Work in the Bonanza Creek LTER provides considerable insight into the patterns, causes, and consequences of this belowground C allocation.

Belowground allocation has a number of important ecosystem consequences beyond the simple fact that C allocated belowground comes at the expense of aboveground growth. Belowground and aboveground tissues differ substantially in the rates of C and nitrogen (N) incorporation into new tissue, the ratio of growth to respiration, and the rate of tissue decay. For example, despite the small biomass of fine roots relative to aboveground tissues in forest ecosystems, disproportionate amounts of C and N cycle annually through fine roots, which grow, die, and decompose very rapidly and have high N concentrations (Hendrick and Pregitzer 1992, Ruess et al. 1996, 2003).

The objectives of this chapter are to (1) summarize our understanding of the structure and function of fine-root systems in forest types within the Bonanza Creek Experimental Forest, (2) compare our findings with the results of studies of other boreal and temperate ecosystems in order to develop a broader understanding of fine-root function, and (3) identify critical research gaps in our understanding of the role of fine-root systems in boreal ecosystem function.

Fine-Root Morphology and Demography

Fine roots grow more rapidly than the rest of the root system in a forest and are responsible for the bulk of nutrient and water acquisition. Until recently, fine roots were defined rather arbitrarily as roots less than 1–2 mm in diameter, while roots larger than this were considered coarse roots. Only one data set for fine and coarse root biomass has been published for interior Alaskan forests (Ruess et al. 1996), which shows (1) live fine-root biomass ranging from 221 g m^{-2} in floodplain white spruce stands to 832 g m^{-2} in upland birch-aspen stands, (2) a positive correlation between fine-root and coarse-root biomass, with coarse-root biomass averaging 50% greater than fine roots, and (3) no relationship between aboveground biomass and fine or coarse root biomass. Our recent discovery that a large proportion of total fine-root biomass is made up of very fine roots with diameters less than 0.350 mm challenges the accuracy of early fine-root biomass assessments for several reasons: (1) very fine roots are typically ignored or lost during sieving, (2) distinguishing live from dead fine roots is subjective and requires tedious microscopic examination of nearly every fine-root segment, and (3) the smallest fine roots are very fragile and are easily broken and fragmented during the washing and sorting processes. For example, a limited but thorough estimate of live fine-root biomass from floodplain black spruce stands that accounted for these factors (1780 g m^{-2}; Ruess et al. 2003) was more than five times greater than previous estimates from the same site that did not (221 g m^{-2}; Ruess et al. 1996).

There is mounting evidence that the position of a fine-root segment on a branching system is directly related to its function and life history. For example, in a detailed characterization of fine-root demography across nine North American forests, including balsam poplar and white spruce stands from interior Alaska, Pregitzer et al. (2002) concluded that the fine-root systems of woody plants are modular in nature, with strikingly similar demographic patterns across very different forest ecosystems. They dissected fine-root segments by order, using a descriptive scheme similar to that used for streams, and found that first-order roots (the most distal root segments) had the smallest diameters but the highest ratios of length to mass (specific root length, SRL = m g^{-1}; Fig. 12.1 top). These very fine-root segments also had the highest N concentrations (Fig. 12.1 bottom), and are thought to be the most active in terms of growth, respiration, mycorrhizal infection, and nutrient uptake. Although the appearance of the branching pattern varied among the forest ecosystems, first-order roots accounted for >75% of the total number of roots and >50% of the total fine-root length. Forests from interior Alaska had among the highest SRL values, with diameters closely matching the median diameter of roots digitized from minirhizotron photos (clear plastic tubes permanently inserted into the soil that allow the seasonal patterns of root growth and mortality to be documented by repeated photography) installed across the successional floodplain sequence along the Tanana River (Fig. 12.2). From the Alaskan minirhizotron data sets, roots <0.30 mm diameter accounted for an overwhelming percentage of all fine roots present in early-successional shrub-dominated stands (88%), midsuccessional balsam poplar stands (76%), and late-successional white spruce (75%) and black spruce (80%) stands.

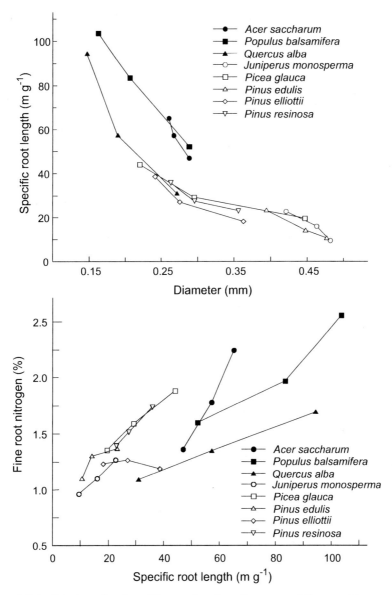

Figure 12.1. Relationship of specific root length to diameter and fine-root N to specific root length for eight forest types. Redrawn with permission from the Ecological Society of America (Pregitzer et al. 2002). Data points within forest types show first-, second-, and third-order roots, distributed left to right (top) and right to left (bottom).

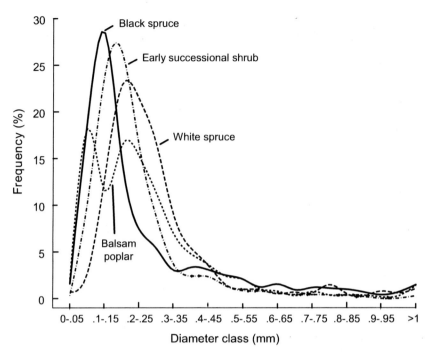

Figure 12.2. Diameter distribution of fine roots digitized from minirhizotron images in four ecosystem types along the Tanana River floodplain. Values represent means across all tubes and time periods.

Of the ecosystems studied by Pregitzer et al. (2002), balsam poplar and white spruce forests from interior Alaska had the highest root N concentrations among angiosperm and gymnosperm forests, respectively. This may be due in part to high concentrations of chitin (C:N = 12:1 vs. root biomass = 25:1) resulting from high infection of ectomycorrhizal fungi in the Alaskan forests, because greater than 95% of all first-order roots are infected with ectomycorrhizae (Lansing, personal communication). High root N may also be linked with acclimation of growth and respiration to low soil temperatures. Burton et al. (2002) found that, when adjusted for temperature, N content explained 94% of the variation in fine-root respiration among 11 forested ecosystems. The Alaskan stands had slightly higher root respiration rates than would be predicted by the cross-site relationship, but respiration response to increasing temperature was very similar among sites (Q_{10} range = 2.4–3.1).

Fine-Root Production and Phenology

The recent development of minirhizotron technology has provided detailed data on seasonal patterns of fine-root production and mortality for locations throughout the boreal forest (Steele et al. 1997, Ruess et al. 1998, 2003). At Bonanza Creek, minirhizotron tubes installed across a 250-year successional forest sequence along the Tanana River show pronounced cross-ecosystem patterns in fine-root processes and provide striking comparisons with similar studies in temperate forests. An important feature that appears to distinguish fine-root dynamics in interior Alaskan forests from temperate and more southerly boreal ecosystems is the approximately six-week time lag between leaf-out and maximum rates of fine-root growth, which typically occur in mid-May and in mid-June to mid-July, respectively (Tryon and Chapin 1983, Ruess et al. 1998). Maximum rates of fine-root growth coincide with canopy leaf-out during spring in northern hardwood ecosystems (Hendrick and Pregitzer 1992, Burke and Raynal 1994, Fahey and Hughes 1994) but become progressively more delayed toward higher latitudes, where root elongation rates are closely tied with delayed soil warming (Tryon and Chapin 1983, Steele et al. 1997, O'Connell et al. 2003, Hendrick and Ruess, unpublished). Fine-root production in black spruce stands peaks approximately one month later than in early-successional, shrub-dominated floodplain forests. This is likely a function of slower soil warming during spring and early summer in black spruce stands that is the result of a continuous moss cover and the presence of permafrost (Viereck 1970). All stands show an asynchrony between monthly rates of fine-root production and mortality within a year. This results from a pronounced seasonal peak of fine-root production during midsummer, coupled with mortality of most of these roots toward fall and over winter (Fig. 12.3; Ruess et al. 1998, 2003). This pattern is not dissimilar to that found for many temperate forests subject to periodic drought during the growing season (Price and Hendrick 1998, Shan et al. 2001, King et al. 2002). In contrast, there is a positive correlation between root production and mortality among years.

Fine roots in boreal forests are adapted to function at low soil temperatures, and fine-root growth rates in interior Alaskan forests are higher than those from many temperate systems when measured at low temperatures (Table 12.1; Tryon and Chapin 1983). For example, N uptake rates measured in situ in balsam poplar stands at 8°C are comparable to rates from temperate forests at 15°C (McFarland et al. 2002) and coincide with very rapid turnover rates of mineral and organic pools of soil N (Chapter 16). Although fine-root production in Alaskan forests peaks during midsummer, we believe that most of the fine-root production measured "over winter" (between October and June) occurs in the spring. However, exactly when root growth is initiated is unknown because our floodplain sites are inaccessible during snowmelt and river ice thaw in late April and early May. We do know that the initiation rate and cessation of root growth are closely coupled to soil temperature (Deans 1979, Teskey and Hinckley 1981, Kuhns et al. 1985, McMichael and Burke 1998), and we found that soil heat sum (days >5 °C) was a very good predictor of cumulative white spruce fine-root growth (unpublished data). Active root function early in spring has implications for plant growth beyond nutrient and water supply to growing tissues. For example, water flux derived from positive root pressure prior to leaf-out in some species plays a significant role in refilling xylem vessels with water columns broken by winter freezing. Prior to leaf-out in early May, Sperry et al. (1994) measured positive root pressure exceeding 30 kPa in paper birch (*Betula neoalaskana*, previously treated as *B. papyrifera*) and green alder (*Alnus viridis*

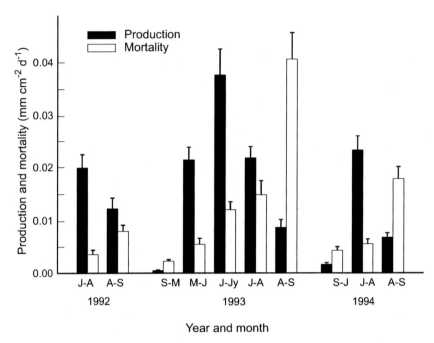

Figure 12.3. Representative seasonal patterns of fine-root production and mortality measured over ten time intervals from 1992 through 1994. Data are from early-successional shrub-dominated stands, averaged across browsed and unbrowsed sites (J = June; Jy = July). All values are expressed as mm root length [cm minirhizotron tube] $^{-2}$ d^{-1} ((SE). Redrawn with permission from the Ecological Society of America (Ruess et al. 1998).

subsp. *fruticosa*, previously treated as *A. crispa*) growing near Fairbanks when soil temperatures rose to 0° at 5 cm depth and rose above freezing at 20 cm depth. They also found that interruption of this water flux inhibited recovery of hydraulic conductivity in early spring.

Fine-root production is concentrated closer to the soil surface in cold boreal forests along the Tanana River than has been reported for many temperate forests. For example, the top 30 cm accounted for an average of 87 ± 4% of annual production in the 1 m profile among mature floodplain stands (Fig. 12.4). Comparative proportions for temperate forests include 56% for sugar maple stands in Michigan (Hendrick and Pregitzer 1993), <50% in loblolly pine (Crocker and Hendrick, personal observation), 50% in white oak (upper 22 cm, Joslin and Henderson 1987). Data from other boreal forests suggest that patterns in the Tanana River forests are typical for cold forests. For example, Safford and Bell (1972) reported similar values for a 39-year-old white spruce plantation in Maine, where 93% of fine-root mass was located in the upper 15 cm and 61% in the upper 5 cm alone. Greater than 80% of black spruce fine roots growing in a Canadian peat soil occurred in the moss layer and the upper 10 cm peat layers (Bhatti et al. 1998).

The positive correlation between average depth distribution of fine-root production and mean growing season soil temperature among Tanana floodplain ecosystem

Table 12.1. Rates of root length production measured across a number of biomes using minirhizotrons.[1] Some values are derived from a whole-tube analysis or represent a selected soil profile; others are growing-season averages or describe peak periods of production.

Ecosystem type	Entire minirhizotron tube[1]		Selected profile		Reference
	Mean	Peak	Mean	Peak	
Angiosperms					
Balsam poplar (*Populus balsamifera*); Alaska	0.068±0.007				Ruess (unpublished)
Mixed willows (*Salix* spp.); Alaskan (browsed)	0.022±0.004				Ruess et al. (1998)
Mixed willows (*Salix* spp.); Alaskan (unbrowsed)	0.047±0.007				Ruess et al. (1998)
Oak shrub (*Quercus* spp.); Florida	0.02–0.05				Dilustro et al. (2002)
Sweetgum (*Liquidambar styraciflua*); Georgia				0.033 (<25 cm)	Price and Hendrick (1998)
Mixed oak (*Quercus* spp.); Tennessee				0.015–0.049 (0–30 cm)	Joslin et al. (2001)
Sugar maple (*Acer saccarum*); Michigan				0.026 to 0.036 (<30 cm)	Hendrick and Pregitzer (1997)
Sugar maple (*Acer saccarum*); Michigan	0.033–0.067				Burton et al. (2000)
Hybrid poplar (*Populus tristis* X *Populus balsamifera*), Wisconsin	0.024–0.045				Coleman et al. (2000)
Northern Hardwoods; New Hampshire	0.039±0.004				Tierney et al. (2003)
Gymnosperms					
Norway spruce (*Picea abies*); Sweden				0.022	Majdi (2001) (0–10 cm)
Longleaf pine (*Pinus palustris*); Alabama				0.026	Pritchard et al. (2001) (0–32.5 cm)
White spruce (*Picea glauca*); Alaska	0.024±0.005				Hendrick and Ruess (unpublished)
Black spruce (*Picea mariana*); Alaska	0.051±0.014		0.063±0.014	0.153±0.040 (10–20 cm)	Ruess et al. (2003)
Slash pine (*Pinus elliotti*); Florida			0.016 (<25 cm)		Schroeer et al. (1999)
Loblolly pine (*Pinus taeda*); North Carolina				0.079 (0–30 cm)	King et al. (2002)
Red pine (*Pinus resinosa*); Wisconsin	0.002 to 0.004				Coleman et al. (2000)

[1] All values are expressed per cm^2 minirhizotron tube (mm fine root cm^{-2} d^{-1}; ± SE where available).

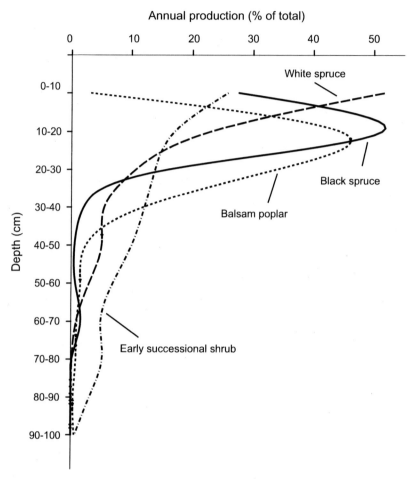

Figure 12.4. Percentage of total annual fine-root production distributed by 10 cm soil horizon increments for four Tanana floodplain ecosystem types. Data are means (± SE) across all minirhizotron tubes within each ecosystem type.

types suggests that the soil temperature profile has a strong influence on root depth distributions. In addition to these among-site patterns, we also observe a progressive increase in fine-root production into deeper soil layers within all forest types as the soil warms through the growing season. This is a function of increased proliferation of roots into deeper profiles, as well as increases in the growth rates of individual roots in those layers. This pattern is similar to that reported for temperate ecosystems (Fernandez and Caldwell 1975, Hendrick and Pregitzer 1993, Persson et al. 1995, Day et al. 1996). Exploration of deeper soil profiles is most dramatic in stands of early-successional shrubs growing contiguous to the Tanana River, where significant increases in fine-root production below 75 cm during mid- to late summer occur at the expense of surface production. One possible explanation for this pattern is fine-root access to subsurface water—perhaps containing significant concentrations of soluble organic and inorganic nutrients—which typically rises in early July as the snowmelt in the Alaska Range recharges river levels in the Tanana River basin (Ruess et al. 1998).

One of the most striking features of fine-root growth in interior Alaskan forests is that rates of daily root length production (RLP) in these systems are very similar to those reported for temperate forests. For example, when expressed as production per unit surface area of minirhizotron tube per day, midgrowing-season values for both deciduous and coniferous forests along the Tanana floodplain are as high as if not higher than values reported for temperate ecosystems using similar techniques (Table 12.1). This is surprising given the low temperature of Alaskan soils and indicates substantial physiological adaptation to low temperature.

Fine-Root Longevity

Until recently, the rates of death and decay of fine roots were inferred from estimates of "fine-root turnover," calculated as the ratio of fine-root production to biomass. Although these estimates were subject to errors associated with destructive cores and root sorting methods, they provided evidence that fine-root systems have substantially higher turnover rates than aboveground tissues. For example, using soil cores from a series of successional forests along the Tanana River, we found an average fine-root turnover rate of 0.90 ± 0.06 year^{-1} (Ruess et al. 1996). This finding indicated that (1) fine-root turnover rates in boreal forests are very similar to those found for temperate forests, (2) fine-root systems are turning over more than an order of magnitude faster than aboveground litter in boreal forests, and (3) overall C turnover in boreal forests is much greater than is suggested by studies of aboveground production.

Minirhizotron technology has provided new insights into belowground dynamics beyond traditional approaches to fine-root turnover because it allows the simultaneous monitoring of production, mortality, and disappearance of individual roots and direct measures of fine-root longevity and decomposition. Because the condition and fate of individual roots are followed across sampling periods, the survival of individuals can be estimated in much the same manner as is done for marked animal populations, without disturbing the root/mycorrhizal association.

Analysis of fine-root longevity among four ecosystem types in interior Alaska reveals mean fine-root lifespans during the growing season that ranged from 52 ± 2 days for balsam poplar stands to 108 ± 4 days for black spruce stands (Table 12.2). In early successional stands, aboveground herbivory by moose and snowshoe hares reduces fine-root lifespan in browsed stands (69 ± 2 days) by approximately 9% relative to unbrowsed stands (76 ± 2 days; Ruess et al. 1998). Fine-root survival decreases substantially over winter in all stands except balsam poplar (Table 12.2). Other patterns of survival common to early shrub stands, balsam poplar, and black spruce, but not white spruce, included highest survival rates during midseason and highest survival among cohorts born early in the growing season.

Although these Alaskan values are at the low end of the range of fine-root lifespans reported from several temperate systems (Hendrick and Pregitzer 1992,

Table 12.2. Fine root survival and decomposition estimates (± SE) for growing season ($\Phi_{growing\ season}$) and over winter (Φ_{winter}) periods generated for four Tanana River stand types using mark-recapture analyses. Mean lifespan (MLS = $|1/\ln(\Phi)|$ × 30) is estimated for growing season only due to uncertainties of exact time of death during winter. Decomposition rates ($\Phi_{decomposition}$) were estimated as 1- Φ for live to dead transitions. Decomposition times are MLS values associated with growing season estimates only (Ruess et al. 1998).

	Early shrub	Balsam poplar	White spruce	Black spruce
Fine root survival				
$\Phi_{growing\ season}$ (mo^{-1})	0.647 ± 0.008	0.563 ± 0.008	0.725 ± 0.005	0.758 ± 0.007
Φ_{winter} (winter^{-1})	0.204 ± 0.014	0.626 ± 0.012	0.541 ± 0.014	0.469 ± 0.012
MLS$_{growing\ season}$ (d)	69 ± 2	52 ± 1	93 ± 3	108 ± 4
Fine root decomposition				
$\Phi_{growing\ season}$ (mo^{-1})	0.453 ± 0.009	0.734 ± 0.008	NE[1]	0.460 ± 0.01
Φ_{winter} (winter^{-1})	0.886 ± 0.008	0.697 ± 0.012	NE[1]	0.725 ± 0.014
Decomposition time$_{growing\ season}$ (d)	50 ± 1	23 ± 3	NE[1]	49 ± 2

[1]NE = not estimated.

Burton et al. 2000, Coleman et al. 2000, Eissenstat et al. 2000, Wells and Eissenstat 2001, King et al. 2002), the lack of common analytical approaches among studies prevents any legitimate cross-biome conclusions at this stage. Nevertheless, a general characteristic of woody plant root systems appears to be that the smallest diameter, first-order roots have lifespans typically less than a year, and, in some cases, substantially less than a year (Wells and Eissenstat 2000, King et al. 2002, Ruess et al. 2003). Pregitzer et al (2002) argued that physiological function and life history traits of fine roots are related to morphology and position along a lateral branch. Several authors have noted that these first-order roots would be relatively inexpensive to build because of their high SLR and low structural content but costly to maintain because of their high N content and respiration rates (Pregitzer et al. 1997, 2002, Eissenstat and Yanai 1997). This suggests that fine-root lifespan is associated with other physiological and morphological traits of fine roots in much the same manner as has been observed for aboveground tissues (Reich et al. 1992, Baruch and Goldstein 1999). Strong support for this comes from an inverse relationship between fine-root diameter and lifespan found by several studies (Eissenstat et al. 2000, Wells and Eissenstat 2001, King et al. 2002). Alaskan forests provide further evidence for the growth/lifespan hypothesis, showing that among Tanana floodplain ecosystem types, the average maximum growth rate of fine roots is inversely correlated with the mean lifespan of those roots (Fig. 12.5). Although this relationship is driven by a taxonomic distinction between angiosperms and gymnosperm-dominated stands, which may have a strong environmental component, we predict that mean lifespan is inversely correlated with fine-root growth and respiration when scaled at the taxon, species, or individual stand level.

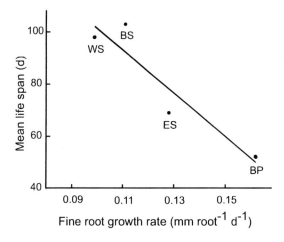

Figure 12.5. Relationship between mean growing season lifespan of roots and maximum fine-root production averaged across all roots and minirhizotron tubes at four Tanana floodplain ecosystem types (ES = early successional shrub, BP = balsam poplar, WS = white spruce, BS = black spruce).

King et al. (2002) recently found that mycorrhizal roots lived nearly twice as long as nonmycorrhizal roots in an eight-year-old loblolly pine plantation in North Carolina. We found a similar response for white spruce fine roots, where roots infected with ectomycorrhizae lived slightly longer (235 days) than uninfected roots (180 days; Hendrick and Ruess unpublished). Eissenstat and Yanai (1997) argued that increased lifespan should maximize nutrient acquisition efficiency, or the ratio of nutrient uptake per unit C cost associated with all physiological activities of the root/mycorrhizal system. Given that ectomycorrhizae can substantially increase nutrient uptake capacity, albeit at a poorly understood cost (Rygiewicz and Anderson 1994), it seems reasonable that mycorrhizal infection will increase mean fine-root lifespan. The fact that most fine roots in ecosystems along the Tanana floodplain are heavily infected with ectomycorrhizae but also have very short lifespans presents an interesting puzzle for further study (see discussion later in this chapter).

Fine-Root Decomposition

One of the most significant contributions of the Bonanza Creek root studies has been the use of minirhizotrons to determine the decomposition rate of intact fine roots (Ruess et al. 1998, 2003). A distinct advantage of this approach is that it follows the decay rate in situ of those roots that likely contribute most to total fine-root production. Estimates from Tanana floodplain forests indicate a greater than 70% chance that first-order roots found dead in the fall will completely decompose over winter and that decomposition time of first-order roots during the growing season is on the order of one to two months (Table 12.2). Thus, it appears that most first-order roots, even in black spruce forests, are growing, dying, and completely disappearing within the growing season—a radical departure from the notion that boreal forests are systems characterized by slowly growing vegetation and slowly decaying soil organic matter.

Previous estimates of root decomposition from other ecosystems have been derived primarily from the disappearance of roots from litterbags or tethered as bundles and placed in or beneath the forest floor. While these studies have led to valuable predictive relationships for fine-root decomposition based on substrate quality and climate (Silver and Miya 2001), they typically include size cohorts (1–2 mm diameter) that are substantially larger than those now thought to be the most active physiologically (Pregitzer et al. 2002). For example, if we predict fine-root decomposition for Tanana floodplain ecosystems from the inverse relationship between fine-root decay and latitude established from a literature review of litterbag studies (Silver and Miya 2001), we would estimate a value (14.6 years) that is two orders of magnitude slower than what we have found using minirhizotrons. Our minirhizotron estimates are not necessarily in conflict with these predictions; they just emphasize that fine-root systems are a modular continuum of morphology, function, and life history that is manifested in differences in decomposition rates among cohorts of different function (Zak et al. 2000). We recently found that longer-living fine roots decompose more slowly after death (Ruess et al. 2003), suggesting that the influence of substrate quality on decomposition rate found for larger fine-root size classes (Silver and Miya 2001) and aboveground tissues (Flanagan and Van Cleve 1983) also applies to the most rapidly growing root cohorts.

Insights regarding rapidly growing first-order roots gained from minirhizotron studies notwithstanding, we know from biomass harvests that larger-size classes of both live and dead roots represent large biomass pools (Ruess et al. 1996) about which we know very little. Pregitzer et al. (2002) estimated that first-order roots constitute 75% of all roots less than 1 mm diameter. In terms of production, however, this value is surely an underestimate, given the increase in lifespan with root order. Larger-size classes, which are longer lived and much slower to decompose, would thus contribute substantially less to total fine-root production and/or the annual fluxes of N and C through the fine-root system (Ruess et al. 2003). Some of the most interesting data on the age and decomposition rate of larger, functionally older cohorts of fine roots come from radiocarbon methods that take advantage of the large pulse of ^{14}C injected into the atmosphere from thermonuclear weapons testing during the 1960s. Using this approach, Gaudinski et al. (2001) estimated a mean age of 3–18 years for live fine roots and 10–18 years for dead fine roots for three temperate forests (cohorts <2 mm). A plausible hypothesis regarding C cycling in boreal systems is that most of the rapidly cycling labile fraction is derived from the growth activities and decay products of first-order roots, while stored soil C is derived from decay products of more recalcitrant above- and belowground tissues. Given the apparent magnitude of belowground allocation in boreal forests, the decomposition rate and eventual fate of C from all fine-root age cohorts remain key questions in the dynamics of C and element cycling in these ecosystems.

Linking Fine Roots to Ecosystem Carbon and Nutrient Cycling

Translating minirhizotron root length production values (mm cm^2 day^{-1}) to units of biomass production (kg biomass ha^{-1} yr^{-1}) involves a number of assumptions, none of which has been rigorously validated (Ruess et al. 2003). In addition to determining appropriate SRL values for roots observed in minirhizotron images, calculations require an estimate for the focal depth of the camera, and an assumption that fine roots are evenly distributed relative to the tube (Merrill and Upchurch 1994). Recent reviews emphasize that all approaches to measuring fine-root growth, including soil coring (Nadelhoffer 2000), ecosystem C balance (Davidson et al. 2002), ecosystem N balance (Ruess et al. 1996), and minirhizotrons (Ruess et al. 2003), have limitations. Nonetheless, the data derived from minirhizotrons and other methods are highlighting that ecosystem patterns in C and nutrient cycling are difficult to interpret without some understanding of ecosystem C allocation to roots.

The relative amounts of C allocated to above- and belowground production vary among the vegetation types in interior Alaska, and determining the controls over this variability has been an active area of research at Bonanza Creek. Low nutrient availability or soil moisture often correlates with an increase in the proportional allocation belowground. Belowground production data for the dominant ecosystem types along the Tanana floodplain show that fine roots constitute between 28% (white spruce) and 58% (both early-successional shrubs and late-successional black spruce) of total production (including nonvascular plants), with mature balsam poplar stands (38%) somewhere in the middle of these values (Table 12.3). Previous research indicated that boreal conifers allocate proportionately more C to fine roots than do boreal deciduous forests (Ruess et al. 1996, Gower et al. 1997, Steele et al. 1997), which has been ascribed to greater nutrient availability in deciduous forests (Ruess et al. 1996). Among our floodplain stands, black spruce but not white spruce supports this deciduous-conifer contrast. However, no clear environmental or nutrient availability trend that we have studied appears to explain this disparity. An important control over belowground allocation in boreal forests may be soil temperature. For example, Hendrick and Ruess (unpublished) found that, among floodplain white spruce stands, higher root mortality was correlated with lower soil temperatures, perhaps explaining greater belowground allocation on colder soils. Similarly, Vogel et al. (2005) reported that mature black spruce forests allocated more C belowground where decomposition was slower and soil temperature colder but also stated that the primary influence on allocation may have been lower moisture or N availability at the slow-decomposition site. Ultimately, variability in aboveground versus belowground C allocation has pronounced influences on patterns and rates of C and nutrient cycling through the boreal ecosystem.

Large pools of soil C have accumulated in Bonanza Creek and in the rest of the boreal forest (Chapter 15; Van Cleve et al. 1983). We believe that a readily decomposable C fraction derived from the growth activities and decay products of first-order roots controls the majority of annual C and N cycling. Roots account for a greater fraction of litter input to the forest floor than aboveground detritus (Ruess et al. 1996), and first-order fine roots decompose at a much faster rate than litter in boreal forest types (Ruess et al. 2003, Steele et al. 1997). The reduced difference in decomposition between fine roots of deciduous and coniferous forests supports the observation that white spruce needles can decompose faster than either aspen or balsam poplar litter (Fig 12.6). We have re-examined earlier estimates of C dynamics in light of our new understanding of fine-root processes and have found that the

Table 12.3. Net annual primary production (kg biomass ha^{-1} yr^{-1}; ±SE) of fine roots and aboveground components for four forest types along the Tanana River floodplain (n = 3).

	Early shrub	Balsam poplar	White spruce	Black spruce
Aboveground				
Cryptogams[1]	0	0	784 ± 369	730 ± 144
Shrubs[2]	958 ± 179	1182 ± 291	324 ± 226	394 ± 15
Trees[3]	0	4054 ± 129	3433 ± 1047	498 ± 139
Belowground				
Fine roots[4]	1366 ± 191	3036 ± 429	1814 ± 605	2284 ± 746
Total	2324 ± 126	8272 ± 482	6530 ± 1009	3906 ± 901

[1]Cryptogam production was assessed by measuring site specific green fall cryptogam biomass; measuring seasonal growth of the two major cryptogam species, the feathermosses *Hylocomium splendens* and *Pleurozium schreberi*; allocating the biomass of other cryptogams to these two feathermoss species in the proportion in which they occur; and converting length growth and moss density to biomass accretion.

[2]Shrub production in black spruce is from Ruess et al (2003). At other sites, shrub production was estimated from litterfall measured across at least two years, assuming that 50% of AG growth was allocated to leaf production and that leaves lost 10% mass during senescence. The shrub increment in balsam poplar stands does not include contributions from a dense cover of *Rosa acicularis*.

[3]AG tree production in balsam poplar stands was also estimated from litterfall because DBH/biomass relationships established for younger balsam poplar trees (Yarie and Van Cleve, unpublished data) proved unreliable in estimating aboveground biomass of these older stands. Tree production in white and black spruce was the sum of annual wood increments and litterfall. Wood increments were estimated from DBH/biomass relationships (Yarie and Van Cleve, unpublished), using diameter increments from tree cores (black spruce) and band dendrometers (white spruce), averaged over the 1990s. Litterfall was estimated as a percentage of total AG annual increment, using values from Van Cleve et al. (1983). Increments at all sites were measured across at least two years, but not always the same years.

[4]Fine root production was determined using minirhizotrons as described by Ruess et al. (2003).

estimated rates of C and nutrient cycling through the soil system increase and species differences decrease when belowground allocation is included. A common index of decomposition rate, forest floor turnover time, is estimated as the forest floor mass divided by the C deposited annually as litter, usually only aboveground. In theory, the litter C is equivalent to the amount of C emitted from the forest floor through heterotrophic respiration. The lower the index (measured in years), the shorter the turnover time, the greater the decomposition rate, and, generally, the greater the N availability to plants. Van Cleve et al (1983) used only aboveground litter and calculated forest floor turnover times of 178 years for black spruce and 20 years for deciduous forests. They ascribed these differences to both poorer litter quality and an unfavorable soil environment in black spruce forests. But if we include fine-root litter in forest floor residence time estimates, these values decrease by a factor of 6 for black spruce (28 years) and 3 (7 years) for deciduous forests (Ruess et al. 1996). Vogel et al. (2005) also found that forest floor turnover time is much greater in black spruce when heterotrophic respiration is substituted for aboveground litter in turnover time estimates. The faster forest floor turnover times

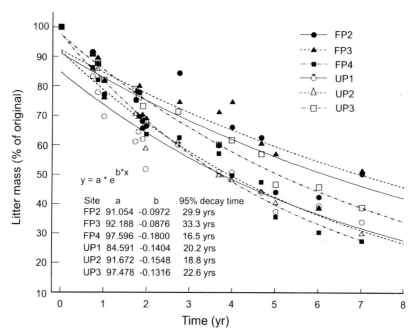

Figure 12.6. Change in organic matter content of litter in upland (UP1 = early successional, UP2 = birch/aspen, UP3 = white spruce) and floodplain (FP2 = alder/poplar, FP3 = balsam poplar, FP4 = white spruce) stands from an LTER litter bag experiment initiated in 1989 (http://www.lter.uaf.edu/Data_catalog_detail.cfm?dataset_id=70). Each data point represents the mean of approximately 30 litter bags.

derived with more complete C balance approaches reconciles the extremely fast root decomposition rates observed with minirhizotrons and what had previously been assumed to be very slow C turnover in boreal ecosystems.

Research Gaps and Conclusions

Because we have intensively studied fine-root dynamics in only a few stands within each of the forest types along the Tanana River, we have a limited understanding of the patterns and controls over belowground allocation at the landscape scale. At this spatial scale, aboveground production in black spruce, our most widely distributed forest type, can vary by nearly an order of magnitude. But how do fine-root growth and respiration rates, longevity, and decomposition scale to aboveground processes across these production gradients? What determines these patterns? Answering these questions for any of our forest types would require an intensive study across a range of stands that varied in tree age and density, topography, soil nutrient and water status, and perhaps disturbance history. We now recognize that belowground C allocation can constitute a large fraction of annual production, but connecting fine-root lifespan and decomposition rate to the eventual fate of fine-root C remains a key uncertainty in understanding the dynamics of C and nutrient cycling of boreal forests.

Among the most significant gaps in our knowledge of belowground function in boreal forests is a relatively poor understanding of the role of mycorrhizae in regulating C and nutrient cycling and, potentially, community composition and dynamics. The paucity of people working with mycorrhizae in the North American boreal forest and the added difficulties of studying mycorrhizal roots and fungi are responsible in part. But the problem is compounded by the fact that all three dominant mycorrhizal types (ectomycorrhizae or ECM, arbuscular mycorrhizae or AM, and ericoid mycorrhizae or ERM) occur in the boreal forest biome, often within the same stands. The reason for this is that, although all three mycorrhizal types improve host nutrient acquisition, they do so in a variety of ways that impact both the form and the amount of nutrients available to plants. All three increase host access to N and/or P via fungal hyphal exploration but differ in their ability to access otherwise unavailable forms of N or P. For example, many AM, ECM, and ERM fungi can access various forms of P enzymatically (Pearson and Read 1975, Alexander and Hardy 1981, Mitchell and Read 1981, Dighton 1991, Joner and Johansen 2000,) but the most complex enzymes are largely confined to ERM fungi (Leake and Miles 1996). Some ERM and ECM fungi can enzymatically access organic forms of N (Leake and Read 1990, Bending and Read 1996, Colpaert and Van Laere 1996, Burke and Cairney 1998, Tibbett et al. 1999), but there can be substantial differences in the amount of enzymes produced by both types (Bending and Read 1996).

These functional differences among mycorrhizal fungi may play a role in determining boreal forest species assemblages. Read (1984) predicted that soils characterized by low mineral N and P availability and high organic matter (OM) content (and thus organic forms of N and P), such as those supporting black spruce, should be dominated by ERM plants. This is in fact often the case in the understory of black spruce forests, where Ericaceous shrubs such as *Ledum* and *Vaccinium* are often abundant. Soils characterized by intermediate levels of mineral N and P availability are presumed to be predominantly ECM, as is true for midsuccessional floodplain balsam poplar and upland birch or aspen forests, as well as late-successional white spruce forests. The ECM fungi in these environments can potentially supply N and P to the host from both organic and inorganic sources, depending on the fungal partner and the OM content of the soil (Dinkelaker and Marschner 1992, Turnbull et al. 1995, Bending and Read 1996,). Conversely, AM fungi enhance primarily mineral P, rather than N uptake, and are likely to be abundant only in early-successional environments dominated by mineral forms of both N and P. The fact that some of the genera dominating both early- and midsuccessional boreal forests, notably willow and poplar, are capable of forming both AM and ECM roots lends support to the notion that accumulating OM might be associated with a reduction in mineral-form-dependent AM species and increase in ECM species (or roots). However, we have not observed any significant levels of AM colonization in balsam poplar roots sampled from midsuccessional floodplain forests (personal observation). Van Cleve et al. (1983) suggested that the immobilization of P in woody debris and the forest floor may be responsible for measured decreases in total soil

P during succession in floodplain forests along the Tanana River. If this is true, then the production of P mineralizing enzymes by mycorrhizae, especially ERM, may play an especially important role in determining the composition and production of late-successional forests.

We and others have proposed that nutrients transferred directly from decaying fine roots to live fine roots via mycorrhizal connections would be an important plant adaptation for "short-circuiting" losses to and uptake from nutrient pools produced from soil organic matter decomposition by saprophytic microbes (Entry et al. 1992, Finlay and Soderstrom 1992, Ruess et al. 1996). Using minirhizotrons, we have observed extensive fine-root/mycorrhizal proliferation surrounding dead roots and decomposing forest floor litter (personal observation). Ectomycorrhizal-mediated transfer of C and nutrients between individual tree species has been demonstrated (Simard et al. 1997), although the exact nature of these transfers is poorly understood (i.e., live-to-live and/or dead-to-live).

One of the most overlooked aspects of root/mycorrhizal associations is the influence of mycophagous soil microfauna on plant growth and nutrient cycling processes. Mycophagous animals such as collembolans can reach high densities in boreal forests (Cowling and MacLean 1981). Our casual observations using minirhizotrons suggest that collembolans are more abundant in interior Alaskan boreal forests than in several other North American forests we have studied, such as sugar maple (northern Michigan), tulip poplar (North Carolina), oak-hickory (Georgia), and loblolly pine plantations (Florida). We have also observed active collembolan populations during winter in black spruce forests when ambient air temperature was -35 °C and soil temperature was -5 °C (Vogel, personal observation). Although these organisms are known to graze VAM and ECM mycorrhizae (Fitter and Sanders 1992, Ek et al. 1994), effects on plant growth are not always negative (Setala 1995). Mycophagous grazers can directly enhance nutrient cycling rates or, through selective feeding, modify the competitive balance among mycorrhizal species or between mycorrhizal and saprophytic hyphae (Clarholm 1985, Shaw 1988, Setala and Huhta 1991, Klironomos et al 1999). These foodweb relationships are complicated by the fact that ectomycorrhizae have recently been shown to "feed" on collembolans. Klironomos and Hart (2001) found that *Laccaria bicolor*, an ectomycorrhizal species common to boreal forests, appears to paralyze live collembolans and subsequently transfer N from the dead microfauna to the host plant.

References

Alexander, I. J., and K. Hardy. 1981. Surface phosphatase activity of Sitka spruce mycorrhizas from a Serpentine site. Soil Biology and Biochemistry 13:301–305.
Baruch, Z., and G. Goldstein. 1999. Leaf construction cost, nutrient concentration, and net CO_2 assimilation of native and invasive species in Hawaii. Oecologia 121:183–192.
Bending, G. D., and D. J. Read. 1996. Effects of the soluble polyphenol tannic acid on the activities of ericoid and ectomycorrhizal fungi. Soil Biology and Biochemistry 28:1595–1602.

Bhatti, J. S., N. W. Foster, and P. W. Hazlett. 1998. Fine root biomass and nutrient content in a black spruce peat soil with and without alder. Canadian Journal of Soil Science 78:163–169.

Bidartondo, M. I., H. Ek, and W. H. Soderstrom. 2001. Do nutrient additions alter carbon sink strength of ectomycorrhizal fungi? New Phytologist 151:543–550.

Burke, R. M., and J. W. G. Cairney. 1998. Carbohydrate oxidases in ericoid and ectomycorrhizal fungi: A possible source of Fenton radicals during the degradation of lignocellulose. New Phytologist 139:637–645.

Burke, M. K., and D. J. Raynal. 1994. Fine root growth phenology, production, and turnover in a northern hardwood forest ecosystem. Plant and Soil 162:135–146.

Burton, A. J., K. S. Pregitzer, and R. L. Hendrick. 2000. Relationships between fine root dynamics and nitrogen availability in Michigan northern hardwood forests. Oecologia 125:389–399.

Burton, A. J., K. S. Pregitzer, R. W. Ruess, R. L. Hendrick, and M. F. Allen. 2002. Fine root respiration rates in North American forests: Effects of nitrogen concentration and temperature across biomes. Oecologia 131:559–568.

Clarholm, M. 1985. Interactions of bacteria, protozoa and plants leading to mineralization of soil nitrogen. Soil Biology and Biochemistry 17:181–187.

Coleman, M. D., R. E. Dickson, and J. G. Isebrands. 2000. Contrasting fine-root production, survival and soil CO_2 efflux in pine and poplar plantations. Plant and Soil 225:129–139.

Colpaert, J. V., and A. Van Laere. 1996. A comparison of the extracellular enzyme activities of two ectomycorrhizal and a leaf-saprotrophic basidiomycete colonizing beech litter. New Phytologist 133:133–141.

Cowling, J. E., and S. F. MacLean Jr. 1981. Forest floor respiration in black spruce taiga forest ecosystem in Alaska. Holarctic Ecology 4:229–237.

Davidson, E. A., K. Savage, P. Bolstad, D. A. Clark, P. S. Curtis, D. S. Ellsworth, P. J. Hanson, B. E. Law, Y. Luo, K. S. Pregitzer, J. C. Randolph, and D. Zak. 2002. Belowground carbon allocation in forests estimated from litterfall and IRGA-based soil respiration measurements. Agricultural and Forest Meteorology 113:39–51.

Day, F. P., E. P. Weber, C. R. Hinkle, and B. G. Drake. 1996. Effects of elevated atmospheric CO_2 on fine root length and distribution in an oak-palmetto scrub ecosystem in central Florida. Global Change Biology 2:143–148.

Deans, J. D. 1979. Fluctuations of the soil environment and fine root growth in a young Sitka spruce plantation. Plant and Soil 52:195–208.

Dighton, J. 1991. Acquisition of nutrients from organic sources by mycorrhizal autotrophic plants. Experientia 47:362–369.

Dilustro, J. J., F. P. Day, B. G. Drake, and C. R. Hinkle. 2002. Abundance, production and mortality of fine roots under elevated atmospheric CO_2 in an oak-scrub ecosystem. Environmental and Experimental Botany 48:149–159.

Dinkelaker, B., and H. Marschner. 1992. In vivo demonstration of acid phosphatase activity in the rhizosphere of soil grown plants. Plant and Soil 144:199–205.

Eissenstat, D. M., and R. D. Yanai. 1997. The ecology of root lifespan. Advances in Ecological Research 27:1–60.

Eissenstat, D. M., C. E. Wells, R. D. Yanai, and J. L. Whitbeck. 2000. Building roots in a changing environment: Implications for root longevity. New Phytologist 147:33–42.

Ek, H., M. Sjogren, K. Arnedbrandt, and B. Sodrstrom. 1994. Extramatrical mycelial growth, biomass allocation and nitrogen uptake in ectomycorrhizal systems in response to collembolan grazing. Applied Soil Ecology 1:155–169.

Entry, J. A., C. L. Rose, and K. Cromack Jr. 1992. Microbial biomass and nutrient concentrations in hyphal mats of the ectomycorrhizal fungus *Hysterangium setchellii* in a coniferous forest soil. Soil Biology and Biochemistry 24:447–453.

Fahey, T. J., and J. W. Hughes. 1994. Fine root dynamics in a northern hardwood forest ecosystem, Hubbard Brook Experimental Forest, NH. Journal of Ecology 82:533–548.

Fernandez, O. A., and M. M. Caldwell. 1975. Phenology and dynamics of root growth of three cool semi-desert shrubs under field conditions. Journal of Ecology 63:703–714.

Finlay, R., and B. Soderstrom. 1992. Mycorrhiza and carbon flow to the soil. Pages 134–160 *in* M. F. Allen, editor. Mycorrhizal Functioning. Chapman and Hall, New York.

Fitter, A. H., and I. R. Sanders. 1992. Interactions with the soil fauna. Pages 333–354 *in* M. F. Allen, editor. Mycorrhizal Functioning. Chapman and Hall, New York.

Flanagan, P. W., and K. Van Cleve. 1983. Nutrient cycling in relation to decomposition and organic matter quality in taiga ecosystems. Canadian Journal of Forest Research 13:795–817.

Gaudinski, J. B., S. E. Trumbore, E. A. Davidson, A. C. Cook, D. Markewitz, and D. D. Richter. 2001. The age of fine-root carbon in three forests of the eastern United States measured by radiocarbon. Oecologia 129:420–429.

Gower, S. T., J. G. Vogel, J. M. Norman, C. J. Kucharik, S. J. Steele, and T. K. Stow. 1997. Carbon distribution and aboveground net primary production in aspen, jack pine, and black spruce stands in Saskatchewan and Manitoba, Canada. Journal of Geophysical Research 102:29,029–29,041.

Hendrick, R. L., and K. S. Pregitzer. 1992. The demography of fine roots in a northern hardwood forest. Ecology 73:1094–1104.

Hendrick, R. L., and K. S. Pregitzer. 1993. The dynamics of fine root length, biomass, and nitrogen content in two northern hardwood ecosystems. Canadian Journal of Forest Research 23:2507–2520.

Hendrick, R. L., and K. S. Pregitzer. 1997. The relationship between fine root demography and the soil environment in northern hardwoods forests. Ecoscience 4:99–105.

Joner, E. J., and A. Johansen. 2000. Phosphatase activity of external hyphae of two arbuscular mycorrhizal fungi. Mycological Research 104:81–86.

Joslin, J. D., and G. S. Henderson. 1987. Organic matter and nutrients associated with fine root turnover in a white oak stand. Forest Science 33:330–346.

Joslin, J. D., M. H. Wolfe, and P. J. Hanson. 2001. Factors controlling the timing of root elongation intensity in a mature upland oak stand. Plant and Soil 228:201–212.

King, J. S., T. J. Albaugh, H. L. Allen, M. Buford, B. R. Strain, and P. Dougherty. 2002. Below-ground carbon input to soil is controlled by nutrient availability and fine root dynamics in loblolly pine. New Phytologist 154:389–398.

Klironomos, J. N., and M. M. Hart. 2001. Animal nitrogen swap for plant carbon. Nature 410: 651–652.

Klironomos, J. N., E. M. Bednarczuk, and J. Neville. 1999. Reproductive significance of feeding on saprobic and arbuscular mycorrhizal fungi by the collembolan, *Folsomia candida*. Functional Ecology 13:756–761.

Kuhns, M. R., H. E. Garrett, R. O. Teskey, and T. M. Hinckley. 1985. Root growth of black walnut trees related to soil temperature, soil water potential, and leaf water potential. Forest Science 31:617–629

Leake, J. R., and W. Miles. 1996. Phosphodiesters as mycorrhizal P sources. I. Phosphodiesterase production and the utilization of DNA as a phosphorus source by the ericoid mycorrhizal fungus *Hymenoscyphus ericae*. New Phytologist 132:435–444.

Leake, J. R., and D. J. Read. 1990. Proteinase activity in mycorrhizal fungi. I. The effect of extracellular pH on the production and activity of proteinase by ericoid endophytes from soils of contrasted pH. New Phytologist 115:243–250.

Majdi, H. 2001. Changes in fine root production and longevity in relation to water and nutrient availability in a Norway spruce stand in northern Sweden. Tree Physiology 21:1057–1061.

McFarland. J. M., R. W. Ruess, K. Kielland, and A. Doyle. 2002. Cycling dynamics of NH_4^+ and amino acid N in soils of a deciduous boreal forest ecosystem. Ecosystems 5: 775–788.

McMichael, B. L., and J. J. Burke. 1998. Soil temperature and root growth. Hortscience 33:947–951.

Merrill, S. D., and D. R. Upchurch. 1994. Converting root numbers observed at minirhizotrons to equivalent root length density. Soil Science Society of America Journal 58:1061–1067.

Mitchell, D. T. and D. J. Read.1981. Utilization of inorganic and organic phosphates by the mycorrhizal endophytes of *Vaccinium macrocarpon* and *Rhododendron ponticum*. Transactions of the British Mycological Society 76:255–260.

Nadelhoffer, K. J. 2000. The potential effects of nitrogen deposition on fine-root production in forest ecosystems. New Phytologist 147:131–139.

O'Connell, K. E. B., S. T. Gower, and J. M. Norman. 2003. Net ecosystem production of two contrasting boreal black spruce forest communities. Ecosystems 6:248–260.

Pearson, V., and D. J. Read. 1975. The physiology of the mycorrhizal endophyte of *Calluna vulgaris*. Transactions of the British Mycological Society 64:1–7.

Persson, H., Y. Von Fircks, J. Majdi, and L. O. Nilsson. 1995. Root distribution in a Norway spruce *(Picea abies* (L.) Karst.) stand subjected to drought and ammonium-sulphate application. Plant and Soil 168–169:161–165.

Pregitzer, K. S., M. E. Kubiske, C. K. Yu, and R. L Hendrick. 1997. Relationships among root branch order, carbon, and nitrogen in four temperate species. Oecologia 111:302–308.

Pregitzer, K. S., J. L. DeForest, A. J. Burton, M. F. Allen, R. W. Ruess, and R. L. Hendrick. 2002. Fine root architecture of nine North American trees. Ecological Monographs 72:293–309.

Price, J. S., and R. L. Hendrick. 1998. Fine root length production, mortality and standing root crop dynamics in an intensively managed sweetgum (*Liquidambar styraciflua* L.) coppice. Plant and Soil 205:193–201.

Pritchard, S. G., M. A. Davis, R. J. Mitchell, S. A. Prior, D. L. Boykin, H. H. Rogers, and G. B. Runion. 2001. Root dynamics in an artificially constructed regenerating longleaf pine ecosystem are affected by atmospheric CO_2 enrichment. Environmental and Experimental Botany 48:55–69.

Read, D. J. 1984. The structure and function of the vegetative mycelium of mycorrhizal roots. Pages 215–240 *in* D. H. Jennings and A. D. M. Rayner, editors. The Ecology and Physiology of the Fungal Mycelium. Cambridge University Press, Cambridge, UK.

Reich, P. B., M. B. Walters, and D. S. Ellsworth. 1992. Leaf lifespan in relation to leaf, plant, and stand characteristics among diverse ecosystems. Ecological Monographs 62:365–392.

Ruess, R. W., R. L. Hendrick, and J. P. Bryant. 1998. Regulation of fine root dynamics by mammalian browsers in early successional Alaskan taiga forests. Ecology 79:2706–2720.

Ruess, R. W., K. Van Cleve, J. Yarie, and L. A. Viereck. 1996. Contributions of fine root production and turnover to the carbon and nitrogen cycling in taiga forests of the Alaskan interior. Canadian Journal of Forest Research 26:1326–1336.

Ruess, R. W., R. L. Hendrick, A. J. Burton, K. S. Pregitzer, B. Sveinbjörnsson, M. F. Allen, and G. Maurer. 2003. Coupling fine root dynamics with ecosystem carbon cycling in black spruce forests of interior Alaska. Ecological Monographs 73:643–662.

Rygiewicz, P. T., and C. P. Andersen. 1994. Mycorrhizae alter quality and quantity of carbon allocated below ground. Nature 369:58–60.

Safford, L. O., and S. Bell. 1972. Biomass of fine roots in a white spruce plantation. Canadian Journal of Forest Research 2:169–172.

Schroeer, A. E., R. L. Hendrick, and T. B. Harrington. 1999. Root, ground cover, and litterfall dynamics within canopy gaps in a slash pine (*Pinus elliottii* Engelm.) dominated forest. Ecoscience 6:548–555.

Setala, H. 1995. Growth of birch and pine seedlings in relation to grazing by soil fauna on ectomycorrhizal fungi. Ecology 76:1844–1851.

Setala, H., and V. Huhta. 1991. Soil fauna increase *Betula pendula* growth: Laboratory experiments with coniferous forest floor. Ecology 72:665–671.

Shan, J., L. A. Morris, and R. L. Hendrick. 2001. Soil carbon and fine root dynamics in slash pine (*Pinus elliottii*) plantations under different management intensities. Journal of Applied Ecology 38:932–941.

Shaw, P. J. A. 1988. A consistent hierarchy of fungal feeding preferences of the Collembola *Onychiurus armatus*. Pedobiologia 31:179–187.

Silver, W. L., and R. K. Miya. 2001. Global patterns in root decomposition: Comparisons of climate and litter quality effects. Oecologia 129:407–419.

Simard, S. W., D. A. Perry, M. D. Jones, D. D. Myrold, D. M. Durall, and R. Molina. 1997. Net transfer of carbon between ectomycorrhizal tree species in the field. Nature 388: 579–582.

Sperry, J. S., K. L. Nichols, J. E. M. Sullivan, and S. E. Eastlack. 1994. Xylem embolism in ring-porous, diffuse-porous, and coniferous trees of northern Utah and interior Alaska. Ecology 75:1736–1752.

Steele, S. J., S. T. Gower, J. G. Vogel, and J. M. Norman. 1997. Root mass, net primary production and turnover in aspen, jack pine, and black spruce forests in Saskatchewan and Manitoba. Canada. Tree Physiology 17:577–587.

Teskey, R. O., and T. M. Hinckley. 1981. Influence of temperature and water potential on root growth of white oak. Physiologia Plantarum 52:362–369.

Tibbett, M., F. E. Sanders, J. W. G. Cairney, and J. R. Leake.1999. Temperature regulation of extracellular proteases in ectomycorrhizal fungi (*Hebeloma spp.*) grown in axenic culture. Mycological Research 103:707–714.

Tierney, G. L., T. J. Fahey, P. M. Groffman, J. P. Hardy, R. D. Fitzhugh, C. T. Driscoll, and J. B. Yavitt. 2003. Environmental control of fine root dynamics in a northern hardwood forest. Global Change Biology 9:670–679.

Tryon, P. R., and F. S. Chapin III. 1983. Temperature control over root growth and root biomass in taiga forest trees. Canadian Journal of Forest Research 13:827–833.

Turnbull, M. H., R. Goodall, and G. R. Stewart. 1995. The impact of mycorrhizal colonization upon nitrogen source utilization and metabolism in seedlings of *Eucalyptus grandis* Hill ex Maiden and *Eucalyptus maculata* Hook. Plant Cell and Environment 18:1386–1394.

Van Cleve, K., L. Oliver, R. Schlentner, L. A. Viereck, and C. T. Dyrness. 1983. Productivity and nutrient cycling in taiga forest ecosystems. Canadian Journal of Forest Research 13:747–766.

Viereck, L. A. 1970. Forest succession and soil development adjacent to the Chena River in interior Alaska. Arctic and Alpine Research 2:1–26.

Vogel, J.G., D.W. Valentine, and R.W. Ruess. 2005. Soil and root respiration in mature

Alaskan black spruce forests that vary in soil organic matter decomposition rates. Canadian Journal of Forest Research 35:161–174.

Wells, C. E., and D. M. Eissenstat. 2001. Marked differences in survivorship among apple roots of different diameters. Ecology 82:882–892.

Zak, D. R., K. S. Pregitzer, J. S. King, and W. E. Holmes. 2000. Elevated atmospheric CO_2, fine roots and the response of soil microorganisms: A review and hypothesis. New Phytologist 147:201–222.

13

Mammalian Herbivory, Ecosystem Engineering, and Ecological Cascades in Alaskan Boreal Forests

Knut Kielland
John P. Bryant
Roger W. Ruess

Introduction

The mammalian herbivores of the taiga forests include members of the largest (moose) and smallest (microtines) vertebrates that inhabit North American terrestrial biomes. Their abundance in a particular area fluctuates dramatically due to seasonal use of particular habitats (moose) and external factors that influence demographic processes (microtines). The low visibility of herbivores to the casual observer might suggest that these animals have minimal influence on the structure and the function of boreal forests. On the contrary, seedling herbivory by voles, leaf stripping by moose, or wholesale logging of mature trees by beaver can profoundly change forest structure and functioning. These plant-herbivore interactions have cascading effects on the physical, chemical, and biological components of the boreal ecosystem that shape the magnitude and direction of many physicochemical and biological processes. These processes, in turn, control the vertical and horizontal interactions of the biological community at large. Herbivores act as ecosystem engineers (Jones et al. 1994) in that they reshape the physical characteristics of the habitat, modify the resource array and population ecology of sympatric species, and influence the flux of energy and nutrients through soils and vegetation. Additionally, many herbivores are central to a variety of human activities. Both consumptive and nonconsumptive use of wildlife represents a pervasive aspect of life in the North. In this chapter, we examine the interactions of mammalian herbivores with their environment, with an emphasis on moose, and attempt to delineate the biotic and abiotic conditions under which herbivores influence the phenotypic expression of vegetation. We also examine the role of herbivores, and of wildlife in general, in the context of human perceptions and interactions with their environment. Human-environment interactions are both direct and indirect and pertain to a variety of social expressions. The relationship between humans and wildlife has economic, cultural, and psychological dimensions, which underscore the importance of these animals in a broader social, as well as ecological, context.

Population Densities, Habitat Requirements, and Patterns of Activity

Northern ecosystems such as the boreal forest are characterized by extreme seasonality and pronounced change in resource availability between summer and winter. Not surprisingly, these conditions are reflected in the population dynamics of the animals that inhabit these environments, particularly in smaller-bodied herbivores. Although herbivore groups exhibit huge differences in average population densities per unit area (from 0.1 moose km^{-2} to 10,000 mice km^{-2}), moose, hares, and microtines have nearly the same biomass per unit land area (Chapter 8).

The early stages of primary and secondary succession of taiga forests are dominated by deciduous vegetation (Chapter 7). The principal species include willows and poplar on floodplains and birch and aspen in the upland forests. Herbivores such as moose, beavers, and hares are attracted to these early successional stands, whereas porcupines, red squirrels, and microtines make more extensive use of the later-successional conifer stands. All of these herbivores stay active throughout the year but exhibit seasonal patterns of activity dictated by their energy budget. During winter, forage availability and quality decline, and animals lower their activity correspondingly. Moose may spend as much as 70% of their time resting during winter (Risenhoover 1986). During the summer months, by contrast, the activity budget is divided approximately equally between foraging and resting (Sæther et al. 1992), and moose typically follow a relatively loose schedule of two hours of feeding and two hours resting around the clock.

Foraging Strategies

The dietary composition of subarctic herbivores varies by the season and is dictated to a large extent by forage availability. During the growing season, the animals seek out forage high in protein and energy, primarily green forage such as grasses, forbs, aquatic plants, and the leaves of deciduous woody plants. During the winter, these forage items are largely unavailable for moose and snowshoe hares. Whereas beavers and microtines base their diet largely on food stored in caches under the ice and subnivian space, respectively, moose, hares, and porcupines must rely on food still "on the hoof." The change from herbaceous to woody forage for moose and hares causes these herbivores to face a diet during winter that has roughly half the energy and protein content of their summer diet. The plants available to herbivores include both conifers and broad-leaved species. Porcupines are specialists that exploit the cambium of white spruce (Harder 1980). Hares mix deciduous and evergreen species in their diet (Bryant and Kuropat

1980), whereas moose rely almost exclusively on deciduous browse such as willow, birch, and aspen (Wolff 1976, Schwartz and Renecker 1998). Another important element of their respective foraging ecologies pertains to the handling of food items. Porcupines and hares can high-grade their foods because their dentition and manual dexterity allow them to consume the highly digestible portion of the woody browse (bark/cambium), and discard the largely indigestible portion (wood). By contrast, moose consume the entire woody stem. Thus, plant species-specific traits pertaining to twig morphology (e.g., shoot length and thickness, degree of taper) have greater foraging consequences for moose (Kielland and Osborne 1998) than for the smaller herbivores.

These foraging strategies also influence stand structure and soil processes. The activities of porcupines can increase standing dead material. Hares generally snip off the twigs from which they consume the bark, thus increasing coarse woody debris on the forest floor. Moose grind up entire twigs and leave behind feces high in microbial N but also high in fine woody debris. Thus, herbivores differ qualitatively in their effects on ecosystem function.

Wildlife, Herbivory, and Element Cycling

Moose, hares, and other boreal herbivores affect nutrient cycling in two basic ways: indirectly, by modifying the quality and quantity of plant litter (below- and aboveground) for decomposition, and directly, by accelerating nutrient turnover by consumption of forage and subsequent excretion.

Consequences of Herbivory for Plant Populations in Early Succession

The effects of herbivory on plant population dynamics are distinct from abiotic disturbances because herbivores are selective in their foraging regarding plant species, plant age, and plant part (Bryant et al. 1983). As such, herbivory may enhance certain demographic processes (e.g., vegetative shoot growth) and curtail others (e.g., seed set; Kielland et al. 1997). One major consequence of these differential effects on the forage species is to shift the age distribution of the twig populations on browsed plants toward younger age classes (Fig. 13.1), resulting in vegetation that is architecturally and chemically very different from unbrowsed vegetation.

In herbivore exclosures along the Tanana River, overstory species composition of early successional stands has undergone a substantial change after 15 years of mammal exclusion, as indicated by species-specific changes in leaf litter fall. During the first eight years of mammal exclusion, biomass of willow litter as a proportion of total litter fall increased more than twofold (Kielland et al 1997) because of a shift in the competitive balance between willow and alder in the absence of winter browsing of willows by moose and hares (Fig. 13.2). This signified a temporary reversal of succession. By contrast, the strong dominance of alder in the presence of browsing suggests that browsing significantly facilitates the successional transition from willow to alder thickets over a 15-year period. These findings demonstrate that

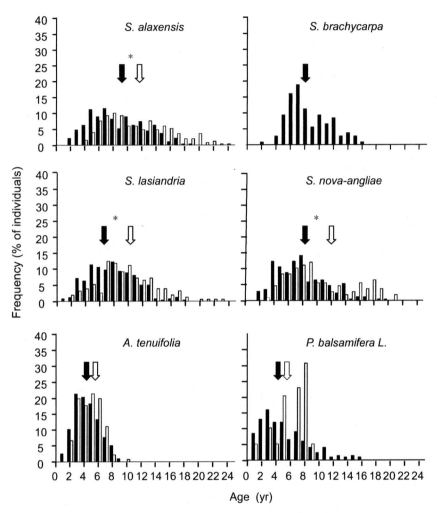

Figure 13.1. Stem age distribution of four willow species, alder, and poplar in willow communities along the Tanana River in areas of high (solid) and low (open) moose density. Arrows indicate the average age. Significant differences are indicated by asterisk. Data from Butler (2003).

herbivory has been as much a force in shaping these stands over the past six years as have "normal" plant recruitment and growth.

The substantial change in the ratio of alder:willow litterfall is closely tied to selective browsing by moose on willows (Bryant and Kuropat 1980, Risenhoover 1989, Wolff 1976) and demonstrates that animals can be important causative agents of successional change. The long-term effect of browsing appears to be the replacement of palatable deciduous species with long-lived unpalatable evergreens (Bryant and Chapin 1986, Pastor et al. 1988), which are characterized by slower carbon and nutrient turnover (Pastor et al. 1993). However, in the short-term (<30 years), mamma-

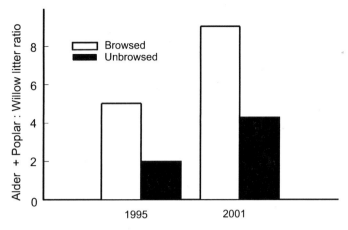

Figure 13.2. Ratio of alder-plus-poplar to willow litter after approximately 8 and 14 years of herbivore exclusion. Modified with permission from *Oikos* (Kielland and Bryant 1998) and Kielland (unpublished).

lian herbivory modifies plant carbon allocation above- and belowground, with marked consequences for element cycling. Winter browsing results in up to a twofold increase in production of current annual growth stems (Bryant 1987) but at the same time reduces fine-root production by 30% (Ruess et al. 1998). This reallocation of biomass production shifts carbon turnover increasingly aboveground, causing the 50% consumption to represent a smaller proportion of total plant biomass than it might appear (Fig. 13.3). Under sustained levels of heavy browsing, however, this production scenario breaks down. Browsing eventually removes sufficient buds and energy reserves that mycorrhizal colonization on roots declines (Rossow et al. 1997) and aboveground stems experience increased mortality (Butler 2003). In the nutrient-deficient soils of early-successional stands (Van Cleve and Viereck, 1981), this reduction of fine-root biomass and associated mycorrhizae adversely affects the competitive capacity of willows. Moreover, the shading of willows and the enhanced soil nutrient availability that accompany the establishment of alders in primary succession improves the nutritional quality of herbivore-favored willow species, such as feltleaf willow, making them more susceptible to additional heavy browsing (Bryant 1987). These factors, combined with the low shade tolerance of willows (Walker et al. 1986), provide a likely set of mechanisms by which browsing accelerates the replacement of willows by alder and underscores the importance of competitive (i.e., biotic) interactions early in succession (Bryant and Chapin 1986, Walker and Chapin 1986). This transition is particularly important in the development of these boreal forests because the alder stage of succession accrues approximately 60% of total ecosystem soil nitrogen (Chapter 15) and is associated with order-of-magnitude increases in productivity, litter fall, soil nutrient stocks, as well as significant changes in soil organic matter chemistry (Van Cleve et al. 1993, Viereck et al. 1993, Kielland et al. 1997). Herbivory exerts both quantitative and qualitative effects on forest ecosystems by augmenting and accelerating these changes (Fig. 13.4).

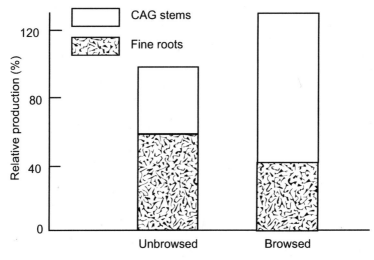

Figure 13.3. Relative production of current annual growth (CAG) stems and fine roots of early-successional vegetation in the absence and presence of mammalian herbivory. Relative production is root + CAG production relative to unbrowsed stands. Data from Bryant (1987) and Ruess et al. (1998).

Herbivore Interactions with Plant Chemistry

The interaction between herbivores and their food supply is complex because the effects differ depending on time scale. For example, winter browsing removes both growing points and depletes energy reserves that alter the chemical makeup of the leaves produced the following growing season. Generally, herbivory on a given species results in litter with higher nitrogen concentrations and higher rates of decomposition (Fig. 13.5), thus speeding up element cycling in both the terrestrial and the aquatic environments (Irons et al. 1991). Moreover, experiments with snowshoe hares have found increased leaf nitrogen concentration and higher digestibility of previously browsed feltleaf willow (Bryant unpublished data), which mirror the observations of moose browsing (Kielland and Bryant 1998). These findings suggest that browsing causes a short-term increase in litter decomposition and element cycling, within the successional stage that browsing occurs. On the Tanana River floodplain, browsing by moose and hares results in increased turnover potentials of both leaf litter and soil organic matter, due to increases in plant litter concentrations of mineralizable carbon and nitrogen following browsing (Kielland et al. 1997).

However, the chemical consequences of herbivory differ among plant parts. Whereas browsing improves the chemical quality of *foliage* for consumers, be they moose or microbes, browsing induces chemical defenses that render the woody tissue (i.e., *the bark*) less attractive to herbivores. Species such as alder, birch, balsam poplar, and quaking aspen respond to herbivory by producing adventitious shoots that have significantly higher concentrations of terpenes and phenolic resins that repel browsers like snowshoe hares (Bryant 1981). Thus,

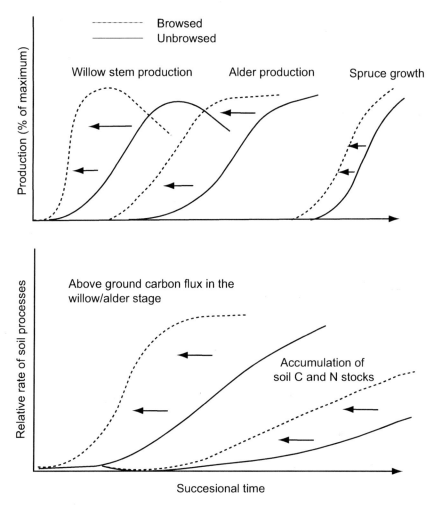

Figure 13.4. Generalized diagram of browsing effects on vegetation and soil processes during primary succession on the Tanana River floodplain. Arrows indicate the direction of changes in production caused by browsing. Browsing delays the successional changes in production by all plant species that would otherwise occur in the absence of browsing.

increases in nutrient cycling due to leaf chemistry are somewhat offset by opposing effects on bark chemistry. But, because the proportion of bark on current annual growth is relatively low compared to that of foliage, the negative feedback of plant bark chemistry probably has greater effect on browsing behavior than on nutrient cycling per se.

Over the long term, herbivory causes a significant reduction in nutrient cycling through changes in species composition (Pastor et al. 1993), especially when it involves a shift from deciduous to evergreen species. Litter of evergreen species has lower nitrogen and phosphorus concentrations but higher concentrations of

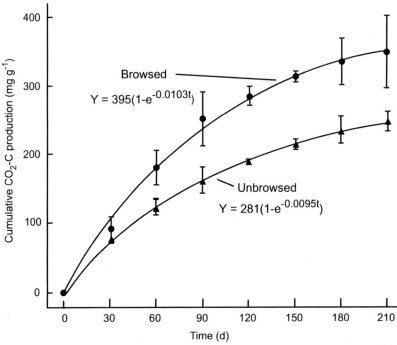

Figure 13.5. Effects of browsing on decomposition potentials (CO_2 release during decomposition in a laboratory incubation) of feltleaf willow litter (Mean ± S.E). Redrawn with permission from *Oikos* (Kielland et al. 1997).

recalcitrant/toxic compounds such as lignin, tannin, and resin, compared to deciduous species such as willows (Bryant and Kuropat 1980). These chemical characteristics cause evergreen litter to decompose more slowly than deciduous litter (Fox and Van Cleve 1983, Berg and Staaf 1987). Thus, high concentrations of indigestible substances and chemical defenses that deter herbivory (Bryant and Kuropat 1980) are correlated with intrinsic rates of litter decomposition (Chapter 14). Over successional time, mammalian herbivory accelerates succession to a dominance of conifers, resulting in decreasing rates of organic matter turnover and element cycling (Pastor et al. 1993).

Direct Consumption and Defecation Effects on Element Cycling

The landscape pattern of food consumption by herbivores depends on their population abundance and distribution. Thus, an assessment of herbivore effects on forests must entail an understanding of both population biology and nutritional ecology across temporal and spatial scales. The effects of these species-specific characteristics are further modified by migratory behavior and external drivers such as climate and predation.

The diel activity pattern of moose is characterized by alternating periods of feeding and resting/rumination (Risenhoover 1986). Moose typically consume upwards of 3% of their body weight daily during the summer and perhaps about half that much in the winter (Schwartz and Renecker 1998). Thus, during a typical winter day, about 4–5 kg of woody browse goes down the gullet of a moose, and at the end of the day approximately 2–3 kg of feces has come out the other end. Depending on moose densities and habitat characteristics, these feeding activities can greatly influence the carbon flux of aboveground vegetation early in succession as well as the movement of energy and nutrients through the boreal ecosystem in general. In high-quality habitats along parts of the Tanana and Koyukuk Rivers, moose may annually clip more than 90% of all current annual growth twigs of the dominant willow species (Kielland 1996, Seaton 2002, Butler 2003). This translates into a biomass removal of nearly 50% (Kielland and Osborne 1998, Seaton 2002). This level of biomass removal represents a significant acceleration of aboveground carbon turnover and a change in stand structure (see later discussion). Many of these plant species, such as willows, have the capacity to increase stem production following herbivory (compensatory growth; Bryant 1987), causing the standing crop (biomass accumulation) to fluctuate tremendously between the dormant and growing seasons (Fig. 13.6). The net result is increased herbivore-mediated carbon flux

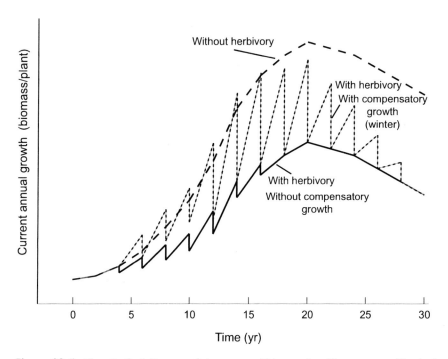

Figure 13.6. Hypothetical diagram of aboveground biomass in willow communities in the presence and absence of herbivory. Biomass removed by browsing is estimated at 50% yr^{-1} (Kielland unpublished, Seaton 2002), assuming regrowth after browsing (Bryant 1987).

from vegetation to the soil, demonstrating the multiplier effects that herbivores have on element cycling. If we assume that current annual stem production is about half of total net aboveground primary productivity (Ruess et al. 1996), moose through their feeding activities alone account for the turnover of 25% of this biomass!

In addition to accelerating carbon and nitrogen turnover through direct consumption, herbivory confers a whole suite of indirect effects on vegetation. For example, browsing by moose and snowshoe hares increases litter decomposition in the field by approximately 20% and increases nitrogen flow through the same pool by more than 40%, due to elevated N concentrations in plants subject to browsing compared to unbrowsed plants (Kielland et al. 1997). These multiplier effects again clearly demonstrate how small changes in one process translate into large effects on other processes (White 1983). The flux of materials through ecosystems is influenced by trophic interactions from decomposer organisms up to the top consumer (often humans). Though we generally think of photosynthesis and decomposition as the major conduits of energy and nutrient flow in ecosystems, higher trophic processes are equally important in defining the dynamics of the system. Thus, the consumption of 50% of the current annual willow growth by moose represents an important avenue of carbon flux in these northern ecosystems.

The combined effects of climate, social interactions, and food availability reduce the maximum lifespan of animal species to actual life expectancy of perhaps no more than 20–30% of their potential lifespans. For example, bear/wolf predation on moose and caribou in interior Alaska may amount to 20–25% annually (R. Boertje, pers. comm.). Moose can live up to 15 years, but these levels of predation result in a population turnover of four to five years under stable conditions. Thus, an entire wildlife population turns over in the span of an LTER grant renewal cycle. Life in the woods is tough, and the entrances and exits of the players occur rapidly (Chapter 8). These population fluxes drive nutrient cycling, not so much through their own processing of materials but as through their mediation of material flux in taiga forests.

Herbivores as Ecosystem Engineers

Mammalian herbivores in boreal forests affect a wide array of environmental components, including physical, chemical, and biological factors, as direct consequences of animal behavior and as indirect effects of interactions with other species.

Microclimate and Environmental Geometry

Exclosure studies reveal many significant herbivore effects on the physical structure of vegetation, as well as effects on plant species composition. After more than a decade of herbivore exclusion, canopy height of the dominant willow species was 12–50% less in stands inside exclosures than in stands exposed to browsing. In addition, large reductions in twig density resulted in a much more open canopy outside the exclosures. These canopy changes significantly altered the physical environment. For example, we found lower relative humidity, higher light intensity, lower soil moisture, and increased soil temperature outside the exclosures (Kielland and Bryant 1998). Thus, moose act as ecosystem engineers by altering the habitat and environmental conditions for other biota. These effects translate into changes in the distribution and abundance of other organisms, as described later.

Soil Biogeochemistry and Hydrology

Herbivore impacts on the chemistry of early successional surface soils illustrate well the many subtle and interactive processes that are set in motion by plant-herbivore interactions. Herbivory reduces total soil organic carbon largely by lowering fine-root production, but increases soil respiration (Kielland et al. 1997), in part due to input of more decomposable leaf litter (Fig. 13.5). The enrichment of soil $\delta^{13}C$ signature associated with herbivory is attributable primarily to changes in physical processes brought about by browsing-induced changes in vegetation structure. Browsing opens up the canopy, which results in higher temperatures and lower relative humidity (Kielland and Bryant 1998). These conditions enhance evaporation from the soil, which in turn increases capillary rise of water and thus deposition on the soil surface of calcite and gypsum salts (Dyrness and Van Cleve 1993; Chapter 7). Consequently, willow stands that are heavily browsed exhibit much higher levels of soil carbonate (calcite) as a proportion of total soil carbon (Fig. 13.7) than those that are not browsed. Because calcium carbonate is highly enriched in ^{13}C ($\delta^{13}C$ -4‰; Marion et al. 1991), compared to the total pool of soil organic carbon ($\approx$-20‰), these soils are correspondingly enriched in ^{13}C (Fig. 13.7). Thus, herbivory sets in motions a cascade of biological, physical, and chemical processes that ultimately produce the ecological attributes we observe. To our knowledge, this is the first example from a high-latitude ecosystem of how fundamental geochemical processes, in addition to vegetation changes, are modulated by the foraging behavior of a large mammalian herbivore.

Higher-Order Interactions

We have seen how browsing by moose and other mammalian herbivores changes a variety of ecological parameters, including soil chemistry, nutrient cycling, species composition, and the climate near the ground. These changes in the physical and biological environment have cascading effects on biodiversity and community structure of invertebrate guilds living on the forest floor. In general, browsing tends to increase the abundance of insects and the richness of higher taxa (Suominen et al. 1999). These changes are correlated with higher biomass of forbs, grasses, and mosses, as observed in other studies of boreal forests (McInnes et al. 1992, Suominen et al. 1999). Moreover, changes in leaf chemistry of browsed plants affect their interaction with other herbivores. For example, winter-browsed individuals of willow and poplar on the Tanana River floodplain support almost twice as many leaf-galling insects as do individuals of the same species protected from browsing (Roininen et al. 1997). Mammalian browsing results in the production of longer

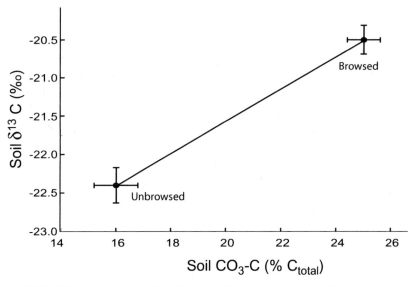

Figure 13.7. Changes in inorganic carbon chemistry of surface soils in the absence and presence of browsing. Data from Kielland and Bryant (1998).

shoots, which enhance the density of insects. Thus, the process of herbivory serves as a nexus for the interactions between the largest vertebrates and the smallest invertebrates in these ecosystems (Fig. 13.8).

The river floodplains of interior Alaska represent an important habitat for breeding passerine birds. Bird species diversity and density are greater in lowland habitats along the Tanana River than in similar upland habitats, due to higher primary productivity in the lowlands (Spindler and Kessel 1980). Out of 25 species of breeding birds that occupy territories on the Tanana River floodplain near the Bonanza Creek Experimental Forest, more than 95% of territories were held by migrant song birds. Species richness and the number of territories were positively correlated with vegetation productivity, both doubling across the gradient from black spruce forests to the willow communities in early succession (Johnson 1999). This observation is diametrically opposite those from studies in more temperate climates, where bird species diversity peaks in later succession (Johnson 1999).

The cascading effects of herbivory, as with most ecological phenomena, depend on both temporal and spatial scale, and the observed trajectory of one response may change radically over time. Plant-herbivore interactions are not just about animals eating vegetation and perhaps some response on part of the plants. Through a multiplicity of unexpected feedbacks, these interactions reshape the physical characteristics of habitats and modify the resource array and population ecology of sympatric species. These higher-order interactions, in turn, produce different expressions of ecosystem energy and nutrient flow, species composition, and productivity of both plants and animals.

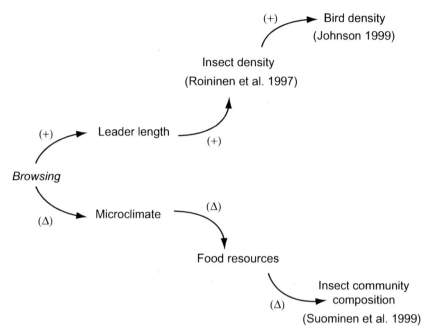

Figure 13.8. Higher-order interactions in community dynamics set in motion by mammal browsing on early-successional vegetation. These scenarios are based on data of Roininen et al. (1997), Souminen et al. (1999), and Johnson (1999).

Beyond Science: Services of Wildlife to Society

The value of wildlife can be viewed in terms of consumptive and nonconsumptive use. These values may not be mutually exclusive, but in Alaska, as in other parts of the United States, emotions often run high regarding competing interests of different users. Because human interactions with wildlife take on many facets—esthetic, recreational, economic, and spiritual—decisions that are targeted at one dimension of our relationship with wildlife invariably result in effects on other dimensions.

Alaska is a pluralistic society with mixed subsistence-cash economies existing side by side with a large industrial capital economy (Wolfe 1989). Many Alaskans, both urban and rural, engage in certain levels of mixed economies by combining commercial wage employment with harvest of wild resources, including fishing, trapping, and hunting.

Many aspects of nonconsumptive use of wildlife, particularly those associated with tourism, are closely tied to the state's economy. Tourism is the second-largest industry in the state, and the attraction to visitors from "Outside" is in large part a result of the wilderness values that surpass those found in any of the other 49 states. In view of the approximately $1 billion dollars spent annually by visitors (Alaska Department of Commerce and Economic Development 1999), wildlife is a commodity that has tremendous impact on the regional economy.

However, the consumptive use of wildlife is very important for many Alaskan residents. Rural Alaskans on average harvest more meat, fish, and waterfowl per capita annually than most Americans buy in grocery stores (Alaska Fish and Game 1989, p. 14). The amount of wild foods harvested per year varies considerably from urban centers like Anchorage and Fairbanks to rural communities like Minto and Gambell, whose annual per capita harvest is approximately five times greater than the annual purchase of fish, poultry, and red meat by the average household in the lower 48 states (Alaska Fish and Game 1989, p. 14). From an analytical perspective, the harvest of wild foods is a vital economic sector in rural Alaska. By way of comparison, the monetary value of moose meat derived from hunting in Alaska alone exceeds $12 million annually, which is more than the total stumpage value for timber harvested in the boreal region of the state (Wurtz and Zasada 2001).

In rural Alaska, fish and wildlife are important because they provide food and other goods that otherwise require substantial expenditure of cash. Moreover, the use of these resources in subsistence-related activities gives local residents more choices that result in greater stability, security, and control than any single economic sector could provide. Last, fishing, hunting, and trapping are activities that people value, independent of their economic dimension; these activities connect them to the natural world and give them identity and a sense of context of where they are and how they are living (Chapter 20; Wolfe 1989).

Conclusions

The ecological interactions set into motion by trophic exchanges discussed in this chapter demonstrate the potential array of both clear, direct effects and more subtle indirect effects that cascade through the ecosystem. Putting together the picture of interactions across microbial, plant, and megafaunal levels is akin to piecing together a jigsaw puzzle, except there is no guarantee that the pieces have fixed shapes or are of a definite number. Mammalian herbivores in the boreal forest interact with humans along many cultural dimensions; perhaps their most important influence is in shaping the ways we have come to perceive the natural world.

References

Alaska Department of Commerce and Development. 1999. Alaska Visitor Industry Economic Impact Study. Prepared by the McDowell Group, Juneau, AK.
Alaska Department of Fish and Game. 1989. Alaskan's per capita harvests of wild foods. Subsistence: Adapting ancient ways to modern times. Alaska Department of Fish and Game November–December:14–15.
Berg, B., and H. Staaf. 1987. Release of nutrients from decomposing white birch leaves and Scots pine needle litter. Pedobiologia 30:55–63.
Bryant, J. P. 1981. Phytochemical deterrence of snowshoe hare browsing by adventitious shoots of four Alaskan trees. Science 213:889–890.
Bryant, J. P. 1987. Feltleaf willow-snowshoe hare interactions: plant carbon/nutrient balance and floodplain succession. Ecology 68:1319–1327.

Bryant, J. P., and F. S. Chapin III. 1986. Browsing–woody plant interactions during boreal forest plant succession. Pages 213–225 *in* K. Van Cleve, F. S. Chapin III, L. A. Flanagan, L. Viereck, A., and C. T. Dyrness, editors. Forest Ecosystems in the Alaskan Taiga: A Synthesis of Structure and Function. Springer-Verlag, New York.

Bryant, J. P., and P. Kuropat. 1980. Selection of winter forage by subarctic browsing vertebrates: The role of plant chemistry. Annual Review of Ecology and Systematics 11:261–285.

Bryant, J. P., F. S. Chapin III, and D. Klein, R. 1983. Carbon/nutrient balance of boreal plants in relation to vertebrate herbivory. Oikos 40:357–368.

Butler, L. G. 2003. The role of mammalian herbivores in primary succession on the Tanana river floodplain, interior Alaska. M.S. Thesis, University of Alaska Fairbanks.

Dyrness, C. T., and K. Van Cleve. 1993. Control of surface soil chemistry in early-successional floodplain soils along the Tanana River. Canadian Journal of Forest Research 23:979–994.

Fox, J. R., and K. Van Cleve. 1983. Relationships between cellulose decomposition, Jenny's k, forest-floor nitrogen, and soil temperature in Alaskan taiga forests. Canadian Journal of Forest Research 13:789–794.

Franzmann, A. W., P. Arneson, and J. Oldemeyer. 1976. Daily winter pellet groups and beds of Alaskan moose. Journal of Wildlife Management 40:374–375.

Harder, L. D. 1980. Winter use of montane forests by porcupines (*Erethizon dorsatum*) in southwestern Alberta, Canada: Preferences, density effects, and temporal changes. Canadian Journal of Zoology 58:13–19.

Irons, J. G., III, J. P. Bryant, and M. Oswood, W. 1991. Effects of moose browsing on decomposition rates of birch leaf litter in a subarctic stream. Canadian Journal of Fisheries and Aquatic Sciences 48:442–444.

Johnson, A. K. 1999. Breeding bird density and species diversity in relation to primary productivity in the Tanana River floodplain. M.S. Thesis, University of Alaska Fairbanks.

Jones, C. G., J. H. Lawton, and M. Shackak. 1994. Organisms as ecosystem engineers. Oikos 69:373–386.

Kielland, K. 1996. Browse relations of moose along the middle Koyukuk River, Alaska. Alaska Department of Fish and Game, Division of Wildlife Conservation, Fairbanks.

Kielland, K., and J. P. Bryant. 1998. Moose herbivory in taiga: Effects on biogeochemistry and vegetation dynamics in primary succession. Oikos 82:377–383.

Kielland, K., and T. Osborne. 1998. Moose browsing on feltleaf willow: Optimal foraging in relation to plant morphology and chemistry. Alces 34:149–155.

Kielland, K., J. P. Bryant, and R. W. Ruess. 1997. Moose herbivory and carbon turnover of early successional stands in interior Alaska. Oikos 80:25–30.

Marion, G. M., D. S. Introne, and K. Van Cleve. 1991. The stable isotope chemistry of $CaCO_3$ on the Tanana River floodplain of interior Alaska: Composition and mechanisms of formation. Chemical Geology 86:97–110.

McInnes, P., R. Naiman, J. Pastor, and Y. Cohen. 1992. Effects of moose browsing on vegetation and litterfall of the boreal forests of Isle Royale, Michigan, USA. Ecology 73:2059–2075.

Pastor, J., R. Naiman, B. Dewey, and P. McInnes. 1988. Moose, microbes, and the boreal forest. Bioscience 38:770–777.

Pastor, J., B. Dewey, R. Naiman, J. P. McInnes, and Y. Cohen. 1993. Moose browsing and soil fertility in the boreal forests of Isle Royale National Park. Ecology 74:467–480.

Risenhoover, K. L. 1986. Winter activity patterns of moose in interior Alaska. Journal of Wildlife Management 50:727–734.

Risenhoover, K. L. 1989. Composition and quality of moose winter diets in interior Alaska. Journal of Wildlife Management 53:568–577.

Roininen, H., P. W. Price, and J. P. Bryant. 1997. Response of galling insects to natural browsing by mammals in Alaska. Oikos 80:481–486.

Rossow, L., J. P. Bryant, and K. Kielland. 1997. Effects of above-ground browsing by mammals on mycorrhizal infection in an early successional taiga ecosystem. Oecologia 110:94–98.

Ruess, R. W., R. L. Hendrick, and J. P. Bryant. 1998. Regulation of fine root dynamics by mammalian browsers in early successional Alaskan taiga forests. Ecology 79:2706–2720.

Ruess, R., K. Van Cleve, J. Yarie, and L. A. Viereck. 1996. Contributions of fine root production and turnover to the carbon and nitrogen cycling in taiga forests of the Alaskan interior. Canadian Journal of Forest Research 26:1326–1336.

Sæther, B.-E. 1992. The final report from the project "Moose-forest-society." NINA Forskningsrapport 28:1–153.

Schwartz, C. C., and L. A. Renecker. 1998. Nutrition and energetics. Pages 441–478 in A. W. Franzmann and C. C. Schwartz, editors. Ecology and Management of North American Moose. Smithsonian Institution Press, Washington, DC.

Seaton, C. T. 2002. Winter foraging ecology of moose in the Tanana Flats and Alaska Range foothills. M.S. Thesis, University of Alaska Fairbanks.

Spindler, M. A., and B. Kessel. 1980. Avian populations and habitat in interior Alaska taiga. Syesis 13:61–104.

Steele, K. W., and R. M. Daniel. 1978. Fractionation of nitrogen isotopes by animals: a further complication to the use of variation in the natural abundance of ^{15}N for tracer studies. Journal of Agricultural Science (Cambridge) 90:7–9.

Suominen, O., K. Danell, and J. P. Bryant. 1999. Indirect effects of mammalian browsers on vegetation and ground-dwelling insects on an Alaskan floodplain. Ecoscience 6:505–510.

Van Cleve, K., and L. Viereck. 1981. Forest succession in relation to nutrient cycling in the boreal forest of Alaska. Pages 185–210 in D. C. West, H. H. Shugart, and D. B. Botkin, editors. Forest Succession, Concepts and Application. Springer-Verlag, New York.

Van Cleve, K., L. Viereck, and G. M. Marion. 1993. Introduction and overview of a study dealing with the role of salt-affected soils in primary succession on the Tanana River floodplain, interior Alaska. Canadian Journal of Forest Research 23:879–888.

Viereck, L. A., C. T. Dyrness, and M. J. Foote. 1993. An overview of the vegetation and soils of the floodplain ecosystems of the Tanana River, interior Alaska. Canadian Journal of Forest Research 23:889–898.

Walker, L. R., and F. S. Chapin III. 1986. Physiological controls over seedling growth in primary succession on an Alaskan floodplain. Ecology 67:1508–1523.

Walker, L. R., J. C. Zasada, and F. S. Chapin III. 1986. The role of life history processes in primary succession on an Alaskan floodplain. Ecology 67:1243–1253.

White, R. G. 1983. Foraging patterns and their multiplier effects on productivity of northern ungulates. Oikos 40:377–384.

Wolfe, R. 1989. Myths: what have you heard. Alaska Fish and Game November-December:16–19.

Wolff, J. O. 1976. Utilization of hardwood browse by moose on the Tanana floodplain of interior Alaska. Research Note PNW–267. USDA Forest Service, PNW Forest and Range Experiment Station, Portland, OR.

Wurtz, T. L., and J. C. Zasada. 2001. Boreal blending: Timber and moose in Alaska's interior. Pages 1–5 in S. Duncan, editor. PNW Science Findings. USDA Forest Service, PNW Forest and Range Experiment Station, Portland, OR.

14

Microbial Processes in the Alaskan Boreal Forest

Joshua P. Schimel
F. Stuart Chapin III

Introduction

Forest ecosystems typically occur in moderate environments where growing season rainfall is adequate to support tree growth and where nongrowing season conditions are not too extreme. The Alaskan boreal forests, however, occur at the limit of the forest biome, in an environment that is climatically extreme, with strong physical gradients. The seasonal variation in temperature is among the greatest on earth, with winter temperatures as low as $-50°C$ and summer growing season temperatures that can reach $+30°C$ (Chapter 4). The growing season is short, the climate is semi-arid, and growing season rainfall is limited. Forests exist in the region because evapotranspiration is also limited. Steep south-facing slopes can be too dry to support tree growth (Chapter 6). In contrast, in flat, low-lying areas, low evapotranspiration combined with permafrost produces wetlands despite the low rainfall.

Regular drought makes the forest highly susceptible to fires. At large scales (many square kilometers), the boreal forest experiences regular, extensive fires that destroy whole stands, resetting succession (Chapter 17). This regular fire cycle produces a patchwork mosaic of forest stands in different successional stages across the landscape (Dyrness et al. 1986, Kasischke and Stocks 2000; Chapter 7). In large rivers (e.g., the Tanana), the cutting and filling of meander loops washes away some forest stands while depositing new silt bars for colonization and succession (Zasada 1986). At the landscape scale, the biogeochemical cycles in the boreal forest are therefore dominated by landscape structure (e.g., dry uplands vs. wet lowlands) and by disturbance (particularly fire). At smaller scales, however, the strong feedbacks between plant and soil processes control much of the functioning of individual forest stands, and possibly the rate of transition among successional stages.

In this chapter, we discuss how microbial processes in the boreal forest produce unusual patterns of nutrient cycling that drive the overall functioning of boreal forest stands. Figure 14.1 illustrates the linkages between plant and microbial communities that dominate the functioning of the boreal forest soil system.

In the feedbacks between plant and soil processes, plants drive the loop largely through inputs of organic materials. Plant leaves, roots, and dead wood become litter that supplies the bulk of organic matter to soils and build the O (organic) horizon, which has important roles in soil insulation, hydrology, and nutrient cycling. Variation in the quality of the litter inputs regulates the rate of decomposition and nutrient recycling (Flanagan 1986, Bonan et al. 1990, Hobbie et al. 2000). However, plants also supply dissolved chemicals to the soil either as root or as foliar exudates or leachates. These chemicals are important C sources in their own right, but they may also act to modify the processing of litter and soil organic matter (SOM), either by binding with other compounds or through direct effects on soil microorganisms.

Microbes, in their turn, drive the feedback loop by producing the extracellular enzymes that break litter polymers down into small, soluble organics that can be taken up and metabolized, by acting as a sink for dissolved organic carbon (DOC), and by releasing and taking up nutrients. The specific composition of the microbial community is a function of the chemical composition of plant litter (Fierer et al.

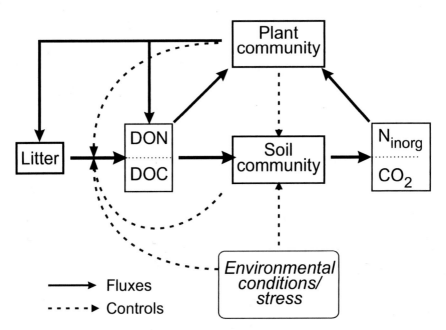

Figure 14.1. Linkages between plant and soil microbial communities in soil C and N cycling, and the environmental controls over those processes. Plants and microbes act as pools of material within the C and N cycles, while the specific composition of their communities acts as a control on the fluxes within biogeochemical cycles.

2001), so there are complex linkages among plants and microbes at the levels of both community composition and biogeochemical cycling.

Succession

The nature of plant-soil feedbacks changes through the course of succession. Change in the composition of the plant communities alters the litter deposited on and in the soil. There is a strong successional trend toward increasing recalcitrance from early-stage shrubs (willow, alder, etc.), through midsuccession deciduous trees (birch, aspen, balsam poplar), to late-successional white spruce and, in some cases, to black spruce. This trend contributes to an accumulation of the forest floor and to a reduction in the rate of N mineralization and overall nutrient cycling. In this, the boreal forest differs from some other forest ecosystems such as the eastern deciduous forest, where N availability and nutrient cycling rates sometimes increase through succession (Vitousek and Melillo 1979, Robertson and Vitousek 1981). Although overall patterns of change in soil processes in the boreal forest resemble patterns found in other ecosystems, there are many ways in which the specific dynamics of the boreal forest appear quite different from those in most other forest ecosystems.

Temperature

The boreal forest is the coldest and least productive of all forested biomes (Viereck et al. 1986). Within interior Alaska, both productivity and nutrient cycling correlate closely with temperature (Van Cleve et al. 1983), suggesting a key role of temperature in regulating microbial processes. However, temperature exerts a variety of indirect effects (Fig. 14.2) that have an even stronger effect on microbes than does temperature per se. The most important of these are the presence of permafrost, which governs drainage and soil aeration, and plant community composition, which influences the quantity and quality of litter inputs to soil. These indirect effects are addressed in subsequent sections.

When other effects are controlled, microbial respiration increases exponentially with temperature to temperatures well above those that occur in the field (Flanagan and Veum 1974). In the field, however, soil respiration shows a discontinuity in the temperature response at 17°C (Gulledge and Schimel 2000). Below this temperature, soil respiration shows an exponential response to temperature, while, above 17°C, respiration rates are lower and unresponsive to temperature. This discontinuous response appears to result from an interaction with soil moisture. Rainwater is cool, and evaporation acts to keep soils cool until they have dried out. As a result of this, for soils to exceed 17°C, they must first dry enough for root and microbial respiration to become water-limited. Interestingly, given the cold climate, the boreal forest soil microbial community does not appear to be dominated by psychrophiles, that is, microbes that function optimally at low temperature (Flanagan and Veum 1974).

When direct and indirect effects are considered together, temperature has a profound effect on microbial processes. The largest soil carbon (C) accumulations occur

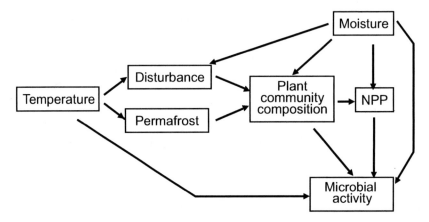

Figure 14.2. Direct and indirect effects of temperature and moisture on soil microbial processes in the boreal forest. Temperature effects on plant community composition and NPP are omitted for clarity and because they likely have more limited effects on microbial processes than the linkages diagrammed.

in the coldest sites, even though these sites have the lowest productivity and litter inputs to soils (Van Cleve et al. 1983; Chapter 15). When these soils are incubated in the laboratory under favorable temperature and moisture conditions, they maintain high respiration rates for at least a year (Weintraub and Schimel 2003), indicating that their slow decomposition in the field reflects "environmental protection," rather than low organic matter quality.

Water

Water affects soil biogeochemical processes in at least two fundamental ways. The first is at the landscape scale, in which flat, low-lying areas are generally dominated by permafrost and often become saturated. In these areas, inundation produces anaerobic conditions. Thus, the state of the water table can act as an on/off switch between aerobic and anaerobic processes. This ecotone from permafrost and saturated meadow soils to drier upland soils can take place over mere meters in some places, such as where the Tanana floodplain runs into the hills of the Tanana uplands. The secondary way that water affects soil processes is through drought stress in the upland forests. Because of low summer rainfall, soil drought is common in upland forests with southern exposure (Chapter 11).

Drought stress is important in regulating both the fragmentation and the mineralization components of boreal forest litter breakdown and decomposition (Wagener and Schimel 1998, Schimel et al. 1999). Fragmentation is carried out primarily by soil fauna, whereas metabolism is predominantly a microbial process, with fungi typically dominating in aerobic forest soils (Wagener and Schimel 1998, Schimel et al. 1999). Moisture and moisture stress play important roles in mediating both processes and the identity of organisms that carry them out in the boreal forest.

Moisture Effects on Forest Floor Structure and Soil Faunal Communities

One way in which moisture regulates decomposition is by controlling the composition and activity of the soil invertebrate community within the forest floor (Wagener 1995). Although the soil fauna are capable of shredding leaf litter and so producing a well-mixed forest floor, their activity is limited to areas of adequate moisture. This produces a moisture effect on decomposition that varies through succession. In early succession, litter decomposes quickly (Van Cleve et al. 1993, Clein and Schimel 1995), producing a relatively thin layer that lacks a distinct structure. Late succession is dominated by needle-leaved spruce, which doesn't form complete litter layers, because the needles intermix with the mosses that develop late in succession. The litter of the broad-leaf deciduous species (birch/aspen in uplands, balsam poplar on the floodplain) that dominate the midsuccessional stages, however, produces distinct layers. Given adequate moisture, litter fauna (chironomid and mycetophilid larvae) are capable of fully shredding this litter and potentially accelerating decomposition (Wagener 1995), but surface litter is typically too dry to support their activity. Thus, litter accumulates in discrete and identifiable layers for roughly three years (Fig. 14.3; Wagener and Schimel 1998). Only deeper in the forest floor profile does material stay moist enough to support the activity of shredders and the discrete layering give way to a more fine textured, mixed organic horizon. Thus, the moisture effect on soil fauna produces zones of distinctly different microbial activity patterns through the forest floor profile (Wagener and Schimel 1998). In the surface intact-leaf layers, litter remains C-rich, and N immobilization is the dominant microbial N-cycling process. It is only in the Oe horizon that microbial activity switches and causes substantial net mineralization (Fig. 14.3; Wagener and Schimel 1998). This is also a layer of intense rooting, whereas roots are largely absent from the shallower litter zones. Thus, moisture interacts with litter chemistry to control the patterns of both faunal and microbial communities and processes within the organic horizon in midsuccessional boreal forest stands.

Moisture Effects on Microbial Communities and Processes

In addition to affecting the activities of the soil fauna that alter or create the fine structure of the forest floor, moisture also has strong effects on the activity of the microbes that carry out decomposition. These effects can involve relatively long time lags and complex interactions through moisture-induced changes in the composition of the microbial community. The fundamental response of microbial activity to moisture in boreal forest organic soils appears to be asymptotic (Gulledge and Schimel 2000). That is, respiration is strongly responsive to increases in moisture at low moisture contents but reaches a maximal value at a moisture level somewhat above the average soil moisture content. In mineral soils, there is a strong

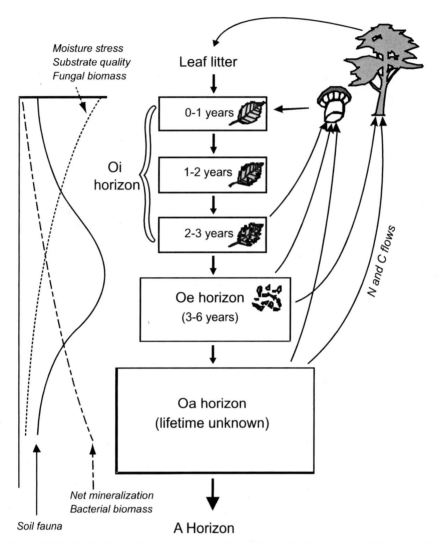

Figure 14.3. Vertical transfer processes in the forest floor of midsuccessional forest stands in the Alaskan boreal forest. Root uptake and fungal translocation move material up the forest floor. Interactive gradients of resource quantity/quality and environmental stress regulate the biotic processes that occur within each layer.

threshold for microbial respiration at roughly –0.1 to –0.2 MPa, while more organic-rich lowland meadow soils show a much flatter response, retaining significant activity below –2 MPa (Gulledge and Schimel 1998). The moisture response appears to interact strongly with temperature to produce the discontinuity in the temperature response mentioned earlier (Gulledge and Schimel 2000).

An additional factor that appears important in regulating microbial activity in the litter layer is the moisture history of the microbial community. Almost all biogeochemical models drive microbial activity in accordance with the instantaneous soil moisture; in other words, there is assumed to be no effect of moisture history on microbial activity or communities. Yet, increasingly, research suggests that microbial community composition responds to soil moisture and to changes in soil moisture (Schimel et al. 1999, Denef et al. 2001, Fierer and Schimel 2002, Fierer et al. 2003). Rapid changes, particularly increases, in soil moisture can impose a substantial stress on soil microorganisms because they must remain in water-potential equilibrium with their environment. As soils dry, microbes accumulate solutes within their cells to maintain a low intracellular water potential. Bacteria typically use amino compounds such as proline and glycine betaine (a quaternary amine; Harris 1981, Boncompagni et al. 1999), while fungi more typically use polyols as osmotic agents. When a dry soil is moistened, the microbes must rapidly release cellular contents to increase their intracellular water potential (Halverson et al. 2000, Fierer and Schimel 2003) or water will flood into the cells, potentially causing them to rupture (Harris 1981, Kieft et al. 1987). If microbes are killed either by intense drought or by the stress of being rewet, then community composition will be altered, and microbes need to either regrow, if any survive, or recolonize materials to be able to process them. Recolonization rates may therefore be an important factor in controlling longer-term responses to major stresses (Schimel and Clein 1996).

In the boreal forest, communities of litter decomposers are sensitive to drought and rewetting stresses. In the lab, even a mild, one-day drying and rewetting treatment can cause a long-term (two-month) reduction in microbial respiration sufficient to reduce total litter decomposition by as much as 25% over two months (Clein and Schimel 1994). Moisture effects were tested in a field study in which moisture was controlled through rain-out shelters and watering (Schimel et al. 1999); four treatments were set up: watered daily, watered weekly, constantly dry, and natural conditions (which were roughly biweekly rains). Birch litter samples were collected before and after rewetting events over one month, and microbial respiration was measured over 10-day periods in the lab to assess the ability of the community to recover and respire. Litter samples that were rewet daily or weekly showed very similar relationships between moisture in the sample, respiration, and microbial biomass. Samples in the "natural conditions" treatment that experienced longer droughts, on the other hand, showed significantly reduced respiration and microbial biomass, even after sample moisture was accounted for (Fig. 14.4).

At the low moisture levels common for surface litter, the samples from the natural moisture regime had about half the respiration of rate of samples that were watered more regularly. Respiration was reduced by drying more strongly than was microbial biomass, possibly as a result of the death of the most active members of the microbial community. These drought-sensitive organisms appeared to include both bacteria and some members of the fungal community that dominates the litter layer (Schimel et al. 1999).

All these pieces of work suggest that moisture limitation and variability have important roles in controlling the composition of the soil communities that are active in litter decomposition and nutrient recycling. The effects of water stress and water potential shock in the boreal forest may be more like those that occur in other

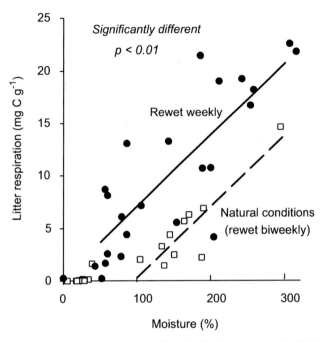

Figure 14.4. Effect of moisture history on soil respiration in decomposing birch litter. Litter was decomposed in the field under varying moisture conditions. It was periodically collected and brought back to the lab, where respiration rates were measured over 10 days at the moisture the litter was collected at. Data from Schimel et al. (1999). Field samples were rewet weekly (solid circles) or biweekly (open squares), a more natural wetting cycle.

semi-arid and arid ecosystems than like those found in the more common mesic conditions of "normal" forests.

Microbial Processes in Wetland and Permafrost Soils

The bulk of the work on microbial processes that drive overall ecosystem function in the Alaskan boreal forest to date has focused on upland soils. Work in wetland systems has focused on CH_4 fluxes and cycling (Funk et al. 1994, Moosavi et al. 1996). Boreal wetlands are a major player in the global CH_4 budget (Matthews and Fung 1987). However, relatively little has been done on overall organic matter and nutrient cycling, or on plant-soil feedbacks in these areas, and black spruce muskeg has been perhaps the most neglected part of the boreal forest ecosystem in terms of microbial processes.

Litter Quantity and Quality

Microbial activity responds not only to the physical environment (moisture and temperature) but also to the quantity and quality of plant litter that serves as the substrate for microbial growth. In general, soil respiration parallels NPP when boreal forest sites are compared. There is higher soil respiration in midsuccessional sites where NPP is higher than in the very-early- or late-successional sites (Viereck et al. 1986, Gulledge and Schimel 2000). Soil respiration is also greater in productive floodplain and on south-facing slopes than in unproductive black spruce sites (Vance and Chapin 2001). Roots account for a larger proportion of total respiration in black spruce than in more productive sites (Ruess et al. 1996; Chapter 12), so microbial activity must be responsible for this high soil respiration in productive sites. In other words, microbial activity is greater where microbes receive the most litter inputs to grow on.

Several chemical properties govern the quality of litter as a resource for soil invertebrates and microbes. These include the size and complexity of molecules, the strength of chemical bonds, and the nutrient content of the material. Two indices of litter quality (C:N ratio and lignin:N ratio) show clear patterns of variation across sites, with the highest litter quality (low C:N and lignin:N) in the most productive sites. Thus, the high microbial activity in productive sites reflects both the high quality and the large quantity of litter inputs.

The high microbial activity in productive sites explains why the annual soil C accumulation rates (15–100 g C $m^{-2}yr^{-1}$) are only a small proportion of annual inputs when both above and belowground inputs are accounted for (100–400 C $m^{-2}yr^{-1}$; Viereck et al. 1986; Chapter 12). Site variation in microbial respiration is due more to differences in microbial activity than to differences in microbial biomass, which is a relatively constant 1–2% of soil C in most sites (Vance and Chapin 2001). The unproductive black spruce sites have the largest microbial biomass, despite their low productivity and soil respiration.

An additional aspect of litter chemistry that appears to play a role in boreal forest decomposition and nutrient cycling processes is the presence of high concentrations of secondary chemicals (particularly tannins) in the litter of balsam poplar (and possibly other species, as well). The effects of such chemicals on soil processes are important in at least some boreal forest communities, producing complex chemical interactions between plants and soil communities. Thus, as succession proceeds, these feedbacks between plants and soil organisms may lead to a codevelopment of linked plant and microbial communities.

The role of plant secondary chemicals in regulating soil processes is best understood in primary floodplain succession. Although new river bars are initially colonized by willow, alder rapidly becomes dominant, in part because of moose and hare herbivory on willow (Chapters 7 and 13). The alder stage of succession is critical because alder is an N-fixer, and much of the ecosystem N is accumulated during this stage, where estimates of fixation rates range from 30–60 kg $ha^{-1}yr^{-1}$ (Uliassi and Ruess 2002). During this stage, N inputs are rapid, plant N concentrations are high, resorption is limited enough (Uliassi and Ruess 2002) that litter falls green, and decomposition is extremely rapid, causing very high N cycling rates, with extensive nitrification (Clein and Schimel 1995). Alder is replaced by balsam poplar (*Populus balsamifera*), which produces low-N litter that is high in polyphenolics and that decomposes quite slowly. In this successional stage, nitrogen (N) cycling is substantially slower, and N availability is low (Clein and Schimel 1995). The question is, then, what causes the dramatic shift in the nature of N cycling. Poplar produces large quantities of simple phenolics and a range of different condensed tannins. These secondary chemicals have complex interactions with soil nutrient cycling at several stages of the flows of material and controls diagrammed in Fig. 14.1. The simple phenolics and other soluble compounds act simply as available DOC. They are substrates for microbial growth, stimulating microbial growth and N-immobilization (Clein and Schimel 1995, Schimel et al. 1996).

The behavior of condensed tannins, on the other hand, is variable with the specific chemical structure of the tannin molecules and with the microbial community. Balsam poplar condensed tannins can be separated into two classes in terms of their effects on soil processes: low-molecular-weight tannins (up through tetramers with molecular weight 1000) and larger tannins (MW>1000; Fierer et al. 2001). The heavy-fraction balsam poplar tannins appear to bind to N-containing substrates and thereby reduce microbial respiration and net N mineralization (Schimel et al. 1996, Fierer et al. 2001). There is also some evidence that they may slightly reduce the activity of extracellular enzymes (Fierer et al. 2001), although this effect is likely less important than binding substrates. The effect of these compounds appears to be similar in soils from both alder- and poplar-dominated stands, as would be expected for chemicals whose effect is extracellular.

The lighter tannins and other simple polyphenolics have more complex effects on microbial activity and appear to act on microbes through uptake and direct physiological effects. First, in poplar litter and in deeper forest floor layers in both alder and poplar stands, these compounds act as C-rich microbial substrates (Fierer et al. 2001) and function like simple phenolics, stimulating microbial growth and N immobilization. The effects of labile compounds appears to be short-lived, however, because when the biomass is starved of the C inputs, N is rapidly released again (Clein and Schimel 1995). Thus, continuous inputs are needed to maintain the immobilization pressure.

The low-molecular-weight tannins, however, are toxic to microbes in alder litter (Fierer et al. 2001). They reduce respiration and N mineralization and inhibit decomposition. Thus, it seems likely that these compounds are involved in causing soil microbial community succession that is associated with the plant succession from alder to poplar domination. The full effects of these microbial community shifts are still, however, unclear.

Finally, poplar tannins are capable of inhibiting N fixation in alder nodules (Schimel et al. 1998). It seems likely that the tannins that are responsible for any inhibition are the low-molecular-weight fraction, which can be taken up by microbes. However, these compounds are rapidly used as substrates by microbes in developed poplar stands and even in deeper alder forest floor layers. Thus, it is possible, if not actually likely, that they are consumed before they can be absorbed by alder nodules and act as inhibitors. This may explain why there does not appear to be any inhibition of N fixation in alder nodules in established poplar stands (Uliassi and Ruess 2002).

The overall effect of the poplar secondary chemicals is to reduce N availability in soil both by reducing decomposition and mineralization and by stimulating immobilization in the soil. The combined effects of these processes, coupled with declines in litter quality through succession, cause a decline in nutrient availability that may increase the competitiveness of spruce relative to poplar and alder and so facilitate plant succession.

Most woody plants in the boreal forest produce tannins that could influence the activity of microbial communities, although we know much less about these effects than in the case of poplar tannins. Nonetheless, most low-molecular-weight tannins likely act as substrates for organisms that are adapted to them and as toxins for those that are not (Fierer et al. 2001). The general net effect of condensed tannins in the boreal forest and elsewhere appears to be a broad-spectrum effect on microbial activity and net N mineralization, reducing NH_4^+ supply, rather than a specific inhibition of nitrification (Schimel et al. 1996, Fierer et al. 2001, Kraus et al. 2004). For example, leaching of these tannins from leaf litter of several boreal species increases aquatic invertebrate use of the litter (Irons et al. 1991). Thus, we suspect that tannin effects, whether inhibitory or stimulatory to microbial activity and N mineralization, may be common in forest ecosystems.

Turnover of DOC/DON

In most ecological studies, the linkage between depolymerization and use of the released DOC and DON has been treated as a black box; decomposition acts on polymeric plant litter and releases CO_2 and mineral nutrients (e.g., Aber and Melillo 2001). Recent studies on terrestrial-aquatic linkages have paid more attention to the generation and consumption of DOC and DON (Aitkenhead and McDowell 2000, Stieglitz et al. 2003). Studies in several ecosystems, including the boreal forest, have begun to pay more attention to direct plant uptake of amino acids (Chapin et al. 1993, Näsholm et al. 1998, Jones and Kielland 2002). Because plants use amino acids and other organic compounds as N sources (Chapin et al. 1993, Schimel and Chapin 1996, Näsholm et al. 1998; Chapter 15) in boreal and other low-nutrient environments, it becomes critical to understand the turnover of DOC and DON in boreal forest soils. Weintraub and Schimel (2003) found that, in most tundra soils, over a year-long incubation, DOC concentrations and the relationship between DOC concentration and respiration remained relatively constant, even though the total amount of respired C was much greater than the original amount of DOC. Thus, DOC was continuously regenerated by exoenzyme and microbial processes, and, although DOC concentrations were the proximal control on respiration rates, the rate-limiting step for overall microbial respiration was the production of DOC. We still know disturbingly little about the rates and controls on DOC generation. Amino acid production and consumption are extremely fast, with turnover times of hours (Jones and Kielland 2002; Chapter 15), but this has not been examined for other chemical forms that may be important to microbial nutrition and possibly to plant nutrition, as well.

The generation of DOC/DON is dominated by extracellular enzymes, while its consumption is dominated by intracellular microbial processes. Although the microbes produce the enzymes, the lifetimes and recovery from stress of the microbes and their enzymes may be different. Thus, in an environment where seasonality is very strong and episodic stress to the microbial communities appears to be common, there is the potential for temporal disconnections to occur in the balance between DOC generation and consumption (Schimel and Weintraub 2003). Developing a better understanding of the role of these compounds, and the microbial role in producing and consuming them, therefore becomes an important area for future research in the boreal forest.

Conclusions

In conclusion, although the broad patterns of biogeochemical processing in the boreal forest are reasonably well documented and understood (Chapter 15), the microbial roles in these processes are not. There are many aspects of boreal forest biogeochemistry where we have shown that the details of microbial dynamics may have substantial large-scale impacts. However, this is only a beginning in understanding the full importance of specific microbial dynamics in controlling the larger functioning of the boreal forest ecosystem.

References

Aber, J., and J. Melillo. 2001. Terrestrial Ecosystems. 2nd ed. Harcourt Academic Press, San Diego.
Aitkenhead, J., and W. McDowell. 2000. Soil C:N ratio as a predictor of annual riverine DOC flux at local and global scales. Global Biogeochemical Cycles 14:127–138.
Bonan, G. B., H. H. Shugart, and D. L. Urban. 1990. The sensitivity of some high-latitude boreal forests to climatic parameters. Climatic Change 16:9–29.
Boncompagni, E., M. Østerås, M.-C. Poggi, and D. Le Rudulier. 1999. Occurrence of choline and glycine betaine uptake and metabolism in the family Rhizobacteriaceae and their roles in osmoregulation. Applied Environmental Microbiology 65:2072–2077.
Chapin, F. S., III, L. Moilanen, and K. Kielland. 1993. Preferential use of organic nitrogen for growth by a non-mycorrhizal arctic sedge. Nature 361:150–153.
Clein, J. S., and J. P. Schimel. 1994. Reduction in microbial activity in birch litter due to drying and rewetting events. Soil Biology and Biochemistry 26:403–406.
Clein, J. S., and J. P. Schimel. 1995. Nitrogen turnover and availability during succession from alder to poplar in Alaskan taiga forests. Soil Biology and Biochemistry 27:743–752.
Denef, K., J. Six, H. Bossuyt, S. Frey, E. Elliott, R. Merckx, and K. Paustian. 2001. Influence of dry–wet cycles on the interrelationship between aggregate, particulate organic matter, and microbial community dynamics. Soil Biology and Biochemistry 33:1599–1611.
Dyrness, C., L. Viereck, and K. Van Cleve. 1986. Fire in taiga communities of interior Alaska. Pages 22–43 in K. Van Cleve, F. S. Chapin III, P. W. Flanagan, L. A. Viereck, and C. T. Dyrness, editors. Forest Ecosystems in the Alaskan Taiga. Springer-Verlag, New York.
Fierer, N., and J. P. Schimel. 2002. Effects of drying-rewetting frequency on soil carbon and nitrogen transformations. Soil Biology and Biochemistry 34:777–787.
Fierer, N., and J. P. Schimel. 2003. A proposed mechanism for the pulse in carbon dioxide production commonly observed following the rapid rewetting of a dry soil. Soil Science Society of America Journal 67:798–805.

Fierer, N., J. P. Schimel, and P. A. Holden. 2003. Influence of drying-rewetting frequency on soil bacterial community structure. Microbial Ecology 45:63–71.

Fierer, N., J. Schimel, R. Cates, and J. Zou. 2001. The influence of balsam poplar tannin fractions on carbon and nitrogen dynamics in Alaskan taiga floodplain soils. Soil Biology and Biochemistry 33:1827–1839.

Flanagan, P. W. 1986. Substrate quality influences on microbial activity and mineral availability in Taiga forest floors. Pages 138–151 in K. Van Cleve, F. S. Chapin III, P. W. Flanagan, L. A. Viereck, and C. T. Dyrness, editors. Forest Ecosystems in the Alaskan Taiga: A Synthesis of Structure and Function. Springer-Verlag, New York.

Flanagan, P. W., and A. K. Veum. 1974. Relationships between respiration, weight loss, temperature and moisture in organic residues in tundra. Pages 249–277 in A. J. Holding, O. W. Heal, S. F. MacLean Jr., and P. W. Flanagan, editors. Soil Organisms and Decomposition in Tundra. Tundra Biome Steering Committee, Stockholm, Sweden.

Funk, D. W., E. R. Pullman, K. M. Peterson, P. M. Crill, and W. D. Billings. 1994. Influence of water table on carbon dioxide, carbon monoxide, and methane fluxes from taiga bog microcosms. Global Biogeochemical Cycles 8:271–278.

Gulledge, J., and J. P. Schimel. 1998. Moisture control over atmospheric CH_4 consumption and CO_2 production in physically diverse soils. Soil Biology and Biochemistry 30:1127–1132.

Gulledge, J., and J. P. Schimel. 2000. Controls over carbon dioxide and methane fluxes across a taiga forest landscape. Ecosystems 3:269–282.

Halverson, L. J., T. M. Jones, and M. K. Firestone. 2000. Release of intracellular solutes by four soil bacteria exposed to dilution stress. Soil Science Society of America Journal 64:1630–1637.

Harris, R. F. 1981. Effect of water potential on microbial growth and activity. Pages 23–95 in J. F. Parr, W. R. Gardner, and L. F. Elliott, editors. Water Potential Relations in Soil Microbiology. American Society of Agronomy, Madison, WI.

Hobbie, J. E., J. P. Schimel, S. E. Trumbore, and J. R. Randerson. 2000. Controls over carbon storage and turnover in high-latitude soils. Global Change Biology 6:196–210.

Irons, J. G., III, J. P. Bryant, and M. W. Oswood. 1991. Effects of moose browsing on decomposition rates of birch litter in a subarctic stream. Canadian Journal of Fisheries and Aquatic Science 48:442–444.

Jones, D., and K. Kielland. 2002. Soil amino acid turnover dominates the nitrogen flux in permafrost-dominated taiga forest soils. Soil Biology and Biochemistry 34:209–219.

Kasischke, E., and B. Stocks. 2000. Fire, Climate Change, and Carbon Cycling in the Boreal Forest. Springer-Verlag, New York.

Kieft, T. L., E. Soroker, and M. K. Firestone. 1987. Microbial biomass response to a rapid increase in water potential when dry soil is wetted. Soil Biology and Biochemistry 19:119–126.

Kraus, T. E. C., R. J. Zasoski, R. A. Dahlgren, W. R. Horwath, and C. M. Preston. 2004. Carbon and nitrogen dynamics in a forest soil amended with purified tannins from different plant species. Soil Biology and Biochemistry 36:309–321.

Matthews, E., and I. Fung. 1987. Methane emission from natural wetlands: Global distribution, area, and environmental characteristics of sources. Global Biogeochemical Cycles 1:61–86.

Moosavi, S., P. Crill, E. Pullman, D. Funk, and K. Peterson. 1996. Controls on CH_4 flux from an Alaskan boreal wetland. Global Biogeochemical Cycles 10:287–296.

Näsholm, T., A. Ekblad, A. Nordin, R. Giesler, M. Högberg, and1 P. Högberg. 1998. Boreal forest plants take up organic nitrogen. Nature 392:914–916.

Robertson, G. P., and P. M. Vitousek. 1981. Nitrification potentials in primary and secondary succession. Ecology 62:376–386.

Ruess, R. W., K. Van Cleve, J. Yarie, and L. A. Viereck. 1996. Contributions of fine root production and turnover to the carbon and nitrogen cycling in taiga forests of the Alaskan interior. Canadian Journal of Forest Research 26:1326–1336.

Schimel, J. P., and F. S. Chapin III. 1996. Tundra plant uptake of amino acid and NH4+ nitrogen in situ: Plants compete well for amino acid N. Ecology 77:2142–2147.

Schimel, J. P., and J. S. Clein. 1996. Microbial response to freeze-thaw cycles in tundra and taiga soils. Soil Biology and Biochemistry 28:1061–1066.

Schimel, J. P., and M. N. Weintraub. 2003. The implications of exoenzyme activity on microbial carbon and nitrogen limitation in soil: A theoretical model. Soil Biology and Biochemistry 35:549–563.

Schimel, J. P., R. G. Cates, and R. Ruess. 1998. The role of balsam poplar secondary chemicals in controlling soil nutrient dynamics through succession in the Alaskan taiga. Biogeochemistry 42:221–234.

Schimel, J. P., J. M. Gulledge, J. S. Clein-Curley, J. E. Lindstrom, and J. F. Braddock. 1999. Moisture effects on microbial activity and community structure in decomposing birch litter in the Alaskan taiga. Soil Biology and Biochemistry 31:831–838.

Schimel, J. P., K. Van Cleve, R. G. Cates, T. P. Clausen, and P. B. Reichardt. 1996. Effects of Balsam Poplar (*Populus balsamifera*) tannins and low molecular weight phenolics on microbial activity in taiga floodplain soil—Implications for changes in N cycling during succession. Canadian Journal of Botany 74:84–90.

Stieglitz, M., J. Shaman, J. McNamara, V. Engel, J. S. Hanley, and G. Kling. 2003. An approach to understanding hydrological connectivity on the hillslope and the implications for nutrient transport. Global Biogeochemical Cycles 17:Article 1105.

Uliassi, D. D., and R. W. Ruess. 2002. Limitations to symbiotic nitrogen fixation in primary succession on the Tanana River floodplain. Ecology 83:88–103.

Van Cleve, K., C. T. Dyrness, G. M. Marion, and R. Erickson. 1993. Control of soil development on the Tanana River floodplain, interior Alaska. Canadian Journal of Forest Research 23:941–955.

Van Cleve, K., L. Oliver, R. Schlentner, L. A. Viereck, and C. T. Dyrness. 1983. Productivity and nutrient cycling in taiga forest ecosystems. Canadian Journal of Forest Research 13:747–766.

Vance, E. D., and F. S. Chapin III. 2001. Substrate limitations to microbial activity in taiga forest floors. Soil Biology and Biochemistry 33:173–188.

Viereck, L., K. Van Cleve, and C. Dyrness. 1986. Forest ecosystem distribution in the taiga environment. Pages 22–43 *in* K. Van Cleve, F. S. Chapin III, P. W. Flanagan, L. A. Viereck, and C. T. Dyrness, editors. Forest Ecosystems in the Alaskan Taiga. Springer-Verlag, New York.

Vitousek, P. M., and J. M. Melillo. 1979. Nitrate losses from disturbed forests: Patterns and mechanisms. Forest Science 25:605–619.

Wagener, S. M. 1995. Ecology of birch litter decomposition and forest floor processes in the Alaskan taiga. Ph.D. Dissertation, University of Alaska Fairbanks.

Wagener, S. M., and J. P. Schimel. 1998. Stratification of soil ecological processes: A study of the birch forest floor in the Alaskan taiga. Oikos 81:63–74.

Weintraub, M. N., and J. P. Schimel. 2003. Interactions between carbon and nitrogen mineralization and soil organic matter chemistry in arctic tundra soils. Ecosystems 6:129–143.

Zasada, J. 1986. Natural regeneration of trees and tall shrubs on forest sites in interior Alaska. Pages 44–73 *in* K. Van Cleve, F. S. Chapin III, P. W. Flanagan, L. A. Viereck, and C. T. Dyrness, editors. Forest Ecosystems in the Alaskan Taiga. Springer-Verlag, New York.

15

Patterns of Biogeochemistry in Alaskan Boreal Forests

David W. Valentine
Knut Kielland
F. Stuart Chapin III
A. David McGuire
Keith Van Cleve

Introduction

As the northernmost forest on Earth, boreal forests endure a combination of environmental challenges common only in subalpine forests elsewhere: extremely cold winters, short growing seasons, cold soils, and limited nutrient availability. Consequently, decomposition has lagged plant production, making circumpolar boreal forest soils one of the largest terrestrial reservoirs of carbon (C). Soil organic matter also constitutes a major source of nutrients, particularly nitrogen (N), that promote plant productivity when released during decomposition. If current trends in high-latitude warming continue (Chapter 4), how will accelerated soil C losses from decomposition compare to the C gains from enhanced plant productivity? This remains an open question of great interest to climate modelers seeking to incorporate biological feedbacks into future generations of general circulation models. This chapter builds on earlier chapters on plants (Chapters 11 and 12), herbivores (Chapter 13), and soil microbes (Chapter 14) to describe the patterns and processes of C and N dynamics in Alaska's boreal forest, paying particular attention to responses of these processes to the interacting influences of disturbance and climatic variations that occur across the landscape and through time. Other nutrients have received less attention in Alaskan research, and that data gap is reflected in this chapter.

Landscape Context and Permafrost

Interior Alaska's boreal forest is a patchwork of successional forest types. The major physiographic zones into which we categorize them reflect the contrasting influences of two major disturbance types: fire in upland and lowland areas results in multiple secondary successional pathways, while a more ordered array of forest types results from a combination of primary succession and variation in flooding frequency during succession on active floodplains (Chapter 7). Within each general physiographic zone (uplands and lowlands, floodplains), differences in the postdisturbance environment further influence vegetation establishment, plant species composition, and, ultimately, element cycling.

The state factor approach has proven useful in understanding landscape variation in biogeochemistry (Chapter 1; Van Cleve et al. 1991). As with other aspects of ecosystem function, element cycling reflects control exerted by major state factors: climate, parent material, potential vegetation, topography, and time since the most recent disturbance event. At one end of the spectrum, black spruce ecosystems occupy the coldest soils, exhibit the slowest growth, and have the least demand for nutrients to support growth. Organic materials produced by black spruce and bryophytes tend to display the greatest resistance to decomposition and therefore the slowest resupply of elements for plant uptake (Van Cleve and Harrison 1985, Van Cleve et al. 1986). Production of slowly decomposing organic matter initiates a positive feedback loop that results in low element supply and increased forest floor thickness, reducing soil temperatures that allow relatively labile organic compounds to persist. More productive forest types (birch, aspen, poplar, and some white spruce) generally display the opposite characteristics, producing less decay-resistant organic matter that recycles elements faster, and do not accumulate thick, insulating forest floors. Early-successional communities, dominated by deciduous vegetation, also tend to display more rapid element cycling.

Small differences in microclimate have more profound influences on biogeochemistry in Alaska's boreal forest than in other biomes because these small temperature differences govern the distribution of permafrost across the landscape. Permafrost, in turn, restricts internal soil drainage and root penetration in the soil profile, which further restricts decomposition and focuses C accretion near the soil surface (Chapter 12). Local climate and vegetation, particularly mosses, interact strongly to govern permafrost characteristics (Chapter 4). The accumulation of an insulating organic surface horizon slows heating in the summer, restricting the depth of the active layer (that portion of the soil profile that thaws seasonally). In extreme cases, the active layer becomes thinner than the O horizon, and the underlying mineral soil no longer participates meaningfully in element cycling.

Correspondence between landscape position, permafrost presence or depth of the active layer, and biogeochemical processes is complex, however. In areas with wide seasonal temperature fluctuation (e.g., valley bottoms that experience the coldest temperatures during winter atmospheric inversions), annual average temperatures may be below freezing and thus be underlain by permafrost. Yet their surface soils may be warmer during the growing season than soils that have higher annual mean temperatures but narrower seasonal temperature amplitudes, such as higher elevation areas that are above winter inversions yet are cooler during summer months). This can result in the counterintuitive situation in which soils in permafrost areas are warmer than those higher in elevation with no permafrost (Vogel 2004) and thus experience more rapid rates of decomposition during the summer months.

Major Element Cycles

Productivity in Alaska's boreal forest is constrained more directly by availability of moisture and nutrients than by low temperatures (Chapter 11). Nutrient supply rates in turn are governed by litter decomposition and C mineralization rates, so the dynamics and fate of decomposing residues and soil C are of primary concern in any treatment of biogeochemistry.

Carbon

The patterns of C storage, release, and internal dynamics in boreal forest soils have captured the attention of earth scientists owing to the enormous C reservoir that these soils contain and the potential for its release as CO_2 into the atmosphere. Because C balance is the relatively small difference between two much larger and highly variable rates—photosynthesis versus multiple pathways of loss—its assessment entails a challenging integration of complex processes and controls occurring over multiple temporal and spatial scales.

Soil C Stocks

Most authors agree that the boreal forest contains large stocks of soil C, which is most usefully expressed per unit area (e.g., Mg ha^{-1} = megagrams per hectare). However, such measurements must be interpreted carefully because two questions are not always carefully answered: "Are peatlands included" and "To what depth are measurements made?" Summaries differ in whether they include the surface O horizon, consisting of layers of recent litter (the Oi horizon; Chapter 3) and increasingly decomposed organic matter (Oe and Oa horizons), in their assessments of C stocks. In nearly all biomes (including the boreal forest), most soil C is in the mineral soil. The O horizon accounts for less than 5% of the total soil C stock in most biomes, but in temperate and boreal forests the percentage is two or three times as large, respectively (Schlesinger 1997).

One recent review of global soil C reservoirs (Jobbágy and Jackson 2000) considered only the surface 1 meter of mineral soil horizons in estimating a pool size of 93 Mg C ha^{-1} in boreal forest soils. This figure is lower (per unit area) than all other surveyed biomes except deserts and sclerophylous shrublands. Additional C also exists below the surface meter of soil that is the basis of the estimates described earlier, so Jobbágy and Jackson (2000) also expanded the depth to the surface 3 m of soil. This raised the total boreal forest mineral soil C inventory to 125 Mg C ha^{-1} but also increased total C stocks in the soils of all other biomes even more in absolute terms, so boreal forests ranked ahead of only deserts in total C stored in mineral soil.

Most other estimates include the surface O horizon and are much larger. One early estimate (Schlesinger 1977, repeated in Schlesinger 1997) put average boreal forest soil C content at 149 Mg C ha^{-1}, of which 20 Mg C ha^{-1} (13.4%) is in the O horizon. This is in relatively good agreement with the estimate from Jobbágy and Jackson (2000). Others have arrived at larger figures. Anderson (1991) (also cited in Van Cleve and Powers 1995) estimated C stored in boreal forest soils at about 170 Mg C ha^{-1}, and Apps et al. (1993) estimated a still larger C inventory of 185 Mg C ha^{-1}, larger than that for nearly any other biome.

These issues dramatically highlight the unique distribution of C in boreal forest soils. The O horizon is larger in boreal forests than in any other biome in terms of global total and per unit area and as a fraction of total soil C, so the difference between C in the mineral soil and C in the organic horizons is least in boreal forest soils (Schlesinger 1997). This top-heavy distribution reflects differences both in vectors and in rates of vertical movement of C within the soil profile and in the vulnerability of soil C to changes in moisture and temperature at the soil surface. As such, the quantities of surface soil C may vary over time more rapidly than total soil C in other biomes, belying the notion that overall boreal forest soil C inventories may not change rapidly in response to a changing climate.

The foregoing discussion does not include the vast northern peatlands embedded within the boreal forest. On average, these wetlands store more than 1600 Mg C ha^{-1} as peat on a land area a fifth the size of the boreal forest biome. This corresponds to a global total nearly twice that of the rest of the boreal forest (419 vs. 231 Pg C; Apps et al. 1993).

In summary, because of their large aerial extent (1.2 billion ha), boreal forest soils represent a total reservoir (179–231 Pg C) that is larger than that in any biome except the tropical forest, which has more than twice the spatial extent (Schlesinger 1997). These figures triple when we include embedded wetlands (Apps et al. 1993). Because of their historic success in sequestering C and their potential responsiveness to changes in climate and disturbance regime, the boreal forest and its associated wetlands have become foci of interest and concern for future changes in C storage and methane release (see later discussion and Chapter 19).

Soil C Dynamics

The rate at which soils accumulate C depends on the relative rates of litter input, soil respiration, and combustion. As in most forests, boreal soils lose C with disturbance but accumulate C between disturbances (Harden et al. 1997, Goulden et al. 1998). Many boreal ecosystems are characterized by frequent disturbances (Chapter 17) and modest aboveground tree productivity (Chapters 11, 12). Why, then, does soil C accumulate so dramatically in many boreal soils? Part of the answer lies in changes in the source of litter—roots and mosses account for a large proportion of litter input in many boreal ecosystems (Chapter 12). However, controls over decomposition, as described in this section, also play a critical role.

Surface litter and the underlying SOM differ substantially in their controls over decomposition. Plant chemistry appears to dominate patterns of foliar litter decomposition rates, whereas environment and organic matter chemistry both contribute to landscape variation in SOM decomposition.

The most dramatic differences in litter decomposition rate observed among Alaskan boreal forests are between black spruce and other forest types. Birch foliar litter decomposes four to five times faster than black spruce foliar litter in the field (Flanagan and Van Cleve 1983). Reciprocal transplants of these litter types demonstrate that the inherent properties of the litter are much more important than environment in explaining this large difference in litter decay rate. Differences in carbon chemistry account for much of the observed variation in foliar litter decomposition rate. Poplar litter, for example, decomposes more slowly than willow or alder litter on the Tanana floodplain in part because of higher concentrations of complex tannins and lignin (Chapter 14; Yarie and Van Cleve 1996). Other patterns of decomposition are less easily explained. White spruce needle litter, for example, decomposes relatively rapidly (Chapter 12), despite its high lignin and tannin concentrations (Yarie and Van Cleve 1996). Litter N concentration also correlates with litter decomposition rate (Flanagan and Van Cleve 1983), as in many biomes, but tests often show that this correlation does not reflect a causal relationship (Prescott 1995, Hobbie 1996). In the boreal forest, for example, enhanced N and P availabilities inhibited wood decomposition relative to both untreated litter and litter to which glucose had been added (Fig. 15.1; Flanagan and Van Cleve 1983). Thus, litter decomposition appears to be constrained more by C chemistry (resistance of C compounds to decay) than by nutrient composition (Chapter 14).

Most litter decomposition studies have focused on the leaf litter or wood produced by vascular plants. The largest proportion of litter input in black spruce forests comes from fine roots and moss (Chapter 12; Ruess et al. 2003), which complicates our understanding of overall litter decomposition dynamics. The role of root litter carbon chemistry in controlling decomposition rates has not been examined, but roots appear and disappear rapidly from serial minirhizotron images, suggesting rapid production and decomposition (Chapter 12; Ruess et al. 1996). The proportion of the root-derived C that is mineralized to CO_2 or transformed into humic substances is unknown. It is apparent, however, that root litter contributes substantially to belowground C dynamics. In the absence of root inputs, Vogel (2004) found that the forest floor mass of three black spruce stands decreased rapidly, suggesting

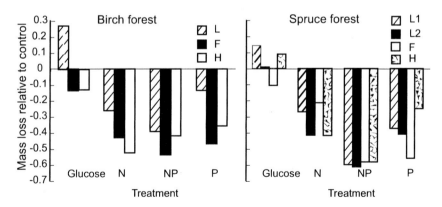

Figure 15.1. Effects of carbon and nutrient additions on decomposition rates of wood substrates placed in different organic horizons (L1 = upper litter layer [undecomposed or Oi]; L2 = lower litter layer [undecomposed or Oi]; F = fermentation layer [partially decomposed or Oe]; H = humus layer [fully decomposed or Oa]. Rates are expressed relative to untreated control plots, set at zero. Data recalculated from Flanagan and Van Cleve (1983).

that a large fraction of the forest floor consists of root litter that turns over rapidly, with much of the decomposition products being lost from the forest floor as either CO_2 or soluble C. Mosses, on the other hand, decompose slowly (Turetsky 2003), perhaps contributing more to long-term SOM accumulation in black spruce forests.

In contrast to litter, the decomposition and turnover of the soil organic mat appear to be controlled by both environment and chemistry. Reciprocal transplants (Flanagan and Van Cleve 1983) and laboratory incubations of soil organic matter (Vance and Chapin 2001) both show that black spruce SOM decomposes more slowly than SOM from other forest types, suggesting an important role of SOM chemistry, just as described for foliar litter. Unlike foliar litter, however, SOM turnover is also strongly influenced by the subsurface environment. Soils on south aspects often warm to temperatures typical of temperate soils during the summer, enabling instantaneous rates of litter decomposition that rival those of their southern counterparts—but over a much shorter time period. Once coniferous forests develop thick moss layers and underlying O horizons, their cold, wet environment strongly limits decomposition. Laboratory incubations of this organic layer under favorable conditions of temperature and moisture demonstrate that most of this organic matter is highly labile and decomposes rapidly (Neff and Hooper 2002, Vogel 2004), just as in arctic tundra (Weintraub and Schimel 2003). Field warming experiments also lead to rapid decomposition of black spruce SOM (Van Cleve et al. 1990). Temperature may also alter the nature of the residual humus and increase the relative quantities of CO_2 evolved per unit of organic matter processed (Thornley and Cannell 2001). Decomposition of SOM accumulated under cold conditions responds sensitively to changes in temperature. For example, soils sampled from cold environments exhibited higher Q_{10}'s (R_{T+10}/R_T where R is microbial respiration at two temperatures differing by 10°C) than did soils sampled from warmer environments (Kirschbaum 1995). Together, these observations suggest that there may be large positive feedbacks between climate warming and C release from the cold soils of boreal forests (Kirschbaum 2000).

Ecosystem carbon turnover is often indexed as the mean residence time (MRT) of the forest floor, which strictly speaking requires measurement of C loss from the forest floor. This is often approximated as litterfall rates by assuming steady state, that is, that the forest floor is neither aggrading nor degrading. This assumption is questionable in most biomes, especially the boreal forest, where the forest floor clearly aggrades during succession. By measuring mass loss of forest floor samples in litterbags placed back into the forest floor, Flanagan and Van Cleve (1983) avoided this assumption. They found that the MRT of spruce litter in forest floor increased with increasing forest floor thickness, while MRT for broad-leafed litter did not differ across a range of deciduous forest floor thicknesses (Fig. 15.2; Flanagan and Van Cleve 1983). These patterns and rates are very similar to the litter MRTs measured across a range of mature black spruce stands in interior Alaska, which ranged from 76–102 years (Vogel 2004).

A recent paper presented a simple model of soil C accumulation with potentially important implications for the geographic distribution of changes in C balance with climate. Because C fixation occurs aboveground and C mineralization is largely a belowground process, Swanson et al. (2000) reasoned that the difference between

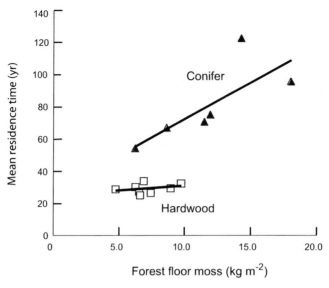

Figure 15.2. Mean residence time of the forest floor, calculated as forest floor mass divided by annual decomposition of forest floor, for hardwood (aspen and birch) and coniferous (white and black spruce) stands. Data recalculated from Flanagan and Van Cleve (1983).

above- and belowground growing degree-days (GDD, Chapter 4; thus AGDD-BGDD) would constitute an "above-ground thermal advantage." Larger advantages would correspond to higher rates of peat accumulation. AGDD and BGDD converge at low latitudes (both are high) and at extreme high latitudes (both are near zero; Fig. 15.3). At intermediate latitudes, AGDD-BGDD reaches a maximum at approximately the 0°C mean annual soil temperature (MAST) isotherm, which corresponds to the major zone of peat accumulation across North America (Swanson et al. 2000). This model also implies that climate warming might differ in its effect on soil C storage across the boreal forest; in the southern boreal forest, there should be a net C loss resulting from a shrinking aboveground thermal advantage; whereas in the northern boreal forest there should be a net C gain resulting from an increasing aboveground thermal advantage. Incorporation of simple models such as this could improve biome-scale assessments of C balance.

Disturbance Impacts

Although boreal forests historically are disturbance-driven ecosystems, Alaska's boreal forests are unusual in the circumpolar north because non-anthropogenic disturbance patterns, especially wildfire, still dominate the landscape. Depending on severity, wildfire transfers varying amounts of organic C from the ground surface, where most fires are carried in coniferous forests, to the atmosphere (Chapter 17). Severe fires, especially those in late summer, burn deep into the O horizon because of chronic drying, releasing more C than would be likely in the

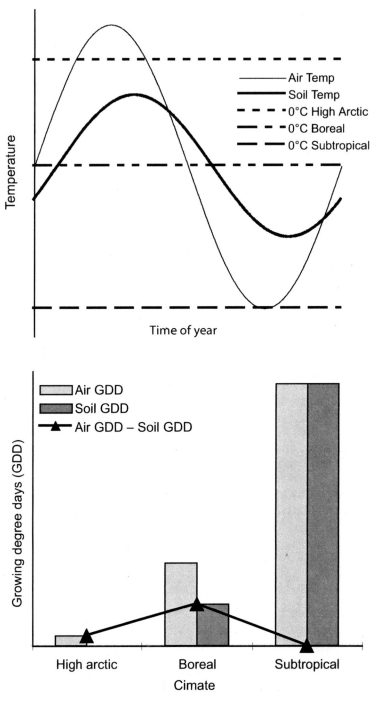

Figure 15.3. Annual cycle of soil and air temperature, soil and air growing degree-days (GDD), and aboveground thermal advantage (air GDD–soil GDD) in high-arctic, boreal, and subtropical ecosystems. Adapted from Swanson et al. (2000).

same stand earlier in the growing season (Kasischke and Johnstone in press). Less severe fires may do little more than scorch the surface of the forest floor, leaving most C in place.

In early July 1999, a major experimental examination of wildfire behavior and the role of wildfire on boreal forest dynamics began. Dubbed "Frostfire" to embrace the contrasts in extremes inherent in the boreal forest, the "prescribed wildfire" achieved moderate severity in about 35% of the 11 km^2 watershed C4 within the Caribou Poker Creeks Research Watershed, primarily in closed-canopy black spruce vegetation. Previous studies had suggested that the warm, moist conditions typical of postfire periods drive a sharp increase in heterotrophic respiration (Richter et al. 2000). However, measurements of surface CO_2 exchange have not provided evidence of such a spike, either at Frostfire (Valentine 2002, unpublished) or elsewhere in interior Alaska (O'Neill 2000). Depending on how slowly productivity recovers, the integrated C losses for postfire years can equal or exceed direct fire losses (Zhuang et al. 2002).

Repeated fires further complicate trends in soil C accumulation owing to the pyrogenic transformation of biomass into charred materials ("black carbon"). Harden et al. (1997, 2000) hypothesized that fires cause not only the single largest loss of C from surface soils but also the single largest input of C into mineral soil (Fig. 15.4), resulting from the movement of small pieces of charred materials deeper

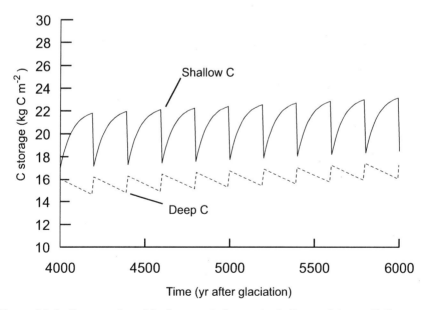

Figure 15.4. Conceptual model of temporal changes in shallow and deep-soil C storage associated with succession and wildfire. During succession, C accumulates in shallow soils and is lost from deep soils, whereas wildfire causes a loss of shallow-soil C to the atmosphere (combustion) and to deep soils (leaching). Redrawn with permission from Blackwell Publishing, Ltd. (Harden et al. 2000).

into the soil profile. Their hypothesis also predicts that, between fires, litter inputs to surface soils exceed respiratory losses, whereas C losses exceed inputs in mineral soils. Black carbon accumulation is potentially a major contributor to the long-term C balance of boreal forests.

The overall pattern after wildfire is C loss from soils until litter inputs exceed microbial respiration, after which rapid regrowth causes rapid net C accretion. Later in stand development, rates of NPP and C accretion slow (Fig. 15.5; Van Cleve et al. 1993), and variability in net C accretion increases among older, otherwise similar stands, even among otherwise similar sites.

Methane

Boreal forests in Alaska and elsewhere contain an abundance of peatlands resulting from poor drainage, often the combined result of permafrost and the insulating and drainage-impeding nature of mosses. Under these conditions, C metabolism proceeds more slowly via anaerobic pathways that generate a variety of reduced trace gases, such as methyl sulfides and, ultimately, methane (CH_4). Although CH_4 emissions are constrained by both slow anaerobic metabolism and partial oxidation in surface sediments and are far lower than CO_2 emissions would be from an aerated environment, CH_4 is 20 times more powerful than CO_2 (per mole) as a "greenhouse gas" (Chapter 19). Northern wetlands may be a major potential source

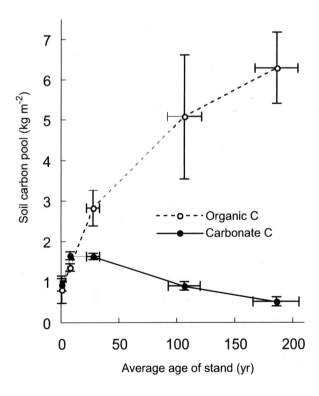

Figure 15.5. Organic C accretion in organic matter and carbonate disappearance during floodplain succession. Redrawn with permission from the *Canadian Journal of Forest Research* (Van Cleve et al. 1993).

of CH_4 to the atmosphere, and a great deal of effort has been expended quantifying CH_4 emissions in northern wetlands (Hamilton et al. 1994, Suyker et al. 1996). Measured CH_4 emissions rates vary substantially depending on water table, substrate quality, temperature, and pH.

Methane fluxes from wetlands in Alaska's boreal forests have not received as much attention as those in Canada, and the importance of methane biogeochemistry remains a gap in our knowledge. Methane fluxes from Alaskan wetlands appear to be comparatively low, however (Bartlett et al. 1992, Martens et al. 1992). This may be the result of temperature or other constraints on the acetate metabolic pathway that is thought to dominate methane production in freshwater wetlands (Hines et al. 2001).

As in most uplands elsewhere, bacteria in well-drained or otherwise aerobic boreal forest soils consume atmospheric CH_4 (Whalen et al. 1990, Gulledge and Schimel 2000). Rates of CH_4 consumption are very low (generally <0.5 mg C m^{-2} d^{-1}) and are inhibited by ammonium in soils (Gulledge et al. 1997). These patterns are similar to those observed in other ecosystems (Steudler et al. 1989, Mosier et al. 1996).

Nitrogen

As with carbon, boreal forests have very large stocks of soil organic N (Van Cleve and Alexander 1981). Nonetheless, N supply rates directly limit plant productivity throughout Alaska's interior forests (Chapter 11). Recent estimates suggest that the annual inputs of inorganic N (dry and wet deposition, N fixation and mineralization) cannot account for observed plant N uptake (Ruess et al. 1996). In black spruce forests, these conventional measures of N flux represent less than 50% of the annual N requirement of the vegetation (Table 15.1). Part of the reason for this discrepancy between N supply and N demand is that boreal trees absorb part of their N in an organic form (Näsholm et al. 1998, Kielland 2001, McFarland et al. 2002). This recent discovery of the central role of organic N in boreal N cycling has led to major changes in our paradigm of how N cycles through ecosystems (Chapin 1995, Atkin 1996).

Table 15.1. Nitrogen budgets for floodplain and upland taiga forests in interior Alaska showing contributions of inorganic nitrogen fluxes and annual nitrogen requirement of vegetation.

Parameter	Alder[1]	Balsam Poplar[1]	Birch[2]	White Spruce[1]	Black Spruce[2]
Supply[3]	4.34	2.40	2.45	1.47	0.38
Requirement[3]	7.46	2.23	5.25	1.35	0.92
S/R*(100)[4]	58	107	47	109	41

Source: Data calculated from Ruess et al. (1996) and Van Cleve et al. (1983).

[1] Floodplain stands.
[2] Upland stands.
[3] Units are in g m^{-2} y^{-1}.
[4] Proportion of annual requirement supplied by inorganic N (%).

N inputs

On floodplains, N cycling is dominated by inputs from thinleaf alder (*A. tenuifolia*), especially in early- to mid-succession. These inputs establish an N legacy that carries through later stages of forest development. Because of the low precipitation and lack of industrial development in interior Alaska, N deposition is relatively minor (Chapter 16), except in the earliest successional stands of willow, which have very low rates of N fixation and N mineralization. Initially, N fixation accounts for nearly 80% of the N that is absorbed by vegetation. In later successional stages, N fixation typically accounts for 10–20% of the N budget (Ruess et al. 1996). Ultimately, in late succession, the amounts of N accumulated in mineral soil in the floodplain and in upland/lowlands may be similar. However, N accumulates more rapidly in floodplain than in upland forests because of the higher rates of N fixation by the floodplain alder (Anderson et al. 2004). This difference in N inputs is reflected in the generally higher foliar N concentrations of deciduous trees on the floodplain than in the uplands (Yarie and Van Cleve 1983), although this difference is not apparent in white spruce.

In upland secondary successional forests, a substantial legacy of N remains after fire because variable burn intensities can consume little of or nearly the entire forest floor. It would be interesting to know the age of organic materials (C) in these postburn ecosystems. Accumulation of decay-resistant organic materials (e.g., lignin, charcoal, and humus) over several fire cycles may also influence supplies of N from mineralization through chemical stabilization, physical protection, and cooling.

Soil Organic N Turnover

Average concentration of total N in the forest floor varies approximately 15-fold through primary floodplain succession in interior Alaska, being lowest in the willow stage and highest in the poplar stage (Walker 1989). This contrasts to a mere twofold variation of total N in the mineral soil. Extractable concentrations of ammonium are typically orders of magnitude greater than nitrate, except in the N-fixing alder stands, where nitrate concentrations approximate those of ammonium. Concentrations of dissolved organic N (DON) in mineral soil are typically two to three times greater than concentrations of ammonium (Walker 1989).

Except in early succession, organic N inputs from plants are the largest N inputs to the soil. The controls over the breakdown of this insoluble organic N to soluble forms is not well understood but may be the rate-limiting step in going from dead organic matter to N forms that are available to plants and microbes. Insoluble organic N is acted upon by a range of proteases, which, in conjunction with the direct release of free amino acids from cell lysis, results in amino acid concentrations in the soil solution on the order of 20–250 μM (Kielland 1995, Raab et al. 1999). The breakdown of proteins to dissolved organic N (DON) is positively correlated with total soil N (Raab et al. 1999) and negatively correlated with soil depth and organic matter quality (Chapin et al. 1988, Jones 1999). Plants are an additional direct source of DON. They can contain up to 30% protein by weight and free amino acids at concentrations between 5–10 mM (Jones and Darrah 1994, Kielland 1994). Rupture of root cells by freeze-thaw cycles, herbivory, or senescence can release substantial DON to the soil.

In boreal surface soils, free amino acid concentrations are about 4–8 µg N g^{-1} dry weight (Kielland 2001, Jones and Kielland 2002), similar to values in arctic tundra soils (Kielland 1995). This DON pool is several-fold (50-fold in black spruce) larger than the ammonium pool (Fig. 15.6; Jones et al. 2002). Because N typically originates as insoluble or soluble organic N and subsequently is transformed to soluble inorganic N forms, the probable gross supply rate must be in this order: soluble organic N >ammonium >nitrate.

Both laboratory and in situ field estimates of amino acid turnover (i.e., complete replacement of the pool) suggest that this pool is highly dynamic, even under the prevailing cold conditions of late-successional boreal ecosystems. In the field, free amino acid turnover times in soils are generally 1–12 hours, depending on soil physicochemical characteristics (Jones et al. 1994), and the N flux through the amino acid pool is large—over an order of magnitude greater than the rate of gross mineralization in black spruce soils (Jones et al. 2002). The turnover of the amino acid pool is approximately three times more rapid in the organic horizon than in the mineral soil (Fig. 15.7 top). This is in part explained by the much more rapid microbial uptake rate of amino acids in the forest floor (Fig. 15.7 bottom). However, in the willow stage of succession, the in situ turnover rates of simple amino acids, such as glycine, alanine, and aspartate, appear to be as rapid as those in the organic horizon in white spruce stands. The large pools and the rapid turnover of free amino acids imply that their production and consumption rates must be very high, both during times of high biological activity and on an annual basis. The rapid consumption

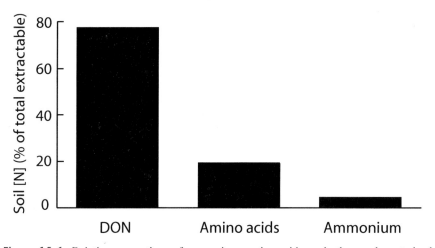

Figure 15.6. Relative proportions of ammonium, amino acids, and other uncharacterized dissolved organic nitrogen (DON) in soil extracts of black spruce forests of the Tanana floodplain. Nitrate was undetectable. Data from Jones et al. (2002).

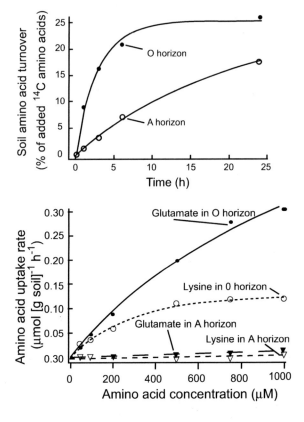

Figure 15.7. Turnover and microbial uptake in organic and mineral soil horizons of a mixture of 15 ^{14}C-labeled amino acids in black spruce soils. Redrawn with permission from *Soil Biology and Biochemistry* (Jones et al. 2002).

of amino acids suggests that the transformation of protein to amino acids (rather than the mineralization of amino acids to NH_4^+) is the major bottleneck in the N cycle in boreal soils. Although organic N turnover has not been thoroughly studied in most ecosystems, we suspect that patterns observed in boreal soils are widely applicable. In both marine and terrestrial ecosystems, the potential and actual turnover of free amino acids are very rapid. For example, the alanine pool in marine sediments turns over in about two to four hours (Jørgensen and Søndergaard 1984, Henrichs and Doyle 1986).

Both proteolysis and N mineralization are positively correlated with total soil N (Marion and Miller 1982, Raab et al. 1999) and negatively correlated with soil depth and organic matter quality (owing to N immobilization potentials; Chapin et al. 1988, Jones 1999). Further, amino acid decomposition appears to exhibit qualitatively different temperature responses than does N mineralization. Whereas amino acid turnover is very sensitive to increases in temperature below 10°C (Jones 1999), N mineralization and nitrification in taiga soils do not respond significantly to temperature until it exceeds 15–20°C (Klingensmith and Van Cleve 1993). Moreover, soil proteolytic activity, unlike N mineralization, is tightly controlled by pH (Leake and Read 1989). In addition, pH strongly affects amino acid uptake in boreal plants (Falkengren-Grerup et al. 2000).

Soil N Mineralization

Short growing season and environmental conditions adverse to decomposition result in a low annual flux of inorganic N in boreal forests. Growing-season (May–September) rates of net N mineralization vary eight-fold across successional stands in interior Alaska (Fig. 15.8). The seasonal dynamics of N mineralization are highly variable, and microbial N immobilization may predominate anytime during the growing season (Gordon and Van Cleve 1983, Boone 1992, Klingensmith and Van Cleve 1993). Van Cleve et al. (1993) found the highest rates of N mineralization early in the season, whereas other studies have found different patterns (Gordon and Van Cleve 1983, Klingensmith and Van Cleve 1993). Moreover, the peak in N mineralization does not always occur synchronously in mineral soil and in the forest floor. Net mineralization rates during the short summer season are much more rapid than over winter (Gordon and Van Cleve 1983). However, cumulative net N production over the winter accounts for nearly 40% of the annual total flux because of the long winter season.

N mineralization in interior forests is strongly affected by substrate quality (Flanagan and Van Cleve 1983), attesting to the importance of vegetation control over ecosystem processes in the north (Chapin et al. 1997). By contrast, temperature appears to have very little effect on N mineralization under ambient soil temperatures between 5–20 °C (Klingensmith and Van Cleve 1993; Viereck et al. 1993). Likewise, soil moisture has negligible effects on N mineralization, except in the white spruce stage of succession, where drought stress significantly reduces tree growth (Chapter 11; Barber et al. 2000).

Though midwinter air and soil temperatures are typically low in boreal forests, the duration of autumn freeze-up can be prolonged, even in these high-latitude ecosystems. High soil moisture, and thus increased soil heat capacity, coupled with

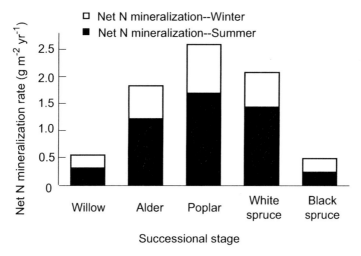

Figure 15.8. Successional changes in summer, winter, and annual net N mineralization in the Tanana River floodplain.

an increasingly insulating snow cover, often leaves the top 10 cm of the soil, where most of the biological activity is concentrated, unfrozen for many weeks after freeze-up has occurred. Indeed, much of the N flux through the soil system occurs after ecologists have returned to their classrooms in the fall. Net N mineralization rates during the growing season from green-up (late May) through freeze-up (late September) accounts for about 60% of the annual inorganic N flux, with the remaining 40% released during the nearly eight months of the apparent dormant season. Similarly, in the Arctic about half of the annual N mineralization occurs during winter (Hobbie and Chapin 1996). Boreal N release during winter occurs primarily from October through January, with negligible N mineralization during early spring in stands of willow, alder, balsam poplar, and white spruce. Conversely, black spruce stands exhibit substantial mineralization after snowmelt during early spring. The high rates of N mineralization in late autumn through early winter coincide with high turnover of fine-root biomass in these stands, suggesting that labile substrate production, rather than temperature, is a major factor in controlling N mineralization in these ecosystems. The results are consistent with the low temperature sensitivity of boreal N mineralization described earlier. These findings suggest that the biogeochemically active season in these ecosystems is much longer than previously recognized. The shoulder seasons of spring and fall may turn out to be as important as the summer growing season for many biogeochemical processes in boreal forests.

Organic Solutes from Plants May Affect N Availability

Some recent studies have suggested that secondary metabolites leached from understory species may have deleterious effects on forest growth. Swedish researchers (Zackrisson and Nilsson 1992, Nilsson et al. 1993, Zackrisson et al. 1996, 1997, Wardle et al. 1998) noted that intensively managed Scots pine (*Pinus sylvestris*) forests in Sweden had declined in productivity and sought an explanation. They found that a ubiquitous understory shrub, crowberry (*Empetrum hermaphroditum*), leached a compound—Batatastin III—that had deleterious effects on germination and early growth of pine, and they suggested that one mechanism to explain this may be depression of N availability. They also noted that charcoal—no longer produced by wildfire in Swedish forests—adsorbed Batatastin III, mitigating its negative effects.

In Alaskan interior forests, Castells et al. (2003) did not find such activity resulting from crowberry but did find evidence that foliar and litter leachates from another common understory shrub in hardwood forests, Labrador tea (*Ledum palustre*), stimulated N immobilization and thereby reduced net N mineralization. The effects of plant secondary metabolites on microbial processes are often complex (Chapter 14).

Plant N Uptake

As described earlier, interior Alaskan forests appear to be dominated by one of two modes of N dynamics. On the one hand, there is strong, direct evidence that addition of inorganic N stimulates the productivity of taiga ecosystems (Chapter 11), a conclusion that is also suggested by significant relationships that link productivity, decomposition, and N mineralization (Van Cleve et al. 1983, 1993). On the other hand, in the N budget of stands such as black spruce, birch, and alder, the combined inorganic N fluxes (mineralization, deposition, and fixation) account for only about 50% of the annual N requirement. This discrepancy suggests that "unconventional" mechanisms of N acquisition, such as direct uptake of amino acids, may play a role in these forests.

Boreal plants exhibit a substantial capacity to absorb organic N, just as do plants from alpine (Raab et al. 1999) and arctic ecosystems (Kielland 2001). In plots labeled with U-^{13}C$_2$^{15}N glycine, 45–90% of the label is taken up as an intact amino acid (Näsholm et al. 1998). Conservative estimates of glycine uptake in interior Alaskan forests show similar results (McFarland et al. 2002). Moreover, estimates of ammonium and glycine uptake in situ show that the forest vegetation takes up these N forms in roughly equal proportions (Fig. 15.9). Thus, despite very rapid microbial turnover of amino acids in the soil, which might compete with uptake by vegetation (Jones 1999), amino acids appear to play a significant role in the N economy of boreal forests.

There are several physiological mechanisms that enable plants to obtain organic N, including direct uptake by roots and association with mycorrhizal fungi that have a high capacity to break down proteins and absorb the resulting amino acids. The direct acquisition of organic N is not an exclusive trait of species in natural ecosystems. Important agricultural species can also take up organic N (Näsholm et al. 2000, Yamagata et al. 2001), suggesting that plant species under radically different edaphic conditions mine the soil for N in both mineral and organic form. It is still unclear how important organic N is to the overall N economy of other ecosystems.

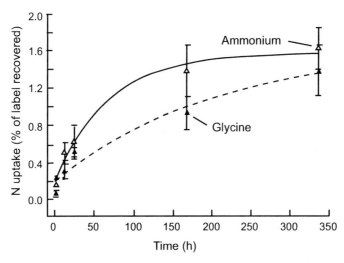

Figure 15.9. Relative plant uptake rates of ^{15}N-labeled NH$_4^+$ and glycine from boreal forest soils. Data are means + SE. Redrawn with permission from *Ecosystems* (McFarland et al. 2002).

Rates of absorption of most nutrients vary predictably among boreal tree species as a function of their intrinsic growth rate, nutrient availability in the soil they occupy, and the ion in question (Chapin et al. 1986). Nutrient uptake rate is typically most temperature-sensitive in species such as aspen that occupy warm soils, compared to species such as black spruce, which occur in very cold soils. Moreover, species from high-fertility soils tend to have higher nutrient absorption rates than species from low-fertility soils because of their greater nutrient demands for growth. The availability/production of mineral N in taiga soils is dominated by ammonium, except in the alder stage of primary succession where nitrate predominates (Van Cleve et al. 1993). Nevertheless, ammonium appears to be the preferred mineral N source for the majority of boreal species, as estimated from laboratory absorption experiments (Chapin et al. 1986).

Isotopic Signatures of Taiga Forest Species

The isotopic signature of N ($\delta^{15}N$) in tissues of northern species also provides evidence of niche differentiation regarding the acquisition of different N forms (Schulze et al. 1994, Michelsen et al. 1996, Nadelhoffer et al. 1996, Kielland et al. 1998). For example, arctic species of different functional types differ substantially in leaf $\delta^{15}N$ values, suggesting differences in the source and possibly the composition of absorbed soil N. Graminoids, which rely primarily on inorganic N and exploit deeper soil horizons, are enriched in the heavy isotope (^{15}N), implying that they are relying on N that has cycled within the ecosystem for a relatively long time. Conversely, mosses and deciduous and evergreen shrubs, which rely more heavily on organic N, are depleted in ^{15}N (Kielland 1997). The patterns of isotope values among boreal species are similar to those in arctic tundra, with $\delta^{15}N$ values ranging from +4‰ in aquatic plants to -10‰ in black spruce (Fig. 15.10; Schulze et al. 1994, Kielland et al. 1998). This variation is consistent with the hypothesis that plants in both ecosystems have evolved diverse strategies for acquiring soil N.

Species in high-fertility soils, such as alder, differ greatly in $\delta^{15}N$ from species such as willow and black spruce, which grow in low-fertility soils (Fig. 15.11). This divergence underscores the differences in N dynamics among ecosystems and suggests variation of N sources across successional gradients. In particular, the very depleted isotope signature of black spruce (*Picea mariana*) is the opposite of what is expected of species growing in century-old peat, where a long history of soil N processing should have enriched soils in ^{15}N via chronic losses of ^{14}N.

Phosphorus

Although most of Alaska's boreal forest soils escaped glaciation (Chapter 2), they have experienced little chemical weathering and hence tend to be relatively "young" (Chapter 3). Along the Tanana River floodplain, for example, total phosphorus (P) content of soils (mass basis) exceeds that of N for nearly the first 200 years of stand development, despite large inputs of alder-fixed N early in succession (Van Cleve et al. 1993). In upland soils, total P contents are similarly high, in some stands rivaling those of N on a mass basis (Van Cleve et al. 1983). Since plants typically

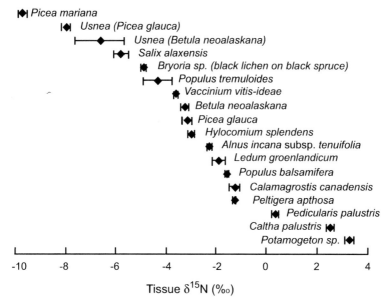

Figure 15.10. Species variation in foliar $\delta^{15}N$ values among boreal plant species. Data are means + SE. Data from Kielland et al. (1998) and Kielland (2001).

require about 14 times more N than P, these observations suggest that P should seldom limit production in these forests (Chapter 11). Three factors may, however, limit P availability in some boreal forest soils.

First, soil pH varies from slightly alkaline (pH 7.5) to strongly acidic (pH 4), largely driven by differences in disturbance regimes and changes through succession (Van Cleve et al. 1993). The higher pH values occur in poorly weathered parent material typical of recently deposited alluvium rich in calcite (Marion et al. 1993), in which calcium phosphates and hydroxyapatite are only sparingly soluble and restrict P availability. Surface soils become mildly acidic later in succession and in upland/lowland areas of all ages in the absence of calcite deposition. Moreover, periodic fire releases P from organic matter and acts as a mildly alkalinizing influence, replenishing soluble P supplies. Greater P availability for plant growth in upland ecosystems is reflected in the substantially higher current foliage P concentrations in both deciduous and white spruce tissue in upland forests compared with floodplain forests. P availability reaches its maximum at pH 6.5, below which precipitation with iron begins to restrict solubility. Surface soil pH values are lowest in the organic horizons under white and black spruce cover types (with associated bryophytes). Available P does not always diminish at low pH in these soils, however. On the Tanana River floodplain, for example, peak concentrations of extractable P occurred at pH 5—well below the theoretical P solubility maximum at pH 6.5 (Van Cleve et al. 1993). In summary, there are several potential mechanisms that might alter the relative availability of N and P in boreal soils, but future research will be required to assess their relative importance.

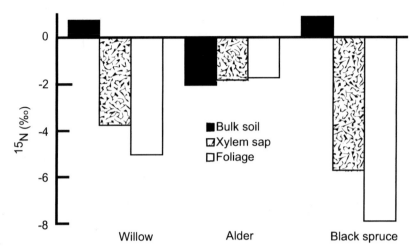

Figure 15.11. The $\delta^{15}N$ values in bulk soil, xylem sap, and foliage in willow, alder, and black spruce stands. In sites with low N availability because they are very young (willow) or very old (black spruce), soil $\delta^{15}N$ values are enriched, while plant tissues are highly depleted in ^{15}N, indicating a less direct pathway from soil into plant.

Second, in excessively well-drained, coarse-textured soils, chemical weathering may be too slow to keep pace with plant demands. In Alaska's boreal forests, these conditions may occur in sandy and gravelly deposits in river floodplains early in succession, in stabilized sand dunes, or anywhere else that parent materials are exposed at the surface.

Third, soils in which restricted drainage or low temperatures limit decomposition accumulate a thick, acidic organic horizon on the soil surface. Over time, this process may become self-reinforcing (paludification) as the accumulating peat not only restricts drainage but also insulates the subsoil from surface heating and enables permafrost to rise, often up into the organic horizon. This severely curtails the rates of mineral weathering and the access of roots to mineral P. Although total P in these organic-rich soils may be relatively high, the poor drainage and low temperatures of the organic horizon restrict mineralization of P from the organic matter, while low soil pH limits its solubility, especially in the presence of soluble iron or aluminum. Although it is not clear which of these processes (restricted mineralization or low solubility) is the dominant mechanism, this pattern frequently occurs in nutrient-poor black spruce stands (Van Cleve et al. 1983).

Conclusions

Biogeochemistry in Alaska's boreal forest is variable both across the landscape and over time. This variability reflects the exceptionally broad dynamic range of conditions imposed by the major state factors on biogeochemical processes, exemplified in the extreme by the interaction of frequent wildfires with underlying permafrost to govern rates of decomposition and nutrient return to vegetation. We have learned a great deal from chronosequence studies by substituting space for time; future repeated measurements within single stands will permit us to verify these patterns.

The major changes that govern boreal forests' productivity and C balance occur rapidly over relatively short time scales. Interannual climatic variability can change the sign of ecosystem C balance by changing both disturbance regimes (fire and insect, Chapters 17 and 9, respectively) and plant root litter inputs (Chapter 12) and decomposition. Rapid root production and turnover rates (Chapter 12) contribute substantially not only to C accretion but also to interannual variation in total surface organic C stocks (Vogel 2004). Because a relatively large fraction of soil C is at or near the soil surface, boreal forest C dynamics track interannual fluctuations in climate and disturbance more closely than in other biomes. This sensitivity, coupled with the large size of the boreal forest soil C reservoir and the predicted magnitude of climate change at high latitudes, reinforces the widely held view that the boreal forest will strongly influence future atmospheric CO_2 levels (Chapter 19). It is nonetheless clear that long-term assessments of the decomposition rates of major litter types, including roots, coarse woody debris, and bryophytes, are needed to complete our understanding of C balance and nutrient cycling.

As with most other terrestrial biomes, the dynamics of N availability generally limit vegetative growth. In the cold soils of Alaska's boreal forest, however, alternative pathways of N availability that do not require mineralization are important, even dominant—especially in late-successional spruce forests. The quantity of N taken up by vegetation as amino acids, as well as the factors that control the uptake rate, are poorly known. Similarly, it remains to be identified when during succession organic N becomes an important N uptake pathway. These basic questions underscore the fundamental shift in our understanding of the N cycle brought about by the discovery of the dominance of organic N in boreal forest nutrition.

References

Anderson, J. M. 1991. The effects of climate change on decomposition processes in grassland and coniferous forests. Ecological Applications 1:326–347.

Anderson, M. D., R. W. Ruess, D. D. Uliassi, and J. S. Mitchell. 2004. Estimating N_2 fixation in two species of *Alnus* in interior Alaska using acetylene reduction and $^{15}N_2$ uptake. Ecoscience 11:102–112.

Apps, M. J., W. A. Kurz, R. J. Luxmoore, L. O. Nilsson, R. A. Sedjo, R. Schimdt, I. G. Simpson, and T. S. Vinson. 1993. Boreal forests and tundra. Water, Air and Soil Pollution 70:39–53.

Atkin, O. K. 1996. Reassessing the nitrogen relations of Arctic plants: A mini-review. Plant, Cell and Environment 19:695–704.

Barber, V. A., G. P. Juday, and B. P. Finney. 2000. Reduced growth of Alaskan white spruce in the twentieth century from temperature-induced drought stress. Nature 405:668–673.

Bartlett, K. B., P. M. Crill, R. L. Sass, R. C. Harriss, and N. B. Dise. 1992. Methane emissions from tundra environments in the Yukon-Kuskokwim Delta, Alaska. Journal of Geophysical Research 97:16645–16660.

Boone, R. D. 1992. Influence of sampling date and substrate on nitrogen mineralization:

Comparison of laboratory-incubation and buried-bag methods for two Massachusetts forest soils. Canadian Journal of Forest Research 22:1895–1900.

Castells, E., J. Peñuelas, and D. W. Valentine. 2003. Influence of the phenolic compound bearing species *Ledum palustre* on soil N cycling in a boreal hardwood forest. Plant and Soil 251:155–166.

Chapin, F. S., III. 1995. New cog in the nitrogen cycle. Nature 377:199–200.

Chapin, F. S, III, P. M. Vitousek, and K. Van Cleve. 1986. The nature of nutrient limitation in plant communities. American Naturalist 127:48–58.

Chapin, F. S, III, N. Fetcher, K. Kielland, K. R. Everett, and A. E. Linkins. 1988. Productivity and nutrient cycling of Alaskan tundra: Enhancement by flowing soil water. Ecology 69:693–702.

Chapin, F. S, III, B. H. Walker, R. J. Hobbs, D. U. Hooper, J. H. Lawton, O. E. Sala, and D. Tilman. 1997. Biotic control over the functioning of ecosystems. Science 277:500–504.

Falkengren-Grerup, U., K. F. Mansson, and M. O. Olsson. 2000. Uptake capacity of amino acids by ten grasses and forbs in relation to soil acidity and nitrogen availability. Environmental and Experimental Botany 44:207–219.

Flanagan, P. W., and K. Van Cleve. 1983. Nutrient cycling in relation to decomposition and organic matter quality in taiga forest ecosystems. Canadian Journal of Forest Research 13:795–817.

Gordon, A. M., and K. Van Cleve. 1983. Seasonal patterns of nitrogen mineralization following harvesting in the white spruce forests of interior Alaska. Pages 119–130 *in* R. W. Wein, R. R. Riewe, and I. R. Methven, editors. Resources and Dynamics of the Boreal Zone, Thunder Bay, Ontario. Association of Canadian Universities for Northern Studies, Ottawa.

Goulden, M. L., S. C. Wofsy, J. W. Harden, S. E. Trumbore, P. M. Crill, S. T. Gower, T. Fries, B. C. Daube, S. Fan, D. J. Sutton, A. Bazzaz, and J. W. Munger. 1998. Sensitivity of boreal forest carbon balance to warming. Science 279:214–217.

Gulledge, J., and J. P. Schimel. 2000. Controls on carbon dioxide and methane fluxes in a variety of taiga forest stands in interior Alaska. Ecosystems 3:269–282.

Gulledge, J., A. P. Doyle, and J. P. Schimel. 1997. Different NH_4^+ inhibition patterns of soil CH_4 consumption: A result of distinct CH_4-oxidizer populations across sites? Soil Biology and Biochemistry 29:13–21.

Hamilton, J. D., C. A. Kelly, J. W. M. Rudd, R. H. Hesslein, and N. T. Roulet. 1994. Flux to the atmosphere of CH_4 and CO_2 from wetland ponds on the Hudson Bay lowlands (HBLs). Journal of Geophysical Research 99(D1):1495–1510.

Harden, J. W., K. P. O'Neill, S. E. Trumbore, H. Veldhuis, and B. J. Stocks. 1997. Moss and soil contributions to the annual net carbon flux of a maturing boreal forest. Journal of Geophysical Research-Atmospheres 102(D24):28805–28816.

Harden, J. W., S. E. Trumbore, B. J. Stocks, A. Hirsch, S. T. Gower, K. P. O'Neill, and E. S. Kasischke. 2000. The role of fire in the boreal carbon budget. Global Change Biology 6(Suppl. 1):174–184.

Henrichs, S. M., and A. P. Doyle. 1986. Decomposition of ^{14}C-labeled substrates in marine sediments. Limnology and Oceanography 31:765–378.

Hines, M. E., K. N. Duddleston, and R. P. Kiene. 2001. Carbon flow to acetate and C-1 compounds in northern wetlands. Geophysical Research Letters 28:4251–4254.

Hobbie, S. E. 1996. Temperature and plant species control over litter decomposition in Alaskan tundra. Ecological Monographs 66:503–522.

Hobbie, S., and F. S. Chapin III. 1996. Winter regulation of tundra litter carbon and nitrogen dynamics. Biogeochemistry 35:327–338.

Jobbágy, E. G., and R. B. Jackson. 2000. The vertical distribution of soil organic carbon and its relation to climate and vegetation. Ecological Applications 10:423–436.

Jones, D. L. 1999. Amino acid biodegradation and its potential effects on organic nitrogen capture by plants. Soil Biology and Biochemistry 314:613–622.

Jones, D. L., and P. R. Darrah. 1994. Simple method for ^{14}C-labelling root maternal for use in root decomposition studies. Communications in Soil Science and Plant Analysis 25:2737–2743.

Jones, D. L., and K. Kielland. 2002. Soil amino acid turnover dominates the nitrogen flux in permafrost-dominated taiga forest soils. Soil Biology & Biochemistry 34:209–219.

Jones, D. L., A. G. Owen, and J. F. Farrar. 2002. Simple method to enable the high-resolution determination of total free amino acids in soil solutions and soil extracts. Soil Biology and Biochemistry 34:1893–1902.

Jones, D. L., A. C. Edwards, K. Donachie, and P. R. Darrah. 1994. Role of proteinaceous amino acids released in root exudates in nutrient acquisition from the rhizosphere. Plant and Soil 158:183–192.

Jørgensen, N. O. G., and M. Søndergaard. 1984. Are dissolved free amino acids free? Microbial Ecology 10:301–316.

Kasischke, E. S., and J. F. Johnstone. In press. Variation in post-fire organic layer thickness in a black spruce forest complex in interior Alaska and its effects on soil temperature and moisture. Canadian Journal of Forest Research.

Kielland, K. 1994. Amino-acid-absorption by arctic plants: Implications for plant nutrition and nitrogen cycling. Ecology 75:2373–2383.

Kielland, K. 1995. Landscape patterns of free amino acids in arctic tundra soils. Biogeochemistry 31:85–98.

Kielland, K. 1997. Role of free amino acids in the nitrogen economy of arctic cryptogams. Ecoscience 4:75–79.

Kielland, K. 2001. Short-circuiting the nitrogen cycle: Strategies of nitrogen uptake in plants from marginal ecosystems. Pages 376–398 in N. Ae, J. Arihara, K. Okada, and A. Srinivasan, editors. Plant Nutrient Acquisition: New Perspectives. Springer-Verlag, Berlin.

Kielland, K, B. Barnett, and D. Schell. 1998. Intraseasonal variation in the delta N-15 signature of taiga trees and shrubs. Canadian Journal of Forest Research 28:485–488.

Kirschbaum, M. U. F. 1995. The temperature dependence of soil organic matter decomposition, and the effect of global warming on soil organic C storage. Soil Biology and Biochemistry 27:753–760.

Kirschbaum, M. U. F. 2000. Will changes in soil organic carbon act as a positive or negative feedback on global warming? Biogeochemistry 48:21–51.

Klingensmith, K. M., and K. Van Cleve. 1993. Patterns of nitrogen mineralization and nitrification in floodplain successional soils along the Tanana River, interior Alaska. Canadian Journal of Forest Research 23:964–969.

Leake, J. R., and D. J. Read. 1989. The biology of mycorrhiza in the Ericaceae.13. Some characteristics of the extracellular proteinase activity of the ericoid endophyte Hymenoscyphus-Ericae. New Phytologist 112:69–76.

Marion, G. M., and P. C. Miller. 1982. Nitrogen mineralization in a tussock tundra soil. Arctic and Alpine Research 14:287–293.

Marion, G. M., K. Van Cleve, and C. T. Dyrness. 1993. Calcium-carbonate precipitation dissolution along a forest primary successional sequence on the Tanana River floodplain, interior Alaska. Canadian Journal of Forest Research 23:923–927.

Martens, C. S., C. A. Kelley, J. P. Chanton, and W. J. Showers. 1992. Carbon and hydrogen isotopic characterization of methane from wetlands and lakes of the Yukon-Kuskokwim Delta, western Alaska. Journal of Geophysical Research 97(D15):16689–16701.

McFarland, J. W., R. W. Ruess, K. Kielland, and A. P. Doyle. 2002. Cycling dynamics of NH_4^+ and amino acid nitrogen in soils of a deciduous boreal forest ecosystem. Ecosystems 5:775–788.

Michelsen, A., I. K. Schmidt, S. Jonasson, C. Quarmby, and D. Sleep. 1996. Leaf ^{15}N abundance of subarctic plants provides field evidence that ericoid, ectomycorrhizal, and non- and arbuscular mycorrhizal species access different sources of soil nitrogen. Oecologia 105:53–63.

Mosier, A. R., W. J. Parton, D. W. Valentine, D. S. Ojima, D. S. Schimel, and J. A. Delgado. 1996. CH_4 and N_2O fluxes in the Colorado shortgrass steppe: 1. Impact of landscape and nitrogen addition. Global Biogeochemical Cycles 10:387–399.

Nadelhoffer, K., G. Shaver, B. Fry, A. Giblin, L. Johnson, and R. McKane. 1996. ^{15}N natural abundances and N use by tundra plants. Oecologia 107:386–394.

Näsholm, T., K. Huss-Danell, and P. Högberg. 2000. Uptake of organic nitrogen in the field by four agriculturally important plant species. Ecology 81:1155–1161.

Näsholm, T., A. Ekblad, A. Nordin, R. Giesler, M. Högberg, and P. Högberg. 1998. Boreal forest plants take up organic nitrogen. Nature 392:914–916.

Neff, J. C., and D. U. Hooper. 2002. Vegetation and climate controls on potential CO_2, DOC and DON production in northern latitude soils. Global Change Biology 8:872–884.

Nilsson, M.-C., P. Högberg, O. Zachrisson, and W. Fengyou. 1993. Allelopathic effects by *Empetrum hermaphroditum* on development and nitrogen uptake by roots and mycorrhizae of *Pinus sylvestris*. Canadian Journal of Botany 71:620–628.

O'Neill, K. P. 2000. Changes in carbon dynamics following wildfire in soils of interior Alaska. Ph.D. Dissertation, Duke University, Durham, NC.

Prescott, C. E. 1995. Does nitrogen availability control rates of litter decomposition in forests? Plant and Soil 168–169:83–88.

Raab, T. K., D. A. Lipson, and R. K. Monson. 1999. Soil amino acid utilization among species of the Cyperaceae: Plant and soil processes. Ecology 80:2408–2419.

Richter, D. D., K. P. O'Neill, and E. S. Kasischke. 2000. Postfire stimulation of microbial decomposition in black spruce (*Picea mariana* L.) forest soils: A hypothesis. Pages 197–213 *in* E. S. Kasischke and B. J. Stocks, editors. Fire, Climate Change, and Carbon Cycling in the Boreal Forest. Springer-Verlag, New York.

Ruess, R. W., K. Van Cleve, J. Yarie, and L. A. Viereck. 1996. Contributions of fine root production and turnover to the carbon and nitrogen cycling in taiga forests of the Alaskan interior. Canadian Journal of Forest Research 26:1326–1336.

Ruess, R. W., R. L. Hendrick, A. J. Burton, K. S. Pregitzer, B. Sveinbjornsson, M. F. Allen, and G. E. Maurer. 2003. Coupling fine root dynamics with ecosystem carbon cycling in black spruce forests of interior Alaska. Ecological Monographs 73:643–662.

Schlesinger, W. H. 1977. Carbon balance in terrestrial detritus. Annual Review of Ecology and Systematics 8:51–81.

Schlesinger, W. H. 1997. Biogeochemistry: An Analysis of Global Change. 2nd ed. Academic Press, San Diego.

Schulze, E. D., F. S. Chapin III, and G. Gebauer. 1994. Nitrogen nutrition and isotope differences among life forms at the northern treeline of Alaska. Oecologia 100:406–412.

Steudler, P. A., R. D. Bowden, J. M. Melillo, and J. D. Aber. 1989. Influence of nitrogen fertilization on methane uptake in temperate forest soils. Nature 341:314–316.

Suyker, A. E., S. B. Verma, R. J. Clement, and D. P. Billesbach. 1996. Methane flux in a boreal fen: Season-long measurement by eddy correlation. Journal of Geophysical Research—Atmospheres 101(D22):28637–28647.

Swanson, D. K., B. Lacelle, and C. Tarnocai. 2000. Temperature and the boreal-subarctic maximum in soil organic carbon. Geographie Physique et Quaternaire 54:157–167.

Thornley, J. H. M., and M. G. R. Cannell. 2001. Soil carbon storage response to temperature: An hypothesis. Annals of Botany 87:591–598.
Turetsky, M. 2003. The role of bryophytes in carbon and nitrogen cycling. The Bryologist 106:395–409.
Valentine, D. W. 2002. Persistent decline in soil respiration following forest fire in interior Alaska. Abstract Soil Science Society of America.
Vance, E. D., and F. S. Chapin III. 2001. Substrate limitations to microbial activity in taiga forest floors. Soil Biology and Biochemistry 33:173–188.
Van Cleve, K., and V. Alexander. 1981. Nitrogen cycling in tundra and boreal ecosystems. Ecological Bulletin 33:375–404.
Van Cleve, K., F. S. Chapin III, C. T. Dyrness, and L. A. Viereck. 1991. Elemental cycling in taiga forests: State-factor control. BioScience 41:78–88.
Van Cleve, K., C. T. Dyrness, G. M. Marion, and R. E. Erickson. 1993. Control of soil development on the Tanana River floodplain of interior Alaska. Canadian Journal of Forest Research 23:941–955.
Van Cleve, K., and A. F. Harrison. 1985. Bioassay of forest phosphorus supply for plant growth. Canadian Journal of Forest Research 15:156–162.
Van Cleve, K., O. W. Heal, and O. Roberts. 1986. Bioassay of forest floor nitrogen supply for plant growth. Canadian Journal of Forest Research 16:1320–1326.
Van Cleve, K., W. C. Oechel, and J. L. Hom. 1990. Response of black spruce (*Picea mariana*) ecosystmes to soil temperature modifications in interior Alaska. Canadian Journal of Forest Research 20:1530–1535.
Van Cleve, K., L. Oliver, R. Schlentner, L. A. Viereck, and C. T. Dyrness. 1983. Productivity and nutrient cycling in taiga forest ecosystems. Canadian Journal of Forest Research 13:747–766.
Van Cleve, K., and R. F. Powers. 1995. Soil carbon, soil formation, and ecosystem development. Pages 155–200 *in* W. W. McFee and J. M. Kelly, editors. Carbon Forms and Functions in Forest Soils. Soil Science Society of America, Madison, WI.
Van Cleve, K., J. Yarie, R. E. Erickson, and C. T. Dyrness. 1993. Nitrogen mineralization and nitrification in successional ecosystems on the Tanana River floodplain, interior Alaska. Canadian Journal of Forest Research 23:970–978.
Viereck, L. A., C. T. Dyrness, and M. J. Foote. 1993. An overview of the vegetation and soils of the floodplain ecosystems of the Tanana River, interior Alaska. Canadian Journal of Forest Research 23:889–898.
Vogel, J. G. 2004. Carbon cycling in three mature black spruce (*Picea mariana* [Mill.] B.S.P.) forests in interior Alaska. Ph.D. Dissertation, University of Alaska Fairbanks.
Walker, L. R. 1989. Soil nitrogen changes during primary succession on a floodplain in Alaska, USA. Arctic and Alpine Research 21:341–349.
Walker, T. W., and J. K. Syers. 1976. The fate of phosphorus during pedogenesis. Geoderma 15:1–19.
Wardle, D. A., O. Zackrisson, and M.-C. Nilsson. 1998. The charcoal effect in boreal forests: mechanisms and ecological consequences. Oecologia 115:419–426.
Weintraub, M. N., and J. P. Schimel. 2003. Interactions between carbon and nitrogen mineralization and soil organic matter chemistry in arctic tundra soils. Ecosystems 6:129–143.
Whalen, S. C., W. W. Reeburgh, and K. S. Kizer. 1990. Methane consumption and emission from taiga sites. Global Biogeochemical Cycles 5: 261–274.
Yamagata, M., S. Matsumoto, and N. Ae. 2001. Possibility of direct acquisition of organic nitrogen by crops. Pages 399–420 *in* N. Ae, J. Arihara, K. Okada, and S. Srinivasan, editors. Plant Nutrient Acquisition: New Perspectives. Springer-Verlag, Berlin.

Yarie, J., and K. Van Cleve. 1983. Biomass and productivity of white spruce stands in interior Alaska. Canadian Journal of Forest Research 13:767–672.

Yarie, J., and K. Van Cleve. 1996. Effects of carbon, fertilizer, and drought on foliar chemistry of tree species in interior Alaska. Ecological Applications 6:815–827.

Zackrisson, O., and M.-C. Nilsson. 1992. Allelopathic effects by *Empetrum hermaphroditum* on seed germination of two boreal tree species. Canadian Journal of Forest Research 22:1310–1319.

Zackrisson, O., M.-C. Nilsson, and D. A. Wardle. 1996. Key ecological function of charcoal from wildfire in the boreal forest. Oikos 77:10–19.

Zackrisson, O., M.-C. Nilsson, A. Dahlberg, and A. Jäderlund. 1997. Interference mechanisms in conifer-Ericaceae-feathermoss communities. Oikos 78:209–220.

Zhuang, Q., A. D. McGuire, K. P. O'Neill, J. W. Harden, V. E. Romanovsky, and J. Yarie. 2002. Modeling soil thermal and carbon dynamics of a fire chronosequence in interior Alaska. Journal of Geophysical Research—Atmospheres 108(8147):doi:10.1029/2001JD001244 [printed 108(D1), 2003].

Part IV

Changing Regional Processes

16

Watershed Hydrology and Chemistry in the Alaskan Boreal Forest
The Central Role of Permafrost

Larry D. Hinzman Jeremy B. Jones
W. Robert Bolton Phyllis C. Adams
Kevin C. Petrone

Introduction

Hydrological processes exert strong control over biological and climatic processes in every ecosystem. They are particularly important in the boreal zone, where the average annual temperatures of the air and soil are relatively near the phase-change temperature of water (Chapter 4). Boreal hydrology is strongly controlled by processes related to freezing and thawing, particularly the presence or absence of permafrost. Flow in watersheds underlain by extensive permafrost is limited to the near-surface active layer and to small springs that connect the surface with the subpermafrost groundwater (Fig. 16.1). Ice-rich permafrost, near the soil surface, impedes infiltration, resulting in soils that vary in moisture content from wet to saturated (Fig. 16.2). Interior Alaska has a continental climate with relatively low precipitation (Chapter 4). Soils are typically aeolian or alluvial (Chapter 3). Consequently, in the absence of permafrost, infiltration is relatively high, yielding dry surface soils (Fig. 16.2). In this way, discontinuous permafrost distribution magnifies the differences in soil moisture that might normally occur along topographic gradients.

Conceptual Model of Boreal Forest Watersheds

Watershed Structure and Physical Controls

Hydrological processes in the boreal forest are unique due to highly organic soils with a porous organic mat on the surface, short thaw season, and warm summer

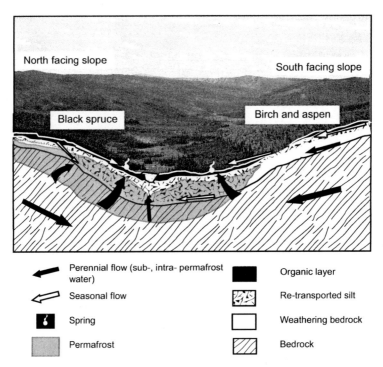

Figure 16.1. The presence or absence of ice-rich permafrost exerts the dominant controlling influence on hydrological processes in northern boreal forests, allowing or preventing infiltration to subsurface groundwater. Soils over ice-rich permafrost tend to be wetter, with thicker organic mats, than soils in permafrost-free areas.

and cold winter temperatures. The surface organic layer tends to be much thicker on north-facing slopes and in valley bottoms than on south-facing slopes and ridges, reflecting primarily the distribution of permafrost. Soils are cooler and wetter above permafrost, which retards decomposition, resulting in organic matter accumulation (Chapter 15). The markedly different material properties of the soil layers also influence hydrology. The highly porous near-surface soils allow rapid infiltration and, on hillsides, downslope drainage. The organic layer also has a relatively low thermal conductivity, resulting in slow thaw below thick organic layers. The thick organic layer limits the depth of thaw each summer to about 50–100 cm above permafrost (i.e., the active layer). As the active layer thaws, the hydraulic properties change. For example, the moisture-holding capacity increases, and additional subsurface layers become available for lateral flow.

Hydrology/Vegetation Interactions

The mosaic of Alaskan vegetation depends not only on disturbance history (Chapter 7) but also on hydrology (Chapter 6). Lowland vegetation must be tolerant of saturated soils in valley bottoms underlain by ice-rich permafrost, where vegetation is

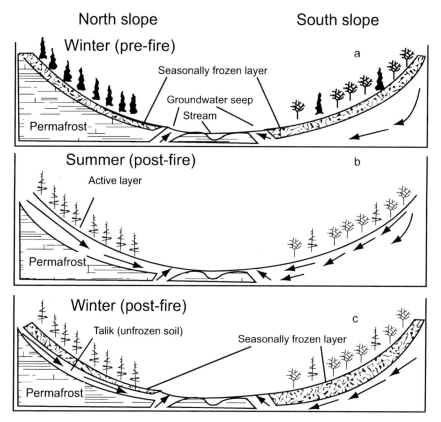

Figure 16.2. Fire effects on active layer dynamics and talik formation. Top: winter refreezing of the active layer in absence of fire; middle: summer active layer depth after fire; bottom: incomplete winter refreezing after fire, forming a talik at depth.

dominated by stunted black spruce, sedge tussocks, and sphagnum mosses. Alder, birch, willow, and white spruce prefer the drier south-facing slopes and hilltops. Vegetation and permafrost distribution are greatly influenced by the local surface energy balance, which in turn governs near-surface thermal processes (Fig. 16.3).

In addition to influencing the energy balance, the species distribution also exerts strong controls on the local water balance (Chapin et al. 2000). The albedo of the deciduous stands is nearly twice that of the darker coniferous forests in winter and in summer (Baldocchi et al. 2000; Chapter 19). Upland stands of deciduous species tend to have higher rates of evapotranspiration (nearly approaching equilibrium evaporation, i.e., the rate of evaporation attained by a freely evaporating wet surface after it saturates the atmosphere with humidity) as compared to upland coniferous stands (about 25–75% of equilibrium evaporation), contributing to the drier soils found in deciduous forests (Baldocchi et al. 2000). In deciduous stands, nearly 90% of summer precipitation returns to the atmosphere via transpiration by the tree canopy, whereas 30–90% of the water loss from coniferous stands occurs

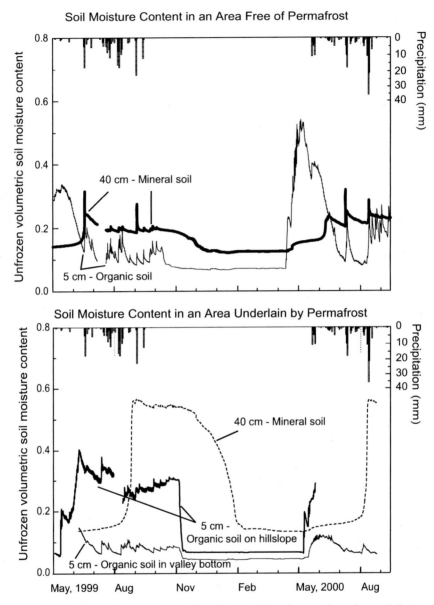

Figure 16.3. Soil moisture variation in a site that is (top) permafrost-free and (bottom) dominated by permafrost.

by evaporation from the moist forest floor (Baldocchi et al. 2000). The higher moisture content of soils under a coniferous stand (larger heat capacity), coupled with the greater insulative capacity of the mosses and the greater cooling associated with greater evaporation from the surface, all contribute to maintenance of cold soils in coniferous stands. Although these differences are largely the result of physiological differences between coniferous and deciduous species, they impart significant differences to the physical environment. Because of their high absorption of net radiation (low albedo) and their limited loss of latent heat (low transpiration), coniferous stands lose most of their energy through sensible heat flux, at rates at least twice those for deciduous stands (Chapin et al. 2000).

Aufeis (icing) exerts a very strong impact on vegetation, geomorphology, and hydrology over limited areas, usually where groundwater springs emerge or along streams (Fig. 16.4). Aufeis usually appears in late winter, especially in February to March. Ice thickness may exceed 10 m but is normally 1–2 m in interior Alaska (Slaughter 1990). Aufeis usually disappears by the middle of June, so that exposure of the ground and the leaf-out of vegetation occurs more than a month later than in snow-covered terrain. Ground-surface temperature remains near 0°C for the entire period of ice cover. Ice also dams stream channels, and new channels may become established temporarily or permanently. Newly developed aufeis kills most perennial (tree) species, dramatically changing the community of the impacted areas.

Watershed Chemistry Dynamics

Wetfall Chemical Deposition

Precipitation in interior Alaska is quite low in solutes, has a relatively high pH, and is relatively uninfluenced by anthropogenic aerosols. Of the base cations in precipitation, sodium is the dominant ion in terms of molar concentration, and calcium and sodium are dominant in terms of charge balance (Table 16.1). For anions, sulfate, chloride, and nitrate have similar molar concentrations, whereas sulfate is the dominant anion in terms of charge. The concentrations of ions exhibit three seasonal patterns. The major cations, with the exception of sodium, have their greatest concentrations in the spring and their lowest concentrations during the autumn and winter. Sodium and chloride have a contrasting pattern, with greatest concentrations in the autumn and lowest concentrations in the spring. Last, nitrate concentrations in precipitation tend to behave independently of the other ions, with winter maximum and summer minimum.

In spite of the low solute concentrations, anthropogenic influences on precipitation chemistry are evident. Sulfate, which is predominantly derived from marine and industrial sources, is quite enriched in precipitation relative to chloride, compared to seawater. Even more pronounced, nitrate and ammonium concentrations are enriched in precipitation by six orders of magnitude relative to seawater, in which nitrogen occurs in trace concentrations. Further, nitrate concentration and enrichment do not exhibit the same seasonality or vary in conjunction with other solutes.

Figure 16.4. Aufeis forms when groundwater is forced to the surface during winter as freezing soil pinches off subsurface flow pathways.

The lack of concordance between nitrate and most other ions in precipitation suggests that nitrate is derived predominantly from an anthropogenic source.

Weathering

Much of interior Alaska is underlain by the Yukon-Tanana metamorphic complex, consisting of polydeformed and polymetamorphosed Upper Paleozoic and older metasedimentary, metavolcanic, and metaplutonic rocks (Newberry et al. 1996) and the Ancestral North America Terrane, consisting of low-grade metamorphic and sedimentary rocks of Late Precambrian to Carboniferous age (Chapters 2 and 3). The geology is moderately to highly weathered because weathering has occurred for millions of years, in the absence of glaciation, without removal of weathering products by glaciers. In some places, one can find ancient erosional surfaces on metamorphic rocks overlain by 55-million-year-old basalt flows. The metamorphic rocks are covered by a thin veneer of loess (wind-blown, glaciallyderived sediment), which was deposited starting about 2 million years ago (Chapter 2). Consequently, surface and ground waters have relatively low weathering-derived ion concentrations and alkalinity. As further evidence of the low degree of current weathering, concentrations of the base cations (calcium, magnesium, potassium, and sodium) in Caribou Creek headwater streams in interior Alaska are not greatly different between soil and deeper ground waters (Table 16.2), contrary to expectation if bedrock weathering were releasing substantial solutes. As a consequence of the low rates of weathering, streams of interior Alaska are not well buffered. For redox-sensitive solutes, the soils of interior Alaska watersheds appear to be an important sink. Nitrate and sulfate are commonly enriched in soils relative to deeper ground

Table 16.1. Precipitation solute concentrations from the National Atmospheric Deposition Program/National Trends Network Poker Creek site in interior Alaska. Concentrations are precipitation-weighted means (with weighted standard errors), 1993–1999.

	Winter	Spring	Summer	Autumn	Annual
Ca^{2+} (μM)	0.54 (0.03)	1.19 (0.11)	0.64 (0.01)	0.46 (0.01)	0.73 (0.01)
Mg^{2+} (μM)	0.20 (0.01)	0.31 (0.02)	0.22 (0.00)	0.21 (0.01)	0.24 (0.00)
K^+ (μM)	0.09 (0.00)	0.66 (0.04)	0.33 (0.01)	0.13 (0.01)	0.35 (0.01)
Na^+ (μM)	1.42 (0.06)	0.94 (0.03)	1.46 (0.03)	1.89 (0.08)	1.40 (0.01)
SO_4^{2-} (μM)	1.07 (0.03)	2.49 (0.05)	1.88 (0.02)	1.03 (0.03)	1.81 (0.01)
Cl^- (μM)	1.48 (0.07)	1.31 (0.03)	1.78 (0.03)	2.49 (0.09)	1.75 (0.01)
NO_3^- (μM)	2.66 (0.08)	2.54 (0.05)	1.94 (0.02)	2.36 (0.06)	2.21 (0.01)
NH_4^+ (μM)	1.06 (0.03)	2.53 (0.11)	1.90 (0.04)	1.10 (0.03)	1.84 (0.02)
H^+ (μM)	3.38 (0.08)	6.57 (0.09)	5.39 (0.05)	3.61 (0.06)	5.20 (0.02)
pH	5.47	5.18	5.27	5.44	5.28

waters due to processes such as atmospheric deposition and organic matter decomposition. In headwater streams of Caribou Creek, however, nitrate and sulfate concentrations are greater in deeper ground water than soil water, suggesting that anaerobic pathways of denitrification and sulfate reduction may consume much of the nitrate and sulfate in soil water (MacLean et al. 1999).

In spite of the presumed loss of nitrate in soils through denitrification, nitrogen export in headwater streams of the Caribou Creek watershed is much greater than input via deposition. Inorganic nitrogen deposition in CPCRW averages 1420 mol N km^{-2} y^{-1} (SE = 175), of which nitrate and ammonium account for 60% and 40% of deposition, respectively. Nitrate export, however, is up to seven times greater, ranging from 1356 to 9499 mol N km^{-2} y^{-1} (Ray 1988, MacLean et al. 1999, Petrone 2005). The export rate of inorganic nitrogen varies inversely with the extent of permafrost underlying watersheds; nitrate flux during 1986 from low (3.5%), medium (18.8%), and high (53.2%) permafrost watersheds averaged 9499, 7677, and 5972 mol N km^{-2} y^{-1}, respectively (Ray 1988). Moreover, dissolved organic nitrogen export is nearly as great as inorganic nitrogen fluxes, with rates ranging from 2239 to 7710 mol N km^{-2} y^{-1} (MacLean et al. 1999, Petrone 2005), and the total export of dissolved nitrogen is upwards of an order of magnitude greater than import by deposition. The magnitude of difference between import and export, in conjunction with greater export from low-permafrost watersheds, is suggestive of a deeper, geologic source of nitrogen or large sources of nitrogen fixation within the watersheds.

Effects of Climate and Permafrost on Streamflow Processes

In nonglacially fed drainages, two periods of high flows usually occur; one during the snowmelt period and the other during late summer or fall rain events (Fig. 16.5). Glacially fed streams or rivers exhibit peak flows during late summer, when large

Table 16.2. Mean soil, ground, and stream water chemistry for Caribou Creek tributaries in interior Alaska (± SE).

	Soil water	Ground water	Stream water	Riparian
Alkalinity (μEq/L)	524.8 (82.9)	740.9 (120.2)	633.9 (16.5)	
Ca^{2+} (μM)	231.0 (37.0)	349.9 (59.6)	291.6 (8.7)	
Mg^{2+} (μM)	84.4 (14.5)	118.5 (14.7)	100.5 (2.4)	
K^+ (μM)	14.2 (2.0)	17.4 (1.7)	15.9 (0.4)	
Na^+ (μM)	48.6 (3.6)	54.9 (5.7)	51.3 (0.9)	
SO_4^{2-} (μM)	50.2 (4.3)	95.1 (17.0)	73.4 (2.9)	
Cl^- (μM)	9.6 (0.3)	9.6 (0.2)	9.8 (0.3)	
Fe^{3+} (μM)	1.7 (0.4)	0.2 (0.2)	1.0 (0.1)	
SiO_2 (μM)	121.7 (6.9)	142.1 (2.7)	131.5 (1.7)	
NO_3^- (μM)	29.6 (5.0)	36.5 (5.2)	33.2 (0.9)	3.9 (1.0)[2]
NH_4^+ (μM)			0.01 (0.01)[1]	3.5 (0.6)[2]
PO_4^- (μM)			0.49 (0.45)[2]	1.3 (0.1)[2]
DON (μM)	86.0 (9.8)		50.0 (5.0)[1]	
DOC (μM)	1157 (68.9)			1928.6 (685.7)[2]

During winter, when soils are frozen, stream discharge is fed by deeper ground waters; ground water solute concentrations were assumed equal to winter stream baseflow concentrations. Soil water concentrations were estimated using Keeling-type plots by solving for the y-intercept of stream water concentrations versus the inverse of discharge and assuming that increase in stream flow is fed by discharge from soils.

[1]From MacLean et al. (1999).
[2]From Petrone, unpublished data.

Source: Soil and ground water means were estimated from stream chemistry data of Ray (1988) except as noted.

inputs of energy from solar radiation accelerate the melting of glaciers (Woo 1986, Adams 1999). In glacially fed systems, solar energy input is the major control over the hydrologic flow pattern, whereas precipitation governs streamflow of nonglacial streams (Woo 1986). Milner and Oswood (1997) present a summary of streamflow patterns for major classes of Alaskan running waters (Chapter 10).

North- and south-facing slopes (permafrost vs. permafrost-free) have different hydrological responses during the snowmelt period. Because of the high ice content of wet soils on north-facing slopes, infiltration of meltwater is limited, resulting in runoff through the near-surface organic layer (with high hydraulic conductivities) or in overland flow (Slaughter and Kane 1979, Woo 1986, Gibson et al. 1993, Carey and Woo 1998, 2000). In interior Alaska, in the areas free of permafrost, surface runoff derived from snowmelt seldom occurs in uplands (Kane and Stein 1983b, Gibson et al. 1993, Carey and Woo 1998) because of the high infiltration rate of nonpermafrost soils relative to the rate water is supplied by snowmelt. In addition, significant rainfall seldom occurs during the snowmelt period. Infiltrating meltwater rapidly raises the temperature of frozen soils through release of latent heat during refreezing (Kane et al. 2001) increasing the infiltration capacity. The infiltration rate of seasonally frozen soils is inversely proportional to the soil moisture before the winter freeze-up (Kane and Stein 1983a). Interstitial ice significantly reduces hydraulic conductivity compared to ice-free, frozen soils. For

Watershed Hydrology and Chemistry in the Alaskan Boreal Forest 277

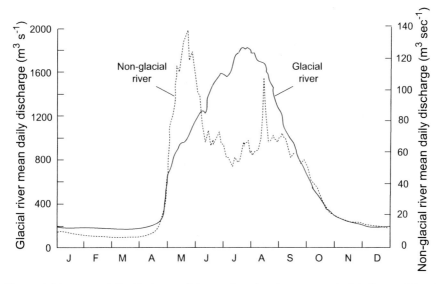

Figure 16.5. Mean daily discharge of rivers whose headwaters are glacial (Tanana River) and nonglacial (Chena River).

example, the infiltration rate of a dry, frozen Fairbanks silt loam is two orders of magnitude greater than that for the same soil that is saturated and frozen.

Although the snowmelt period is usually the major hydrological event of year, the maximum peak stream flow of record occurs during rainstorms in small watersheds because the atmosphere can deliver more rain to these watersheds in a short time period than is physically possible to melt during a similar time period. In very large watersheds, the maximum peak flow is usually caused by snowmelt, because the area contributing to river runoff in a large basin is usually greater than the area of any single rain event. During summer, a soil moisture deficit is created by high evapotranspiration demands. Consequently, the light, showery rainfalls typical of interior Alaska have little effect on stream flow because most of this water is used to meet the soil moisture demand (Gieck and Kane 1986). During large precipitation events, both the soil moisture demand and infiltration rates are met or exceeded, and hydrological responses are observed.

Watersheds underlain by a large proportion of permafrost display higher peak specific discharges (stream discharge normalized by basin area), lower baseflow between storms, and a faster response time to precipitation compared to watersheds with less permafrost (Fig. 16.6; Haugen et al. 1982, Bolton et al. 2000). Water can infiltrate throughout the watershed if there is a low proportion of permafrost, and base flow can be much greater and sustained longer. Water in watersheds with high amounts of permafrost exits hillsides as near-surface runoff, causing a shorter response time to rain and generating stream water with a significantly different chemistry from that found in low-permafrost basins. During storms, streamflow is derived not from the entire watershed but from a smaller region near the stream channel,

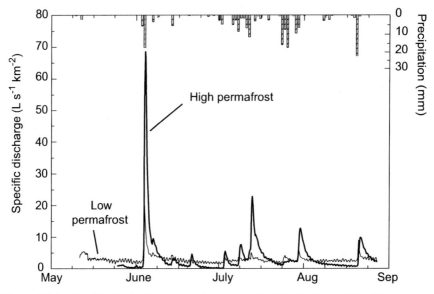

Figure 16.6. Specific discharge (stream flow normalized by basin area) in a high-permafrost and a low-permafrost watershed. Data from Bolton et al. (2000).

called the contributing area. The contributing area in a low-permafrost or permafrost-free watershed is quite small compared with that in high-permafrost watersheds. Consequently, watersheds with large amounts of permafrost may have very flashy flows, with high peaks, but low baseflows.

During both snowmelt and large rainfalls, pipeflow contributes significant amounts of water to streamflow (Roberge and Plamondon 1987, Gibson et al. 1993, Carey and Woo 1998, 2000). Pipeflow occurs at the interface of the organic and mineral soil horizons, providing a preferential runoff mechanism, bypassing the soil matrix (Carey and Woo 2000). During snowmelt and large rainfalls, pipeflow can contribute, respectively, more than 20% and 15% of the total runoff (Carey and Woo 2000).

Nutrient and Water Flowpaths: A Conceptual Model

Stream water chemistry is strongly regulated by discontinuous permafrost in the boreal forest of interior Alaska. At the Caribou Poker Creeks Research Watershed (CPCRW), near Fairbanks, two watersheds are similar in area but differ in permafrost coverage (3% and 50%). Coupled with differences in permafrost and groundwater flows, the high-permafrost watershed has higher DOC and lower dissolved mineral concentrations. MacLean et al. (1999) developed a conceptual model that explains these patterns. During stormflows, permafrost confines water movement to the organic-rich active layer, resulting in the dissolution of DOC and higher concentrations of DOC in stormflow in the high-permafrost watershed compared with the low. Water in the high-permafrost watershed also accumulates fewer minerals from weathering processes because permafrost prevents deep infiltration of groundwater through the cation-rich mineral soil.

Stream Chemistry and Source Delineation

Detailed hydrograph and chemistry observations provide additional insight into water flowpaths and the contributing area for flow during summer storms. During storms in July and September 1999, DOC increased in all watersheds of CPCRW, but the DOC increase was greatest in the high-permafrost watershed (Fig. 16.7). Nitrate revealed opposite patterns for high- and low-permafrost watersheds during these storms. Nitrate was negatively related to discharge in the low-permafrost watershed and positively related to discharge in the high-permafrost watershed (Fig. 16.7). The contributing area for flow, calculated from an end-member mixing model as new water discharge divided by the event precipitation, was 16-fold smaller in

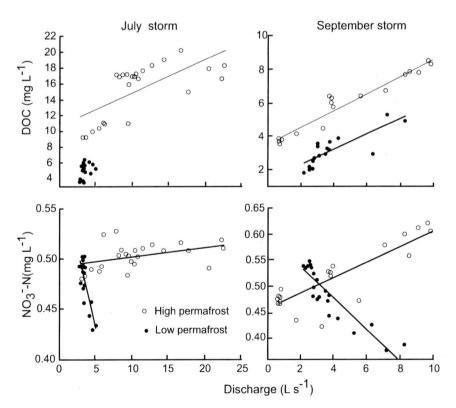

Figure 16.7. Relationships between discharge and stream concentrations of dissolved organic carbon (DOC) and NO_3^- for a July and a September storm in a high-permafrost and a low-permafrost watershed. Linear regressions are significant, $p < 0.05$. Redrawn from Petrone et al. (2000).

the low-permafrost watershed (1.5ha) than in the high-permafrost watershed (23.5 ha) for both storms (Petrone et al. 2000).

Hydrograph and chemistry analyses support the permafrost and water flowpath conceptual model. New water was generated from the permafrost-underlain riparian zone and hillslopes in high-permafrost watersheds, but only in the riparian zone in low-permafrost watersheds (Fig. 16.1). Nitrate was most likely leached in the active layer of the high-permafrost watershed, while the nitrate concentration decreased in the low-permafrost watershed. This is the result of the more direct flowpath in the high-permafrost watershed, where precipitation fell on the saturated valley bottom and entered the stream as saturated overland flow. In the low-permafrost watershed, the water that entered the stream was a mixture of two low-nitrate water sources: precipitation and water in anaerobic soils of the valley bottom, which had probably undergone denitrification. This resulted in a decrease in nitrate concentration during the stormflow peak. For both watersheds, the source area for stormflow was underlain by permafrost, but different flowpaths and source areas (anaerobic riparian soils vs. aerobic hillslope soils) caused contrasting patterns of nitrate chemistry in these streams.

Soil and Groundwater Chemistry

Soil chemistry is affected by permafrost and soil type in interior Alaska (Chapter 15; Yarie et al. 1993, Dyrness and Van Cleve 1993). Black spruce stands underlain by permafrost with cold, organic-rich soils have lower pH and bulk concentrations of N, P, Ca, Mg, Mn, and Zn than permafrost-free deciduous stands (Troth et al. 1976). In addition, soil water DIN and DON are lower in black spruce stands than in deciduous stands (MacLean et al. 1999). These lower nutrient concentrations reflect slower rates of microbial decomposition and nutrient turnover in black spruce than in deciduous stands (Chapter 15). Over time, slow decomposition rates in black spruce result in an accumulation of organic material on the forest floor. A thick moss layer covers the organic layer, reinforcing colder temperatures, and soils begin to thaw only in early summer. In poorly drained sites, the active layer may be less than 30 cm deep. The combined effect of low temperatures due to permafrost and the consequent accumulation of organic material reinforces low productivity in black spruce stands where the rooting zone is limited to the shallow active layer.

Although there are chemical differences between soil types, soils in the rooting zones of black spruce and deciduous forests are consistently lower in Ca, Mg, Na, and SO_4 but higher in DOC compared to their respective well and spring waters (MacLean et al. 1999). It appears that DOC declines as water moves from surface soils into deeper groundwater, and this has been attributed to DOC adsorption by mineral soils in this area (MacLean et al. 1999). Dissolved minerals increase from bedrock weathering through the same flowpath. Stream chemistry reflects the observed soil and groundwater chemistry. Streamflow chemistry in the high-permafrost watershed is more like soil water chemistry during high flows because water is moving through the shallow active later (MacLean et al. 1999, Petrone et al. 2000). In contrast, stream chemistry in the low-permafrost watershed shows this similarity only during the snowmelt period, when the mineral soils are still frozen.

Fluvial Processes in Large Rivers

Ecosystem dynamics on alluvial rivers are closely linked to fluvial processes (Swanson and Lienkaemper 1982, Mason and Begét 1991). These fluvial patterns are determined primarily by the flow regime of the river (Poff et al. 1997), which is, in turn, regulated by climate through control of rate of glacier melt, as well as precipitation (Swanson et al. 1998). Floods occur in spring, when ice jams cause water to back up, and in mid- to late summer, when glacier melt is at its maximum and synoptic storms (i.e., atmospheric systems large enough to affect large regions) produce precipitation (Adams 1999). River flow patterns and flood frequencies regulate silt deposition and the formation and destruction of substrate. Terrace aggradation, along with bank erosion, determine the spatial distribution of successional terraces and the area available for colonization. Primary succession is initiated on those newly formed terraces that continue to aggrade as sediment is deposited by flood events (Yarie et al. 1998; Chapter 7). Subsequent siltation buries nutrient-rich organic layers, provides a mineral soil seedbed required for woody plant establishment, and adds nutrients to the soil (Gill 1973, Bradley and Smith 1986, Krasny 1986, Viereck et al. 1993). Silt deposition also increases productivity through soil warming by burying feathermosses and removing the insulating effects of this organic ground cover (Yarie et al. 1998).

Fluvial processes such as flooding, sedimentation, and erosion interact with biotic processes, including seed dispersal and seedling establishment, to regulate the temporal and spatial variability of forest development on boreal river floodplains (Chapter 7). For example, floods disperse seeds to early-successional sites (Walker et al. 1986), and patterns of silt deposition control the variability in texture of early successional floodplain soils, which influences the colonization of some species (Krasny 1982). Finally, productivity is enhanced in early-successional floodplain stands by the readily accessible groundwater. Surface soil moisture on the Tanana River floodplain depends on water table depth, which is directly related to river level (Viereck et al. 1993). The water-table fluctuation influences the radial growth of the shallow-rooted white spruce trees.

Conclusions and Future Projections

General circulation models predict that the effects of increased ambient temperature will be most pronounced at high latitudes. Discontinuous permafrost in interior Alaska is particularly sensitive to temperature changes because permafrost temperatures are only slightly below freezing (Chapter 4; Osterkamp and Romanovsky 1999). Although permafrost inhibits water infiltration, there is interaction of surface soil water with groundwater through taliks (areas of unfrozen soil within the permafrost), which have upward or downward flow (Woo 1986). Taliks can continuously drain during the winter, allowing ground aufeis to form on toe slopes (Fig. 16.4).

A warming trend could increase the drainage of existing taliks and form new taliks. For instance, higher summer temperatures combined with a changing disturbance regime may increase the active layer thickness each year. Eventually, the thawed soils may not refreeze during the winter months, creating taliks above the permafrost but below the seasonally frozen soil (Fig. 16.1). Drainage of these taliks could increase the interaction of soil water and groundwater. Greater infiltration of precipitation would result in adsorption and mineralization of DOC and DON but increase dissolved ions from weathering processes. Therefore, an apparently simple consequence of climate change in boreal forests—decrease in permafrost—will likely have cascading consequences for soil moisture and soil chemistry, with resulting effects on vegetation and watershed nutrient fluxes.

References

Adams, P. C. 1999. The dynamics of white spruce populations on a boreal river floodplain. Ph.D. Dissertation, Duke University, Durham, NC.

Baldocchi, D., F. M. Kelliher, T. A. Black, and P. Jarvis. 2000. Climate and vegetation controls on boreal zone energy exchange. Global Change Biology 6:69–83.

Bolton, W. R., L. Hinzman, and K. Yoshikawa. 2000. Stream flow studies in a watershed underlain by discontinuous permafrost. Pages 31–36 in D. L. Kane, editor. Proceedings, Water Resources in Extreme Environments, American Water Resources Association, Anchorage, AK.

Bradley, C. E., and D. G. Smith. 1986. Plains cottonwood recruitment and survival on a prairie meandering river floodplain, Milk River, southern Alberta and northern Montana. Canadian Journal of Botany 64:1433–1442.

Campbell, S. Gayon S., and J. M. Norman. 1998. An Introduction to Environmental Biophysics, 2nd ed. Springer-Verlag, New York.

Carey, S. K., and M. K. Woo. 1998. Snowmelt hydrology of two subarctic slopes, Southern Yukon, Canada. Nordic Hydrology 29:331–346.

Carey, S. K., and Woo, M. K. 2000. The role of soil pipes as a slope runoff mechanism, Subarctic Yukon, Canada. Journal of Hydrology 233:306–222.

Chapin, F. S., III, A. D. McGuire, J. Randerson, R. Pielke Sr., D. Baldocchi, S. E. Hobbie, N. Roulet, W. Eugster, E. Kasischke, E. B. Rastetter, A. Zimov, and S. W. Running. 2000. Arctic and boreal ecosystems of western North America as components of the climate system. Global Change Biology 6:211–223.

Dyrness, C. T., and K. Van Cleve. 1993. Control of soil chemistry in early successional floodplain soils along the Tanana River, interior Alaska. Canadian Journal of Forest Research 23: 979–994.

Gibson, J. J., T. W. D. Edwards, and T. D. Prowse. 1993. Runoff generation in a high boreal wetland in Northern Canada. Nordic Hydrology 24:213–224.

Gieck, R. E., and D. L. Kane. 1986. Hydrology of two subarctic watersheds. Pages 283–291 in D. L. Kane, editor. Proceedings, Cold Regions Hydrology Symposium, American Water Resources Association, Anchorage, AK.

Gill, D. 1973. Floristics of plant succession sequence in the Mackenzie Delta. Polarforschung 43:55–65.

Haugen, R. K., C. W. Slaughter, K. E. Howe, and S. L. Dingman. 1982. Hydrology and Climatology of the Caribou-Poker Creeks Research Watershed, Alaska. CRREL Report 82-26, Cold Regions Research and Engineering Laboratory, Hanover, NH.

Kane, D. L., K. M. Hinkel, D. J. Goering, L. D. Hinzman, and S. I. Outcalt. 2001. Non-conductive heat transfer associated with frozen soils. Global and Planetary Change 29:275–292.

Kane, D. L., and J. Stein. 1983a. Field evidence of groundwater recharge in interior Alaska. Pages 572–577 *in* Proceedings Fourth International Conference on Permafrost. National Academy Press, Washington, DC.

Kane, D. L., and J. Stein. 1983b. Water movement into seasonally frozen soils. Water Resources Research 19:1547–1557.

Krasny, M. E. 1982. White spruce (*Picea glauca*) seedling growth and fine root and mycorrhizal activity in four successional communities on the Tanana River floodplain, Alaska. M.S. Thesis, University of Washington, Seattle.

Krasny, M.E. 1986. Establishment of four Salicaceae species on river bars along the Tanana River, Alaska. Ph.D. Dissertation, University of Washington, Seattle.

MacLean, R., M. W. Oswood, J. G. Irons III, and W. H. McDowell. 1999. The effect of permafrost on stream biogeochemistry: A case study of two streams in the Alaskan (USA) taiga. Biogeochemistry 47:239–267.

Mason, O.K., and J. E. Begét. 1991. Late Holocene flood history of the Tanana River, Alaska, USA. Arctic and Alpine Research 23:392–403.

Milner, A. M., and M. W. Oswood. 1997. Freshwaters of Alaska—Ecological Synthesis. Vol. 119. Ecological Studies. Springer-Verlag, New York.

Newberry, R. J., and T. K. Bundtzen (with contributions from K. H. Clautice, R. A. Combellick, T. Douglas, G. M. Laird, S. A. Liss, D. S. Pinney, R. R. Reifenstuhl, and D. N. Solie). 1996. Preliminary Geologic Map of the Fairbanks Mining District, Alaska. State of Alaska Division of Geological and Geophysical Surveys, Public-Data File 96–16.

Osterkamp, T. E., and V. E Romanovsky. 1999. Evidence for warming and thawing of discontinuous permafrost in Alaska. Permafrost and Periglacial Processes 10:17–37.

Petrone, K. C. 2005. Export of carbon, nitrogen, and major solutes from a boreal forest watershed: The influence of fire and permafrost. Ph.D. Dissertation, University of Alaska Fairbanks.

Petrone, K. C., L. D. Hinzman, and R. D. Boone. 2000. Nitrogen and carbon dynamics of storm runoff in three sub-arctic streams. Pages 167–172 *in* Proceedings, American Water Resources Association, Water Resources in Extreme Environments, Anchorage, Alaska.

Poff, N. L., J. D. Allan, M. B. Bain, J. R. Darr, K. L. Prestegaard, B. D. Richter, R. E. Sparks, and J. C. Stromberg. 1997. The natural flow regime. A paradigm for river conservation and restoration. Bioscience 47:769–784.

Ray, S. R. 1988. Physical and chemical characteristics of headwater streams at Caribou-Poker Creeks Research Watershed, Alaska. M.S. Thesis, University of Alaska Fairbanks.

Roberge, J., and A. P. Plamondon. 1987. Snowmelt runoff pathways in a boreal forest hillslope, the role of pipe throughflow. Journal of Hydrology 95:39–54.

Slaughter, C. W. 1990. Aufeis formation and prevention. Pages 433–458 *in* W. L. Ryan and R. D. Crissman, editors. Cold Regions Hydrology and Hydraulics. Technical Council on Cold Regions Engineering Monograph. American Society of Civil Engineers, New York.

Slaughter, C. W., and D. L. Kane. 1979. Hydrologic role of shallow organic soils in cold climates. Pages 380–390 *in* Proceedings, Canadian Hydrological Symposium, National Resources Council, Vancouver, British Columbia.

Swanson, F. J., and G. W. Lienkaemper. 1982. Interactions among fluvial processes, forest vegetation, and aquatic ecosystems, South Fork Hoh River, Olympic National Park. Pages 30–34 *in* E. E. Starkey, J. F. Franklin, and J. W. Mathews, technical coordinators. National Parks of the Pacific Northwest. Oregon State University Forest Research Laboratory, Corvallis, OR.

Swanson, F. J., S. L. Johnson, S. V. Gregory, and S. A. Acker. 1998. Flood disturbance in a forested mountain landscape. Bioscience 48:681–89.

Troth J. L., F. J. Deneke, and L. M. Brown. 1976. Upland aspen/birch and black spruce stands and their litter and soil properties in interior Alaska. Forest Science 22:33–44.

Viereck, L. A., C. T. Dyrness, and M. J. Foote. 1993. An overview of the vegetation and soils of the floodplain ecosystems of the Tanana River, interior Alaska. Canadian Journal of Forest Research 23:889–898.

Walker, L. R., J. C. Zasada, and F. S. Chapin III. 1986. The role of life history processed in primary succession on an Alaskan floodplain. Ecology 67:1243:1253.

Woo, M. K. 1986. Permafrost hydrology in North America. Atmosphere-Ocean 24:201–234.

Yarie, J., K. Van Cleve, C. T. Dyrness, L. Oliver, J. Levinson, and R. Erickson. 1993. Soil-solution chemistry in relation to forest succession on the Tanana River floodplain, interior Alaska. Canadian Journal Forest Research 23:928–940.

Yarie, J., L. Viereck, K. Van Cleve, and P. Adams. 1998. Flooding and ecosystem dynamics along the Tanana River. BioScience 48:690–695.

17

Fire Trends in the Alaskan Boreal Forest

Eric S. Kasischke
T. Scott Rupp
David L. Verbyla

Introduction

Fire is ubiquitous throughout the global boreal forest (Wein 1983, Payette 1992, Goldammer and Furyaev 1996, Kasischke and Stocks 2000). The inter- and intra-annual patterns of fire in this biome depend on several interrelated factors, including the quantity and quality of fuel, fuel moisture, and sources of ignition. Fire cycles in different boreal forest types vary between 25 and >200 years (Heinselman 1981, Yarie 1981, Payette 1992, Conard and Ivanova 1998). Although the increased presence of humans in some regions of boreal forest has undoubtedly changed the fire regime (DeWilde 2003), natural fire is still a dominant factor in ecosystem processes throughout this biome.

Boreal forest fires are similar to those of other forests in that they vary between surface and crown fires, depending on forest type and climatic factors. Surface fires kill and consume most of the understory vegetation, as well as portions of the litter or duff lying on the forest floor, resulting in varying degrees of mortality of canopy and subcanopy trees. Crown fires consume large amounts of the smaller plant parts (or fuels) present as leaves, needles, twigs, and small branches and kill all trees. These fires are important in initiating secondary succession (Lutz 1956, Heinselman 1981, Van Cleve and Viereck 1981, Van Cleve et al. 1986, Viereck 1983, Viereck et al. 1986). Unlike fires in other forest types, smoldering ground fires in the boreal forest can combust a significant fraction of the deep organic (fibric and humic) soils in forests overlying permafrost (Dyrness and Norum 1983, Landhauesser and Wein 1993, Kasischke et al. 2000a, Miyanishi and Johnson 2003). During periods of drought, when water tables are low, or prior to spring thaw, organic soils in peatlands can become dry enough to burn, as well (Zoltai et al. 1998, Turetsky and Wieder

2001, Turetsky et al. 2002). This organic soil consumption is important from several perspectives, because it (a) releases large amounts of trace gases into the atmosphere (Kasischke et al. 1995b, 2005, French et al. 2000, 2003, Kasischke and Bruhwiler 2003); (b) controls the long-term accumulation of carbon in organic soils (Kasischke et al. 1995a, Harden et al. 2000); (c) regulates the moisture and temperature of the ground layer, which in turn influences permafrost formation and loss (Viereck 1983); (d) alters patterns of soil respiration (O'Neill 2000, Richter et al. 2000, O'Neill et al. 2002, 2003, Kim and Tanaka 2003, Bergner et al. 2004); and (e) influences longer-term patterns of forest succession (Viereck 1983, Landhaeusser and Wein 1993, Kasischke et al. 2000b, Johnstone and Chapin in press, Johnstone and Kasischke in press).

During the 1990s, information-gathering and spatial analysis technologies such as satellite-based remote sensors and geographic information systems (GIS) opened new avenues for analyzing patterns of fire in the boreal forest. These technologies are useful not only for producing more accurate and complete maps of fires and their effects (Kasischke and French 1995, Kasischke et al. 1999, Michalek et al. 2000, Isaev et al. 2002, Sukhinin et al. 2004) but also for relating the spatial and temporal patterns of fire occurrence to other important geographic characteristics, such as climate, vegetation cover, and topography (Kasischke et al. 2002). In this chapter, we use several different data sets to investigate the patterns of fire in the Alaskan boreal forest.

Long-Term Perspectives

Although the modern fire record for Alaska dates back only to 1940, inferences about the longer-term fire regime can be made through analysis of lake-bottom sediments and tree rings. Conclusions based on these observations are limited, however, by the relatively small number of studies that have been conducted in the Alaskan and Canadian boreal forest. For example, there are only four published studies from Alaska that utilize fire-scar and/or tree age distributions (Yarie 1981, Mann et al. 1995, Mann and Plug 1999, Devolder 1999), and a similar number of studies analyzing lake bottom sediments (Earle et al. 1996, Hu et al. 1993, 1996, Lynch et al. 2002). The locations of these studies are scattered over a region the size of the state of Montana, spanning a wide range in regional climate, ecosystem type, and topography.

Analysis of lake-bottom sediments provides information on fire occurrence and dominant vegetation cover in Alaska dating back through the Holocene. The onset of the current fire regime occurred in the middle of the Holocene (5600–6500 calendar years before present) when black spruce (*Picea mariana*) replaced white spruce *(P. glauca)* as the dominant forest type (Hu et al. 1993, 1996, Lynch et al. 2002; Chapter 5). Even though the climate was wetter at this time period, the arrival of black spruce (which is more flammable than white spruce) is correlated with significant increases in charcoal and fire frequency. Analysis of charcoal in lake bottom sediments shows that, since black spruce became the dominant tree species, the fire frequencies at one site in interior Alaska averaged 134 years (range 36–301 years; Lynch et al. 2002).

Analysis of tree-ring data to determine stand age distributions has been carried out in only a few locations in Alaska. A study of 371 stands in northeast Alaska in the Porcupine/Black River basins showed that fire cycles in this region were 26 years for hardwood stands (aspen and birch), 36 years for black spruce stands, and 113 years for white spruce stands, with an average of 43 years (Yarie 1981). Finally, on smaller scales, site-specific analysis of tree rings can be used to determine not only the time since the last stand-replacement fire but also the occurrence of surface fires through the presence of fire scars in the tree-ring record (Mann et al. 1995, Mann and Plug 1999).

In summary, our review of the limited historical fire information for Alaska provides only a general picture of past fire regime. It reveals a shifting fire regime in response to climate and vegetation dynamics at millennial time scales and a spatially variable contemporary fire frequency that ranges from 40 to 250+ years.

Contemporary Fire Record

The recent fire record for the Alaskan boreal forest is contained in three databases: (1) a spreadsheet summary of total area burned for a given year since 1940; (2) a spreadsheet summary of the location, ignition source (e.g., natural or human), size, and start and stop times for most individual fire events since 1956; and (3) a spatial database within a GIS containing the boundary of individual fires since 1950, derived from mapped fire perimeters. While this last database is complete back to 1972, there are several years prior to this where fire records are missing, especially in the 1960s (Murphy et al. 2000, Kasischke et al. 2002). This spatial database includes all fires >400 ha (1,000 acres) in size for 1950 to 1987 and all fires >40 ha (100 acres) since 1988. Because it does not contain the boundaries of smaller fires, it is referred to as the Alaskan large-fire database and is available through the Alaska Geophysical Data Clearinghouse at http://agdc.usgs.gov/data/blm/fire/index.html. The exclusion of small fires from this database does not significantly bias analyses of the fire regime because these small fires represent a very small fraction of total area burned (<5%). Kasischke et al. (2002) review the overall accuracy of the fire data archived by the Alaska Fire Service. This review concludes that, although most fires within the state were detected and mapped, the overall accuracy of these maps in terms of fire location and fire area may be poor prior to the 1970s.

A comparison of the seasonal fire areas within the first and third databases mentioned shows an overall good agreement for 80% of the estimates of annual area burned. However, each database contains years for which total area burned is significantly lower than that reported in the other. The large-fire database (3) contains higher estimates of area burned in four years—1950 (by 388,000 ha), 1956 (by 165,000 ha), 1961 (by 551,000 ha), and 1979 (by 89,000 ha). These higher levels of area burned most likely resulted from scaling errors in the approaches used to originally calculate fire areas, which were corrected when the maps were digitized for inclusion in the large-fire database. The fire areas within the large-fire database (3) were significantly lower than the areas reported in annual summary database (1) for six years—1953 (by 143,000 ha), 1957 (by 553,000 ha), 1967 (by 272,000 ha), 1969 (by 862,000 ha), 1971 (by 103,000 ha), and 1974 (by 162,000 ha). In these cases, lower estimates resulted because not all maps of fire boundaries were digitized because of missing data files.

Interannual Fire Patterns

To analyze the interannual patterns of fire, we created a hybrid data set by merging the areas within the official fire records (database 1) and the large-fire database (database 3). We substituted the large-fire database areas in those years when they were >0% larger than the estimates in the official fire records, on the assumption that the official records underestimated the area burned in these years. This procedure resulted in a higher average annual area burned than that provided by either of the two data sources. The average annual area burned between 1961 and 2000 was 268,957 ha yr^{-1} for the official fire statistics (database 1), 232,732 ha yr^{-1} for the large-fire database (database 3), and 286,944 ha yr^{-1} for the combined data set.

Figure 17.1a presents a plot of annual area burned for the years 1961–2000 for the entire state of Alaska. We do not present or discuss data from prior to 1960 because the quality of information is uncertain with respect to fire detection and map accuracy (Murphy et al. 2000, Kasischke et al. 2002). The annual area burned plot presented in Fig. 17.1a exhibits an episodic distribution, with years of high fire activity being interspersed with years of low activity. For the Alaskan boreal forest region, 55% of the total area burned between 1961 and 2000 burned in just six years. The average annual area burned during these episodic fire years was seven times greater than the area burned in the low fire years.

With the observed increase in air temperature and lengthening of the growing season over the past several decades in the North American boreal forest region, a corresponding increase in fire activity between the 1960s and the 1990s might be expected (Stocks et al. 2000). For the Alaska boreal forest region, such an increase in fire activity is not apparent. There was only a small (7%) rise in the annual area burned between 1981 and 2000 (297,624 ha yr^{-1}) compared to the period between 1961 and 1980 (276,624 ha yr^{-1}). The frequency of large fire years has been greater since 1980 (four large fire years) than before 1980 (two large fire years); therefore, the increase in average annual area burned between these periods is largely a result of the increase in frequency of large fire years. The length of the data record for Alaska is insufficient to determine whether the increase in frequency of large fire years is part of natural variability or a response to recent climate warming. Longer-term fire statistics for Alaska show there were two large fire years per decade in the 1940s and 1950s, suggesting that the frequency of large fire years has been relatively constant over the past 60 years.

The number and size distribution of fires in the Alaskan boreal forest region are different during the low fire years and high fire years. According to the large-fire database for the years 1950–1999, during the low fire years, there was an average of 17 fires per year >400 ha, with an average size of 7,800 ha. In contrast, during

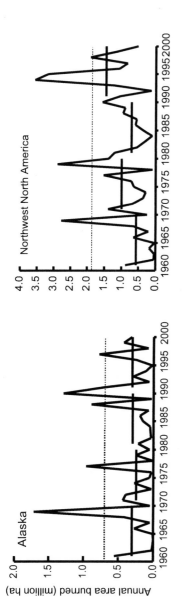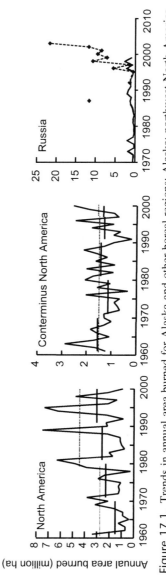

Figure 17.1. Trends in annual area burned for Alaska and other boreal regions: Alaska; northwest North America (Alaska and the Yukon and Northwest Territories); the North American boreal forest region (Canada and Alaska); the conterminous United States; and Russia. The solid horizontal lines represent decadal averages, while the dashed line represents the threshold between low fire and high fire years. The solid diamonds represent satellite burned area estimates for Russia.

high fire years, there were 66 fires >400 ha, with an average size of 20,300 ha. Fig. 17.2 illustrates the difference in percentage of total area burned in different fire size categories between high and low fire years. In low fire years, 73% of the total area burned occurred in fires <50,000 ha in size. In high fire years, 65% of the total area burned occurred in fires >50,000 ha in size. In low fire years, 9% of the total area burned in fires >100,000 ha in size, with no fires affecting areas larger than 200,000 ha. In contrast, during high fire years, 33% of the total area burned in fires >100,000 ha in size.

Sources of Fire Ignitions

There are two sources of ignitions for fires in the boreal forest: lightning and human activities. The importance of lightning as an ignition source in the Alaskan boreal forest was downplayed in earlier studies, with anthropogenic sources of ignitions thought to be more important (Lutz 1959). However, the implementation of lightning detection systems, along with a more thorough review of fire statistics, led to the realization that lightning not only is widespread throughout interior Alaska during the growing season but also is responsible for ignition of the fires that burn most of the area (Barney and Stocks 1983, Gabriel and Tande 1983). Analysis of fire statistics from the Alaska Fire Service shows that, while humans start >61% of all fires, these fires are responsible for only 10% of the total area burned (Table 17.1). The remaining fires are the result of ignitions from lightning.

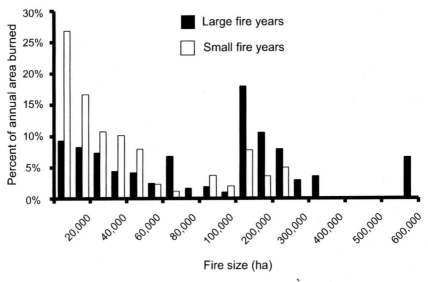

Figure 17.2. Distribution of percent of annual area burned as a function of fire size, showing that, in high fire years, a higher percentage of fires occur in large fire events than in small fire years.

Table 17.1. Summary statistics of human- versus lightning-caused fires in Alaska, 1956–2000.

	Total fires (%)	Area burned (ha)	Area burned (%)
Human causation	61.5%	1,344,979	10.0%
Lightning causation	38.5%	12,060,382	90.0%

Lightning originates in thunderstorm clouds that are formed when and where there is enough convection, vertical instability, and moisture to produce clouds that are high enough to reach temperatures below freezing. A cloud-to-ground lightning flash originates in the water/ice portion of thunderstorm clouds. Convective storms and associated lightning can range in size from individual clouds to synoptic thunderstorms covering thousands of square kilometers. A single thunderstorm may produce most of the annual lightning strikes of an area (Nash and Johnson 1996). In general, lightning strike density increases with the number of convective cells and cloud top height (Pakliam and Maybank 1975). In interior Alaska, there is generally a gradient of increasing lightning density from west to east associated with the cooler-maritime to warmer-continental climate gradient of the region (Chapter 3). However, the fires that originate in lightning strikes are well distributed throughout the interior of the state (Fig. 17.3 top).

The presence of the boreal forest may promote convective thunderstorms. Dissing and Verbyla (2003) found that lightning strike density was consistently highest within boreal forest compared to tundra and shrub zones across the climatic gradient found in central Alaska. There was also a significant relationship between lightning strikes within tundra and distance to the nearest boreal forest edge. From the paleorecord (Chapter 5), we learn that there was an increase in wildfires when black spruce became established in interior Alaska (despite a cooler, moister climate). This may have been due to two factors associated with an increase of black spruce on the landscape: (1) increased landscape heterogeneity and higher sensible heat fluxes leading to increased convective thunderstorms, and (2) increased fuel flammability of black spruce and associated understory vegetation (Lynch et al. 2002).

Fire managers in Alaska recognize two fire seasons. The major fire season typically starts in mid- June, when the earth's surface becomes warm enough to drive convective thunderstorm activity. However, there is an earlier fire season throughout the state that is associated with the extremely dry fuel conditions that occur immediately after snowmelt in late spring (middle to late May). At this time, an extremely dry fire fuel matrix exists because of low precipitation levels, the curing of dead vegetation during winter, and the lack of green vegetation prior to the spring growth period. During this time, human activities cause the majority of fire ignitions. In contrast to lightning ignitions, which are uniformly distributed between the Brooks and Alaska Ranges (Fig. 17.3 top), human fire ignitions are clearly centered around major population centers (Fairbanks and Anchorage), as well as along the major road networks (Fig. 17.3 bottom).

Figure 17.3. Map of fire ignitions for the state of Alaska based on (top) lightning-caused ignitions and (bottom) human-caused ignitions.

Role of Humans in the Alaskan Fire Regime

Humans influence fire activity in the boreal forest in three primary ways: (a) they provide sources of ignitions (Anderson et al. 2000); (b) they carry out fire suppression activities to varying degrees (Ward and Mawdsley 2000); and (c) through deforestation, reforestation, and land clearing activities, they alter the characteristics and distributions of fuels on the landscape.

The degree to which humans are actually changing the fire regime in Alaska is still under study. On the one hand, humans result in increased ignitions (Fig. 17.3 bottom, Table 17.1), but on the other, fire suppression activities tend to be greatest in populated regions because fire management policies place the highest priority on protection of life and property. Analyses of the data within the Alaska large-fire database suggest that fire cycles are longer in areas with higher human population concentrations. The longer fire cycle is likely caused by two factors—higher levels of fire protection and suppression as well as alteration of the fuels. During the early settlement period around 1900, there was significant harvesting of spruce trees to provide fuel for heating, mining, and operation of steamboats (Roessler and Packee 2000). This clearing of spruce forests paved the way for invasion of birch and aspen trees in many sites, which are less flammable than spruce forests. In addition, during the earlier part of the century, there was probably an increase in human-caused fires associated with mining activities and human settlements, leading to an increase in early-successional deciduous tree species in some areas.

Fires deliberately started by Indigenous Peoples can be considered to be part of the natural fire regime of Alaska and include frequent burning of riparian areas to improve habitat for berries, roots for baskets, and wildlife (Lutz 1959, Natcher 2004). In addition, occasional burning of larger areas of forest by Native Americans in Alaska contributed to the fire regime prior to contact with Europeans.

Finally, it should be noted that in dry years and given the right meteorological conditions, large wildfires occur in spite of intensive fire suppression activities. There are numerous examples of large fires near populated areas in Alaska in the past decade—the 40,000–ha fire that was ignited within 1 km of the Tok Forestry Station in 1990; the 17,000-ha Miller's Reach #2 fire that occurred in 1996 near the city of Houston; and the 7,000–ha Donnelly Flats fire that occurred in 1999 on the Fort Greely Military Reserve, near Delta. Each of these fires received immediate and maximum levels of fire suppression yet escaped and burned large areas because the fire and fuel conditions were optimal for rapid fire spread.

Comparisons to Other Regions

Although fire is common throughout the global boreal forest and the temperate forests of the conterminous United States, there are important regional differences. There is considerable geographic variation in dominant tree species throughout the boreal forest, which in turn, influences the fire regime (Wein and McLean 1983, Wirth 2005). Throughout the boreal forest, different species of pines are found on drier sites, with the exception of Alaska, where a complex of white spruce, aspen, and birch dominate these sites. Spruce-fir complexes are common throughout the North American and Eurasian boreal forest on moister sites. In North America, the spruce-fir complexes are found in the eastern part of this region, while black spruce is the dominant wet-site species in the western part of this region, including Alaska. Additionally, larch forests in large parts of Russia dominate poorly drained sites. As reviewed by Payette (1992), the fire return interval in the North American boreal forest ranges between 70 and 180 years. For the Russian boreal region, Conard and Ivanova (1998) report that the fire return interval for southern pine forests is 25 to 50 years, compared to 90 to 130 years for larch forests. Thus, the fire cycles observed in Alaska are similar to those in other boreal regions.

Figure 17.1 presents data on the annual area burned for three different regions of the North American boreal forest, the conterminous United States, and Russia. The northwestern part of North America includes Alaska, the Yukon Territory, and the Northwest Territories, and the North American region includes all of Canada and Alaska. The Canadian fire data are from Amiro et al. (2001) and are based on a digital large-fire database for this region (Stocks et al. 2003). The data for the conterminous United States are from the National Interagency Coordination Center (http://www.nicc.gov/). The Russian data are from official fire statistics reported by Korovin (1996) from the years 1970–1992 and from the Global Fire Monitoring Center for the period 1993–1999 (http://www/uni-freiburg.de/fireglobe/welcome.html). Satellite-based area-burned estimates for Russia are from Cahoon et al. (1994) for 1987, Cahoon et al. (1996) for 1992, and Sukhinin et al. (2004) for 1995–2003.

The annual area burned in the northwest North America (Fig. 17.1b) and the North American boreal (Fig. 17.1c) regions exhibits an episodic pattern similar to that in Alaska (Fig. 17.1a), with large fire years interspersed with small fire years. As in Alaska, area burned in the large years is three to four times greater than in the small years (Figure 17.4). In addition, these regions have exhibited a pronounced increase in fire activity over the past four decades, with at least a doubling of annual area burned between the 1960s and the 1990s. Finally, analyses using the Canadian large-fire database and satellite imagery show that the majority of area burned occurs in large fire events, matching the pattern observed in Alaska. In summary, these data show that the fire regime for Alaska is similar to that found within other regions in North America.

The increases in fire activity throughout the North American boreal forest region as a whole over the past four decades has occurred during a time period when average air temperatures have also been increasing. It is likely that the increased fire activity is a result of the increased warming, especially given the dependence of fire activity and behavior on climate (Stocks et al. 1998, 2000, Gillett et al. 2004). Although there are questions with respect to the overall accuracy of fire records collected prior to 1970 (Kasischke et al. 2002), the increased fire activity in the North American boreal forest appears to be the result of increases in area burned during both low and high fire years. Prior to 1980, there were six large fire years, while after 1980 there were five. The average area burned in large fire years increased from 3.66×10^6 ha yr^{-1} prior to 1980 to 6.45×10^6 ha yr^{-1} after 1980. The average area burned in small fire years increased from 1.13×10^6 ha yr^{-1} prior to 1980 to 1.60×10^6 ha yr^{-1} after 1980. If better data were available, the average values would likely increase for the pre-1980 period, which would lessen the overall level of increases in fire activity. The accuracy of the North American fire record could be improved by searching archives for missing data records or replacing missing maps with information on fire boundaries available from aerial photographs and satellite imagery.

Official fire statistics indicate that the average area burned in the Russian boreal forest region is less than half that in the North American boreal forest region (1.0×10^6 ha yr^{-1}) and that the fire activity increased from 0.7×10^6 ha yr^{-1} during the 1970s to 1.2×10^6 ha yr^{-1} during the 1990s (Korovin 1996). However, the satellite data products developed for Russia indicate that average area burned in Russia averages 8.2×10^6 ha yr^{-1} with the same range of variation between small and large fire years as observed in the North American boreal forest region. The discrepancies between the official fire statistics and the satellite observations are attributed to the fact that (a) only two-thirds of Russia's forests are monitored for fire; and (b) total area burned was purposefully underreported at the district and regional levels

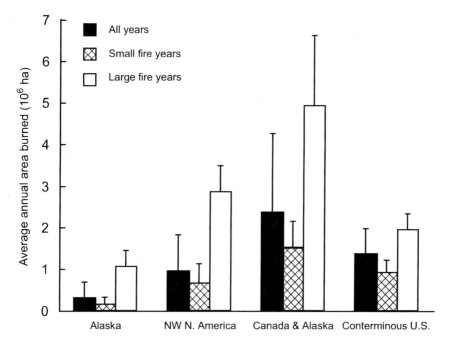

Figure 17.4. Average area burned in different regions of North America in all years between 1961 and 2000 separated into large and small fire years (see Fig. 17.1). The error bar represents one standard deviation.

so that fire managers could qualify for performance bonuses (Conard and Ivanova 1998, Shvidenko and Nilsson 2000).

An important difference between the fire regime in the Russian and the North American boreal forests is the length of the fire season. The North American fire season begins in mid- to late May and continues until early September, whereas the Russian fire season begins in late March or early April and continues into early October. This difference is a result of the fact that the Russian boreal forest covers a greater latitudinal range than the North American boreal forest, and thus snow-free periods during which fires can be started (in the spring) or continue to burn (in the fall) are longer in Russia. As in North America, anthropogenic ignitions are the source of early-season fires in Russia. These early-season, anthropogenic fires were unusually widespread in April and May 2003, when $>16 \times 10^6$ ha of fire occurred in the Lake Baikal region.

In terms of ignition sources, the trends observed for Alaska are also present in Canadian and Russian boreal forests. Murphy et al. (2000) report that in Canada, lightning causes 35% of fires but results in 85% of the total area burned. Korovin (1996) reports that in Russia, humans cause 65% of all ignitions, lightning 15%, and 20% of fires are due to unknown sources.

Finally, the characteristic patterns of burning in the Alaskan and Canadian boreal forests are quite similar. Almost all fires in Alaska and Canada are crown fires that result in nearly 100% mortality of canopy trees. These fires also result in consumption of a significant portion of the organic matter present in litter, lichens, mosses, and organic soils. In contrast, Russian foresters report that in average fire years, only 10% of the fires are stand-replacing crown fires (Shvidenko and Nilsson 2000). All other fires are lower-intensity surface fires, which have much less impact on ecosystem processes than do stand-replacing fires. Analysis of burn scars on satellite imagery, however, indicates that during large fire years in Russia (1987 and 1998), the majority of fires were crown fires, not surface fires. Current research is focused on using satellite imagery to quantify the spatial patterns of burn severity in Alaska, Canada, and Russia.

The temperate forests of North America also experience large fire years; in 2000, in the conterminous United States, 2.5 million ha of forest burned (Fig. 17.1d). Since 1960, an average of 1.35 million ha yr^{-1} has burned, which is 58% of the level of fire activity observed in the boreal forests of North America. Although there is a high degree of interannual variability in fire activity in this region, the difference between high and low fire years is not as pronounced as in the boreal forest, with the ratio being 2.1 between high and low years (Fig. 17.4). The average annual area burned in the conterminous United States has been steady since the 1970s and has decreased slightly since the 1960s (annual area burned was 1.56 million ha yr^{-1} in the 1960s, 1.18 million ha yr^{-1} in the 1970s, 1.28 million ha yr^{-1} in the 1980s, and 1.29 million ha yr^{-1} in the 1990s).

Conclusions

The continuous compilation of fire statistics since the 1940s by the Alaska Fire Service provides information needed to assess important characteristics of the fire regime in the Alaskan boreal forest region. The recent conversion of fire boundary maps to a digital format and the combination of this information with other spatial data sets have provided additional insights. The analyses of these data sets show that large (>50,000 ha) fires occur in all ecoregions of the Alaskan interior. These large fires have taken place in relatively few fire years, which occur with a frequency of every 6 or 7 years. The frequency of large fire years has recently increased dramatically, with four occurring since 1988. The length of the fire records is not sufficiently long to determine whether these recent increases are part of natural variability or represent a response to the recent warming in this region.

The creation of the Alaskan large-fire database represents an important resource for both scientists and resource managers. We are now able to assess spatial and temporal characteristics of these fires and relate them to important landscape features, such as vegetation cover and topography. Kasischke et al. (2002) show that the fire cycle is clearly related to spatial patterns of vegetation cover and climate (precipitation, temperature, and lightning strikes). The dependence of fire cycle on vegetation can be inferred both directly (through comparison to large-scale vegetation cover maps) and indirectly (through analysis of topographic information on aspect and elevation, which, in turn, mediate climate and vegetation distribution).

This database provides the basis for estimating trace gas emissions from fires (Kasischke et al. 1995b, French et al. 2003) and for understanding some of the important sources of variation in vegetation cover. Researchers are just now beginning to incorporate this information into a wide range of biogeochemical and ecosystem process models (Rupp et al. 2000). Equally important are the applications for natural resource managers. Accurate fire location information is important for planning fire management activities. The location and ages of previous fires enables fire managers to more efficiently deploy limited manpower and resources when dealing with large fires in remote regions. In addition, wildlife managers can infer habitat composition and condition from fire maps.

In closing, we note that further research is presently being carried out to improve our understanding of the Alaskan fire regime. Efforts are under way to locate missing fire records and to incorporate them into the database. In some cases, archived aerial photographs or satellite images are being used to map historical fire boundaries. In addition, archived fire records present maps of fire perimeters, and often there are unburned areas or islands within individual fires. Analysis of historical photography and imagery will allow an assessment of the true accuracy of fire perimeter maps for assessing total area burned (French et al. 1996). In addition, important information on the spatial characteristics of fire severity (e.g., levels of tree mortality and consumption of organic soil) can also be inferred through the analysis of satellite imagery (Michalek et al. 2000, Isaev et al. 2002). Such information is critical to determining the ecological and climatological impacts of fire.

References

Amiro, B. D., J. B. Todd, B. M. Wotton, K. A. Logan, M. D. Flannigan, B. J. Stocks, J. A. Mason, D. L. Martell, and K. G. Hirsch. 2001. Direct carbon emissions from Canadian forest fires, 1959–1999. Canadian Journal of Forest Research 31:512–525.

Anderson, K., D. L. Martell, M. D. Flannigan, and D. Wang. 2000. Modeling of fire occurrence in the boreal forest region of Canada. Pages 357–367 *in* E. S. Kasischke and B. J. Stocks, editors. Fire, Climate Change, and Carbon Cycling in the Boreal Forest. Springer-Verlag, New York.

Barney, R. J., and B. J. Stocks. 1983. Fire frequencies during the suppression period. Pages 45–62 *in* R. W. Wein and D. A. MacLean, editors. The Role of Fire in Northern Circumpolar Ecosystems. Wiley, New York.

Bergner, B., J. Johnstone, and K. K. Treseder. 2004. Experimental warming and burn severity alter soil CO_2 flux and soil functional groups in a recently burned boreal forest. Global Change Biology 12:1996–2004.

Cahoon, D. R., Jr., B. J. Stocks, J. S. Levine, W. R. Cofer III, and J. A. Barber. 1996. Monitoring the 1992 forest fires in the boreal ecosystem using NOAA AVHRR satellite imagery. Pages 795–802 *in* J. L. Levine, editor. Biomass Burning and Climate Change. Vol. 2. Biomass Burning in South America, Southeast Asia, and Temperate and Boreal Ecosystems, and the Oil Fires of Kuwait. MIT Press, Cambridge, MA.

Cahoon, D. R., Jr., B. J. Stocks, J. S. Levine, W. R. Cofer III, and J. M. Pierson. 1994. Satellite analysis of the severe 1987 forest fires in northern China and southeastern Siberia. Journal of Geophysical Research 99:18,627–18,638.

Conard, S. G., and G. A. Ivanova. 1998. Wildfire in Russian boreal forests—Potential impacts of fire regime characteristics on emissions and global carbon balance estimates. Environmental Pollution 98:305–313.

Devolder, A. 1999. Fire and climate history of lowland black spruce forests, Kenai National Wildlife Refuge. M.S. Thesis, University of Alaska Fairbanks.

DeWilde, L. 2003. Human Impacts on the Fire Regime of Interior Alaska. M.S. Thesis, University of Alaska Fairbanks.

Dissing, D. D., and D. L. Verbyla. 2003. Spatial patterns of lightning strikes in interior Alaska and their relations to elevation and vegetation. Canadian Journal of Forest Research 33:770–785.

Dyrness, C. T., and R. A. Norum. 1983. The effects of experimental fires on black spruce forest floors in interior Alaska. Canadian Journal of Forest Research 13:879–893.

Earle, C. J., L. B. Brubaker, and P. M. Anderson. 1996. Charcoal in northcentral Alaskan lake sediments: Relationships to fire and late-Quaternary vegetation history. Review of Paleobotany and Palynology 92:85–95.

French, N. H. F., E. S. Kasischke, R. D. Johnson, L. L. Bourgeau-Chavez, A. L. Frick, and S. L. Ustin. 1996. Using multi-sensor satellite data to monitor carbon flux in Alaskan boreal forests. Pages 808–826 *in* J. L. Levine, editor. Biomass Burning and Climate Change. Vol. 2. Biomass Burning in South America, Southeast Asia, and Temperate and Boreal Ecosystems, and the Oil Fires of Kuwait. MIT Press, Cambridge, MA.

French, N. H. F., E. S. Kasischke, B. J. Stocks, J. P. Mudd, D. L. Martell, and B. S. Lee. 2000. Carbon release from fires in the North American boreal forest. Pages 377–388 *in* E. S. Kasischke and B. J. Stocks, editors. Fire, Climate Change, and Carbon Cycling in the Boreal Forest. Springer-Verlag, New York.

French, N. H. F., E. S. Kasischke, and D. G. Williams. 2003. Variability in the emission of carbon-based trace gases from wildfire in the Alaskan boreal forest. Journal of Geophysical Research 107: 8151, doi:10.1029/2001JD000480, 2002 [printed 108(D1), 2003].

Gabriel, H. W., and G. F. Tande. 1983. A Regional Approach to Fire History in Alaska. BLM-Alaska Technical Report No. 9, Bureau of Land Management, Anchorage, AK.

Gillett, N. P., A. V. Weaver, F. W. Zwiers, and M. D. Flannigan. 2004. Detecting the effect of climate change on Canadian forest fires. Geophysical Research Letters 31:18211, doi:10.1029/2004GL02876.

Goldammer, J. G., and V. V. Furyaev, editors. 1996. Fire in Ecosystems of Boreal Eurasia. Kluwer Academic, Dordrecht, The Netherlands.

Harden, J. W., S. E. Trumbore, B. J. Stocks, A. Hirsch, S. T. Gower, K. P. O'Neill, and E. S. Kasischke. 2000. The role of fire in the boreal carbon budget. Global Change Biology 6(Suppl. 1):174–184.

Heinselman, M. L. 1981. Fire intensity and frequency as factors in the distribution and structure of northern ecosystems. Pages 7–57 *in* H. A. Mooney, T. M. Bonnickson, N. L. Christensen, J. E. Lotan, and W. A. Reiners, editors. Fire Regimes and Ecosystem Processes, General Technical Report WO–26. U.S. Department of Agriculture, Forest Service, Washington, DC.

Hu, S. H., L. B. Brubaker, and P. M. Anderson. 1993. A 12,000-year record of vegetation change and soil development from Wien Lake, Central Alaska. Canadian Journal of Botany 71:1133–1142.

Hu, F. S., L. B. Brubaker, and P. M. Anderson. 1996. Boreal ecosystem development in the northwestern Alaska Range since 11,000 yr B.P. Quaternary Research 46:188–201.

Isaev, A. S., G. N. Korovin, S. A. Bartalev, D. V. Ershov, A. Janetos, E. S. Kasischke, H. H. Shugart, N. H. French, B. E. Orlick, and T. L. Murphy. 2002. Using remote sensing for assessment of forest wildfire carbon emissions. Climatic Change 55:231–255.

Johnstone, J. F., and F. S. Chapin III. In press. Effects of soil burn severity on post-fire conifer recruitment in boreal forests. Ecosystems.
Johnstone, J. F., and E. S. Kasischke. In press. Effect of stand-level burn severity on post-fire regeneration in a recently burned black-spruce forest. Canadian Journal of Forest Research.
Kasischke, E. S., K. Bergen, R. Fennimore, F. Sotelo, G. Stephens, A. Janetos, and H. H. Shugart. 1999. Satellite imagery gives a clear picture of Russia's boreal forest fires. EOS—Transactions of the American Geophysical Union 80:141–147.
Kasischke, E. S., and L. P. Bruhwiler. 2003. Emissions of carbon dioxide, carbon monoxide, and methane from boreal forest fires in 1998. Journal of Geophysical Research 107: 8146, doi:10.1029/2001JD000461, 2002 [printed 108(D1), 2003].
Kasischke, E. S., N. L. Christensen Jr., and B. J. Stocks. 1995a. Fire, global warming, and the carbon balance of boreal forests. Ecological Applications 5:437–451.
Kasischke, E. S., and N. H. F. French. 1995. Locating and estimating the areal extent of wildfires in Alaskan boreal forests using multiple-season AVHRR NDVI composite data. Remote Sensing of Environment 51:263–275.
Kasischke, E. S., N. H. F. French, L. L. Bourgeau-Chavez, and N. L. Christensen Jr. 1995b. Estimating release of carbon from 1990 and 1991 forest fires in Alaska. Journal of Geophysical Research 100:2941–2951.
Kasischke, E. S., E. J. Hyer, P. C. Nevelli, L. P. Bruwiler, N. H. F. French, A. I. Sukhinen, J. H. Hewson, and B. J. Stocks. 2005. Influence of boreal fire emissions on northern hemisphere atmospheric carbon and carbon monoxide. Global Biogeochemical Cycles 12:GB1012, doi:10.1029/2004GB002300.
Kasischke, E. S., K. P. O'Neill, N. H. F. French, and L. L. Bourgeau-Chavez. 2000a. Controls on patterns of biomass burning in Alaskan boreal forests. Pages 173–196 *in* E. S. Kasischke and B. J. Stocks, editors. Fire, Climate Change, and Carbon Cycling in the North American Boreal Forest. Springer-Verlag, New York.
Kasischke, E. S., K. P. O'Neill, N. H. F. French, L. L. Bourgeau-Chavez, and D. Richter. 2000b. Influence of fire on long-term patterns of forest succession in Alaskan boreal forests. Pages 214–238 *in* E. S. Kasischke and B. J. Stocks, editors. Fire, Climate Change, and Carbon Cycling in the North American Boreal Forest. Springer-Verlag, New York.
Kasischke, E. S., and B. J. Stocks, editors. 2000. Fire, Climate Change, and Carbon Cycling in the Boreal Forest. Springer-Verlag, New York.
Kasischke, E. S., D. Williams, and D. Barry. 2002. Analysis of the patterns of large fires in the boreal forest region of Alaska. International Journal of Wildland Fire 11:131–144.
Kim Y., and N. Tanaka. 2003. Effect of forest fire on the fluxes of CO_2, CH_4, and N_2O in boreal forest soils, interior Alaska. Journal of Geophysical Research 108: 8154, doi:10.1029/2001JD000663 [printed in 108(D1), 2003].
Korovin, G. N. 1996. Analysis of distribution of forest fires in Russia. Pages 112–128 *in* J. G. Goldammer and V. V. Furyaev, editors. Fire in Ecosystems of Boreal Eurasia. Kluwer Academic Publishers, Dordrecht, The Netherlands.
Landhaeusser, S. M., and R. W. Wein. 1993. Postfire vegetation recovery and tree establishment at the Arctic treeline: Climactic-change–vegetation-response hypothesis. Journal of Ecology 81:665–672.
Lutz, H. F. 1956. Ecological effects of forest fires in interior Alaska. Technical Bulletin Number 1133, U.S. Department of Agriculture, Washington, DC.
Lutz, H. J. 1959. Aboriginal Man and White Man as Historical Causes of Fires in the Boreal Forest, with Particular Reference to Alaska. Yale University Press, New Haven, CT.

Lynch, J. A., J. S. Clark, N. H. Bigelow, M. E. Edwards, and B. P. Finney. 2002. Geographical and temporal variations in fire history in boreal ecosystems of Alaska. Journal of Geophysical Research 107: 8152, doi:10.1029/2001JD000332, 2002 [printed 108(D1), 2003].

Mann, D. H., and L. J. Plug. 1999. Vegetation and soil development at an upland taiga site, Alaska. Ecoscience 6:272–285.

Mann, D. H., C. L. Fastie, E. L. Rowland, and N. H. Bigelow. 1995. Spruce succession, disturbance and geomorphology on the Tanana River Floodplain, Alaska. Ecoscience 2:184–199.

Michalek, J. L., N. H. F. French, E. S. Kasischke, R. D. Johnson, and J. E. Colwell. 2000. Using Landsat TM data to estimate carbon release from burned biomass in an Alaskan spruce complex. International Journal of Remote Sensing 21:323–338.

Miyanishi, K., and E. A. Johnson. 2003. Process and patterns of duff consumption in the mixwood boreal forest. Canadian Journal of Forest Research 32:1285–1295.

Murphy, P. J., B. J. Stocks, E. S. Kasischke, D. Barry, M. E. Alexander, N. H. F. French, and J. P. Mudd. 2000. Historical fire records in the North American boreal forest. Pages 274–288 *in* E. S. Kasischke and B. J. Stocks, editors. Fire, Climate Change, and Carbon Cycling in the North American Boreal Forest. Springer-Verlag, New York.

Nash, C. H., and E. A. Johnson. 1996. Synoptic climatology of lightning-caused forest fires in subalpine and boreal forests. Canadian Journal of Forest Research 26:1859–1874.

Natcher, D. C. 2004. Implications of forest succession on Native land use in the Yukon Flats, Alaska. Human Ecology 32:421–441.

O'Neill, K. P. 2000. Changes in carbon dynamics following wildfire in soils of interior Alaska. Ph.D. Dissertation, Duke University, Durham, NC.

O'Neill, K. P., E. S. Kasischke, and D. D. Richter. 2002. Environmental controls on soil CO_2 flux following fire in black spruce, white spruce, and aspen stands of interior Alaska. Canadian Journal of Forest Research 32:1525–1541.

O'Neill, K. P., E. S. Kasischke, and D. D. Richter. 2003. Seasonal and decadal patterns of soil carbon uptake and emission along an age sequence of burned black spruce stands in interior Alaska. Journal of Geophysical Research 108: 8155, doi:10.1029/2001JD000443 [printed in 108(D1), 2003].

Pakliam, J. E., and J. Maybank. 1975. The electrical characteristics of some severe hailstorms in Alberta, Canada. Journal of the Meteorological Society of Japan 53:363–383.

Payette, S. 1992. Fire as a controlling process in the North American boreal forest. Pages 144–169 *in* H. H. Shugart, R. Leemans, and G. B. Bonan, editors. A Systems Analysis of the Global Boreal Forest. University Press, Cambridge, UK.

Richter, D., E. S. Kasischke, and K. P. O'Neill. 2000. Postfire stimulation of microbial decomposition in black spruce (*Picea mariana* L.) forest soils: A hypothesis. Pages 197–213 *in* E. S. Kasischke and B. J. Stocks, editors. Fire, Climate Change, and Carbon Cycling in the North American Boreal Forest. Springer-Verlag, New York.

Roessler, J. S., and E. C. Packee. 2000. Disturbance history of the Tanana River Basin in Alaska: management implications. Pages 46–57 *in* W. K. Moser and C. F. Moser, editors. Fire and Forest Ecology: Innovative Silviculture and Vegetation Management. Tall Timbers Research Station, Tallahasee, FL.

Rupp, T. S., F. S. Chapin III, and A. M. Starfield. 2000. Response of subarctic vegetation to transient climatic change on the Seward Peninsula in northwest Alaska. Global Change Biology 6:541–555.

Shvidenko, A. Z., and S. Nilsson. 2000. Extent, distribution and ecological role of fire in Russian forests. Pages 132–150 *in* E. S. Kasischke and B. J. Stocks, editors. Fire, Climate Change, and Carbon Cycling in the Boreal Forest. Springer-Verlag, New York.

Stocks, B. J., M. A. Fosberg, B. M. Wotton, T. J. Lynham, and K. C. Ryan. 2000. Climate change and forest fire activity in North American boreal forests. Pages 368–376 *in* E. S. Kasischke and B. J. Stocks, editors. Fire, Climate Change, and Carbon Cycling in the North American Boreal Forest. Springer-Verlag, New York.

Stocks, B. J., M. A. Fosberg, T. J. Lynham, L. Mearns, B. M. Wotton, Q. Yang, J.-Z. Jin, K. Lawrence, G. R. Hartley, J. A. Mason, and D. W. McKenney. 1998. Climate change and forest fire potential in Russian and Canadian boreal forests. Climatic Change 38:1–13.

Stocks, B. J., J. A. Mason, J. B. Todd, E. M. Bosch, B. M. Wotton, B. D. Amiro, M. D. Flannigan, K. G. Hirsch, K. A. Logan, D. L. Martell, and W. R. Skinner. 2003. Large forest fires in Canada, 1959–1997. Journal of Geophysical Research 107: 8149, doi:10.1029/2001JD000484 [printed 108(D1), 2003].

Sukhinin, A. I., N. H. F. French, E. S. Kasischke, J. H. Hewson, A. J. Soja, I. A. Csiszar, E. Hyer, T. Loboda, S. G. Conard, V. I. Romasko, E. A. Pavlichenko, S. I. Miskiv, and O. A. Slinkin. 2004. Satellite-based mapping of fires in Eastern Russia: New products for fire management and carbon cycle studies. Remote Sensing of Environment 93:546–564.

Turetsky, M. R., and R. K. Wieder. 2001. A direct approach to quantifying organic matter lost as a result of peatland wildfire. Canadian Journal of Forest Research 31:363–366.

Turetsky, M. R., K. Wieder, L. Halsey, and D. Vitt. 2002. Current disturbance and the diminishing peatland carbon sink. Geophysical Research Letters 29: doi:10.1029/2001GL014000.

Van Cleve, K., and L. Viereck. 1981. Forest succession in relation to nutrient cycling in the boreal forest of Alaska. Pages 184–211 *in* D. C. West, H. H. Shugart, and D. B. Botkin, editors. Forest Succession, Concepts and Application. Springer-Verlag, New York.

Van Cleve, K., F. S. Chapin III, P. W. Flanagan, L. A. Viereck, and C. T. Dyrness, editors. 1986. Forest Ecosystems in the Alaskan Taiga. Springer-Verlag, New York.

Viereck, L. A. 1983. The effects of fire in black spruce ecosystems of Alaska and northern Canada. Pages 201–220 *in* R. W. Wein and D. A. MacLean, editors. The Role of Fire in Northern Circumpolar Ecosystems. Wiley, Chichester, UK.

Viereck, L. A., K. Van Cleve, and C. T. Dyrness. 1986. Forest ecosystem distribution in the taiga environment. Pages 22–43 *in* K. Van Cleve, F. S. Chapin III, P. W. Flanagan, L. A. Viereck, and C. T. Dyrness, editors. Forest Ecosystems in the Alaskan Taiga. Springer-Verlag, New York.

Ward, P. C., and W. Mawdsley. 2000. Fire management in the boreal forests of Canada. Pages 66–84 *in* E. S. Kasischke and B. J. Stocks, editors. Fire, Climate Change, and Carbon Cycling in the Boreal Forest. Springer-Verlag, New York.

Wein, R. W. 1983. Fire behavior and ecological effects in organic terrain. Pages 81–95 *in* R. W. Wein, and D. A. MacLean, editors. The Role of Fire in Northern Circumpolar Ecosystems. Wiley, New York.

Wein, R. W., and D. A. MacLean. 1983. An overview of fire in northern ecosystems. Pages 1–15 *in* R. W. Wein and D. A. MacLean, editors. The Role of Fire in Northern Circumpolar Ecosystems. Wiley, New York.

Wirth, C. 2005. Fire regime and tree diversity in boreal forests: implications for the carbon cycle. Pages 309–344 in M. Scherer-Lorenzen, C. Körner, and E.-D. Schulze, editors. Forest Diversity and Function: Temperate and Boreal Systems. Springer-Verlag, Heidelberg.

Yarie, J. 1981. Forest fire cycles and life tables: a case study from interior Alaska. Canadian Journal of Forest Research 11:554–562.

Zoltai, S. C., L. A. Morrissey, G. P. Livingstone, and W. J. de Groot. 1998. Effects of fires on carbon cycling in North American peatlands. Environmental Reviews 6:13–24.

18

Timber Harvest in Interior Alaska

Tricia L. Wurtz
Robert A. Ott
John C. Maisch

Introduction

The most active period of timber harvesting in the history of Alaska's interior occurred nearly a century ago (Roessler 1997). The beginning of this era was the year 1869, when steam-powered, stern-wheeled riverboats first operated on the Yukon River (Robe 1943). Gold was discovered in Alaska in the 40-Mile River area in 1886, a find that was overshadowed 10 years later by the discovery of gold in the Klondike, Yukon Territory. By 1898, Dawson City, Yukon Territory, was reported to have 12 sawmills producing a total of 12 million board feet of lumber annually (Naske and Slotnick 1987).

Over the next 50 years, more than 250 different sternwheeled riverboats operated in the Yukon drainage, covering a large part of Alaska and Canada's Yukon Territory (Cohen 1982). This transportation system required large amounts of fuel. Woodcutters contracted with riverboat owners to provide stacked cordwood at the river's edge, at a cost of $7.14 in 1901 (Fig. 18.1; Cohen 1982). Between 100 and 150 cords of wood were required to make the 1400-km round trip from the upper Yukon to Dawson City (Trimmer 1898). Over time, woodcutters moved inland from the rivers' edges, significantly impacting the forest along many rivers of the Yukon drainage (Roessler 1997).

The growth of the town of Fairbanks required wood for buildings and flumes as well as for fuel. In Fairbanks's early days, all electrical generation was by wood fuel at the N.C. Company's power plant. From the founding of the town in 1903 through the 1970s, white spruce harvested in the Fairbanks area was used exclusively by local sawmills, which produced small amounts of green and air-dried lumber. In 1984, however, the Alaska Primary Manufacturing Law was struck down

Figure 18.1. "Wooding up" on the Tanana River, ca. 1910. Photo printed with permission of The Olive Joslyn Bell Collection, Accession number 79-41-165, Archives and Manuscripts, Alaska and Polar Regions Department, University of Alaska Fairbanks.

by the U.S. Supreme Court, removing the legal barrier to round-log export of timber harvested from State lands. During the late 1980s and 1990s, many high-quality logs from State and private land timber sales were exported, primarily to Pacific Rim countries. Declining markets ended this trend in the late 1990s, and there have been no significant exports since the market collapse. The past few years have seen significant shifts in the nations that supply logs to the world market, as well as a much higher level of competition in log supply, conditions that will likely preclude a return of any significant round-log export from interior Alaska to Asia.

Current Situation

Though the contemporary forest products industry in interior Alaska is small and in some cases economically marginal, it is locally important for employment and for providing wood products for local needs. Most of the forest management activity occurring today involves converting unmanaged forest stands to managed stands and addressing forest health issues such as the salvage of trees damaged by insects or fire. Timber harvest occurs primarily in mature white spruce stands with a high component of sawlog material (defined as having a DBH, or diameter at breast height, > 23 cm). A small volume of sawlog-size paper birch and aspen is also harvested for use in both primary (lumber) and secondary (molding, flooring, furniture) manufactured products. All species are used for fuel wood; in many rural areas, wood is often the primary or secondary heating source for individual homes.

Until recently, timber harvested in interior Alaska was typically clearcut because this is efficient and inexpensive and because some of its effects are similar to those of fire. Harvest unit sizes range from 0.4 ha to about 60 ha but are generally 12–16 ha in size. Over the past 10 years, however, aesthetic and environmental concerns have increased the emphasis on partial-harvest methods. Partial-harvest methods used in interior Alaska today include diameter-limit cuts, species-selection cuts, and seedtree cuts, where scattered mature trees are left standing as seed sources. Today, most timber sales involve harvesting the white spruce sawlogs and retaining any white spruce trees smaller than a minimum sawlog diameter. Species such as balsam poplar, paper birch, and aspen are also commonly left standing, as well as large-diameter spruce with either external or internal defects (e.g., broken tops or heart rot) and individual standing dead trees.

An outcome of the thin economic margins of the current forest products industry is that only management costs necessary for timber extraction (timber sale layout, road building, harvest) and, to a lesser extent, reforestation are generally incurred. Intensive silvicultural measures used elsewhere, such as fertilization and precommercial and commercial thinning, have been shown experimentally to improve tree growth in interior Alaska (Van Cleve and Zasada 1976) but when applied operationally can cost more per ha than the value of the harvested timber. Marginal economic returns discourage even those silvicultural treatments that are considered minimal elsewhere, such as scarification of seedbeds, broadcast burning of slash, and application of herbicides to control competing vegetation.

Harvested sites are reforested both by natural and artificial means. Deciduous species—paper birch, quaking aspen, balsam poplar—naturally regenerate by vegetative propagation and seeding. White spruce is regenerated through natural seeding and by the hand planting of containerized nursery stock.

Because white spruce is the primary commercial species in interior Alaska, it has been the focus of most of the silvicultural research conducted there. Early research concentrated on natural regeneration following timber harvest. This work documented that white spruce seed germinates better on mineral soil than on other substrates (Zasada and Gregory 1969, Ganns 1977), and a variety of types of scarification equipment have been tested with a goal of creating mineral soil seedbeds (Zasada 1972, Youngblood and Zasada 1991, Densmore et al. 1999). However, when scarification is immediately followed by a good cone crop, it can result in severe spruce overstocking (Zasada and Wurtz 1990, Wurtz and Zasada 2001). Concerns about the unpredictability of spruce cone production prompted the creation of a State Forest Nursery in Eagle River, Alaska, in the 1980s, to produce containerized seedlings for local reforestation applications. Recent re-examinations of experiments begun in the 1970s and 1980s indicate, however, that spruce seed germinates and becomes established on undisturbed forest floor better than had previously been recognized. It is not necessary for good cone crops to follow logging immediately for them to play a significant role in regenerating the stand. When evaluated 15 or 20 years after harvest, spruce stocking from natural seed rain was as good on unscarified surfaces as on scarified surfaces.

In one case, growth of the young spruce saplings was significantly better on unscarified surfaces, a reversal of the pattern observed on that site shortly after treatment (Wurtz and Zasada 2001). This work documented trends that had already been observed by forest managers. Today, some managers question the need for any artificial regeneration methods at all, and scarification is viewed primarily as a means to promote the regeneration of wildlife browse.

Timber Harvest and Wildlife

The harvest of timber in interior Alaska can affect wildlife habitat and biological diversity in many ways. The impacts associated with the timber harvest can be positive or negative, depending on the wildlife species in question. Some silvicultural practices can mitigate the negative impacts of timber harvest on species that use mature white spruce forests. Examples of practices that can benefit wildlife include the retention of some of the structural features of mature forests, including patches of spruce trees and shrubs, snags, defective trees, and trees of other species. Such features provide cavity trees, resting, nesting and feeding areas for wildlife, and function as a long-term bank of coarse woody debris. For wildlife species that use early-successional vegetation types, timber harvest can have a net positive effect. The conversion of mature spruce to young birch or aspen can increase habitat quality by providing both cover and browse for a variety of animal species (Haggstrom and Kelleyhouse 1996).

Timber management can result in significant habitat enhancement for moose, an economically and culturally important game species in interior Alaska. The fall 2003 population estimate of moose in the state-designated game management unit south of Fairbanks was from 1 to 1.3 moose per km^2 of vegetated habitat (nearly 13,000 km^2; Alaska Department of Fish and Game 2004). The density of moose in this area is managed primarily by regulated hunting, where, since the mid-1990s, between 800 and 1,300 moose have been harvested annually. Silviculture is increasingly recognized as a tool with great potential to help land managers maintain habitat quality and quantity.

The Timber Resource

Interior Alaska has 42 million ha of forested land. About one-tenth of this land produces more than 1.4 m^3 per ha per year of wood and is thus classified by the U.S. Department of Agriculture's Forest Service as capable of sustained commercial wood production (Anonymous 1995). In the 5.8 million ha Tanana Basin, there are approximately 600,000 ha of productive mixed white spruce/hardwood and pure white spruce forest on state lands classified for forestry purposes (Crimp et al. 1997). But, despite the fact that this large land base is available for timber management, the timber industry in interior Alaska is very small. Over the past 30 years, the total harvest from these lands has averaged fewer than 500 ha per year, or less than 5% of the harvest level that is believed to be sustainable (Parsons unpublished).

Why is so little timber harvested in interior Alaska? The answer has many facets, from aspects of the environment to economies of scale. One such component is access. Because of topographic conditions, soil characteristics and permafrost distribution, stands of commercial timber in interior Alaska are typically small and scattered. Productive stands are commonly separated by sizable areas underlain by permafrost, with vegetation ranging from lowland black spruce forest to muskeg. The state's road system, and industrial development in general, is extremely limited. Alaska has only 0.01 km of road per km^2 of land area, compared to California's 0.67 km (Alaska Department of Transportation 2002). Building all-season roads across lowland black spruce forest or muskeg requires expensive measures to insulate the underlying permafrost. Though ice bridges and temporary winter roads are sometimes used to traverse areas where all-season roads would be problematic (Zasada et al. 1987), their use is limited to three to four months per year.

A second obstacle is the distance from interior Alaska to sizable markets. Once harvested logs reach the road or rail system in the Tanana Valley, the closest major market and the closest seaport are more than 550 km away. Such a long haul-distance is a significant economic barrier.

A third reason for the small amount of timber harvested is the small diameter of the logs that make up the timber resource. On even the most productive sites, white spruce trees from interior Alaska rarely exceed 50 cm DBH (Juday and Zasada 1984, Wurtz 1995). Small-diameter timber is more work to harvest and transport and therefore has higher costs (Barbour 1999). The hardwood resources, mainly birch and aspen, are predominantly small in size, with the majority of stems in pure stands falling in the 15–20 cm range at DBH (Crimp et al. 1997). Mixed stands of spruce and hardwood have a higher percentage of hardwood stems in the 20 cm and 25 cm class but often exhibit higher levels of defect because of their older stand ages. This combination of small diameter and high defect limits the competitiveness of the resource on the world log market.

Another reason for the small amount of timber harvested in the interior may be public participation in the forest planning process for state lands. Even a modest proposal for timber harvest is the subject of intense attention by the local public. Because of the limited road system, recreational users and forest harvesters can be concentrated in the same areas, increasing the likelihood of conflict. The state has occasionally deferred timber sales in response to public opposition, leading to concerns among sawmill owners about the predictability of wood supply.

Small-diameter logs, high defect rates in hardwoods, access, infrastructure deficiencies, and long distances to significant markets all suggest that any future wood-processing facilities would need to be large in order to be feasible. Large facilities carry high capital costs and in turn require large volumes of raw material to operate. These large facilities can keep costs down by distributing the cost of production over a high volume of finished product, which translates into low unit costs. Low unit costs can compensate for other factors, such as distance to market and high transportation costs. However, large facilities carry a higher level of risk with regard to long-term viability of the log supply. Issues such as whether the land base can be maintained, whether social, political, and economic conditions are stable, and whether the inventory is accurate are crucial in this regard. These factors collectively contribute to a high level of both risk and uncertainty for commercial timber operations.

Most companies use a rate-of-return calculation to determine the merits of a particular investment against other similar opportunities. Feasibility studies conducted to date for large wood-processing facilities in interior Alaska have not resulted in competitive rate-of-return figures. This situation may change as new markets arise for hardwood resources, as new wood processing technologies are developed, or as road networks expand in Alaska.

Conclusions

Though timber harvest has had only a small impact on the landscape of interior Alaska, it is of interest in the Bonanza Creek LTER program for several reasons. Timber harvest has impacted the forests of interior Alaska over the past hundred years. Today, it is the basis of a small local economy, providing wood products for local needs. Changes in market or political factors may increase the economic return from timber harvesting in interior Alaska, leading to a greater rate of cutting than has occurred to date. Timber harvest proposals and activities are the subject of intense attention by the local public. Logging may be used increasingly in the future as a substitute for fire near human habitation, to improve habitat for certain species of wildlife, and to reduce hazard fuel loads. Finally, the commercial harvest of timber directly adjacent to the Bonanza Creek LTER site provides an opportunity to integrate the scientific understanding of these boreal forest ecosystems with real-world management actions.

References

Alaska Department of Fish and Game. 2004. Personal communication with Don Young, area biologist.
Alaska Department of Transportation. 2002. Personal communication with Jeff Roach, transportation planner.
Anonymous. 1995. Alaska's unreserved productive forestlands. Alaska Cooperative Extension Service: Alaska Natural Resource Management Series, 10600106.
Barbour, R. J. 1999. Diameter and gross product value for small trees. Pages 40–46 *in* Proceedings of the Twenty-seventh Annual Wood Technology Clinic and Show. Sponsored by MillerFreeman, March 24–26, 1999. Oregon Convention Center, Portland.
Cohen, S. 1982.Yukon River Steamboats: A Pictorial History. Pictorial Histories Publishing Co., Missoula, MT.
Crimp, P. W., S. J. Phillips, and G. T. Worum. 1997. Timber resources on state forestry lands in the Tanana Valley. Unpublished report to the State of Alaska, Department of Natural Resources, Division of Forestry.
Densmore, R. V., G. P. Juday, and J. C. Zasada. 1999. Regeneration alternatives for upland white spruce after burning and logging in interior Alaska. Canadian Journal of Forest Research 29:413–423.
Ganns, R. C. 1977. Germination and survival of artificially seeded white spruce on prepared

seedbeds on an interior Alaska floodplain site. M.S. Thesis, University of Alaska Fairbanks.

Haggstrom, D. R., and D. G. Kelleyhouse. 1996. Silviculture and wildlife relationships in the boreal forest of interior Alaska. Forestry Chronicle 72:59–62.

Juday, G. P., and J. C. Zasada. 1984. Structure and development of an old-growth white spruce forest on an interior Alaska floodplain. Pages 227–234 *in* W. R. Meehan, T. R. Merrell, and T. R. Hanley, editors. Fish and Wildlife Relationships in Old-Growth Forests: Proceedings of the Symposium for American Institute of Fisheries Research Biologists, Morehead City, NC.

Naske, C.-M., and H. E. Slotnick. 1987. Alaska—A History of the 49th State. 2nd ed. University of Oklahoma Press, Norman.

Parsons and Associates, Inc. Undated. Tanana state forestry lands: Periodic sustained yield analysis. Unpublished report to the State of Alaska, Department of Natural Resources, Division of Forestry.

Robe, C. F. 1943. The penetration of the Alaskan frontier: The Tanana Valley and Fairbanks. Thesis, Yale University, New Haven, CT.

Roessler, J. S. 1997. Disturbance history in the Tanana River Basin of Alaska: management implications. M.S. Thesis, University of Alaska Fairbanks.

Trimmer, F. M. 1898. The Yukon Territory. Downey and Company, London, UK.

Van Cleve, K., and J. Zasada. 1976. Response of 70-year-old white spruce to thinning and fertilization in interior Alaska. Canadian Journal of Forest Research 6:145–152.

Wurtz, T. L. 1995. Understory alder in three boreal forests of Alaska: Local distribution and effects on soil fertility. Canadian Journal of Forest Research 25:987–996.

Wurtz, T. L., and J. C. Zasada. 2001. An alternative to clearcutting in the boreal forest of Alaska: a twenty-seven-year study of regeneration after shelterwood harvesting. Canadian Journal of Forest Research 31:999–1011.

Youngblood, A. P., and J. C. Zasada. 1991. White spruce artificial regeneration options on river floodplains in interior Alaska. Canadian Journal of Forest Research 21:423–433.

Zasada, J. C. 1972. Guidelines for obtaining natural regeneration of white spruce in Alaska. U.S. Department of Agriculture, Forest Service, Pacific Northwest Forest and Range Experiment Station, Portland, OR.

Zasada, J. C., and R. A. Gregory. 1969. Regeneration of white spruce with reference to interior Alaska: A literature review. U.S. Department of Agriculture, Forest Service Research Paper PNW–79.

Zasada, J. C., and T. L. Wurtz. 1990. Natural regeneration of white spruce on an upland site in interior Alaska. Vegetation Management: An Integrated Approach. Proceedings of the Fourth Annual Vegetation Management Workshop, November 14–16,1989, Forestry Canada Pacific Forestry Centre and British Columbia Ministry of Forests. FRDA Report 109.

Zasada, J. C., C. W. Slaughter, and C. E. Teutsch. 1987. Winter logging on the Tanana River flood plain in interior Alaska. Norwegian Journal of Applied Forestry 4:11–16.

19

Climate Feedbacks in the Alaskan Boreal Forest

A. David McGuire
F. Stuart Chapin III

Introduction

The boreal forest biome occupies an area of 18.5 million km², which is approximately 14% of the vegetated cover of the earth's surface (McGuire et al. 1995b). North of 50°N, terrestrial interactions with the climate system are dominated by the boreal forest biome because of its large aerial extent (Bonan et al. 1992, Chapin et al. 2000b; Fig. 19.1). There are three major pathways through which the function and structure of boreal forests may influence the climate system: (1) water/energy exchange with the atmosphere, (2) the exchange of radiatively active gases with the atmosphere, and (3) delivery of fresh water to the Arctic Ocean. The exchange of water and energy has implications for regional climate that may influence global climate, while the exchange of radiatively active gases and the delivery of fresh water to the Arctic Ocean are processes that could directly influence climate at the global scale (Fig. 19.2). In this chapter, we first discuss the current understanding of the role that boreal forests play in each of these pathways and identify key issues that remain to be explored. We then discuss the implications for the earth's climate system of likely responses of boreal forests to various dimensions of ongoing global change.

Water and Energy Exchange

Most of the energy that heats the earth's atmosphere is first absorbed by the land surface and then transferred to the atmosphere. The energy exchange properties of the land surface therefore have a strong direct influence on climate. Boreal forest

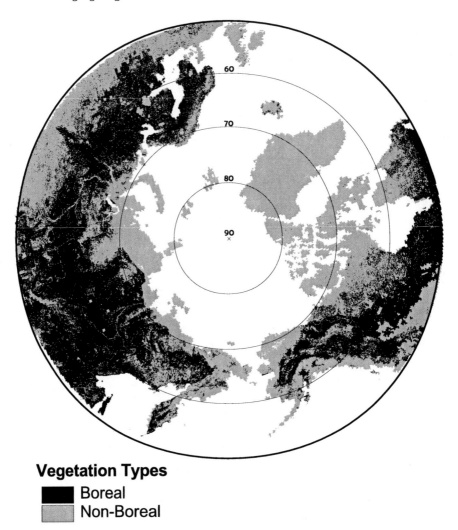

Vegetation Types
■ Boreal
▨ Non-Boreal

Figure 19.1. Polar projection vegetation map indicating the aerial coverage of boreal forest north of 50°N. The vegetation map is courtesy of Catharine Thompson and Monika Calef.

differs from more southerly biomes in having a long period of snow cover, when white surfaces might be expected to reflect incoming radiation (high albedo) and therefore absorb less energy for transfer to the atmosphere. Observed winter albedo in the boreal forest varies between 0.11 (conifer stands) and 0.21 (deciduous stands; Betts and Ball 1997). This is much closer to the summer albedo (0.08–0.15) than to the winter albedo of tundra (0.6–0.8), which weather models had previously assumed to be appropriate for boreal forests. The incorporation of true boreal albedo into climate models led to substantial improvements in medium-range weather forecasting (Viterbo and Betts 1999). There is substantial spatial variability in winter albedo within the boreal forest because of the spatial mosaic of conifer forests,

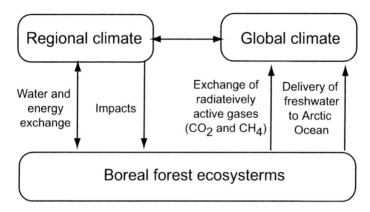

Figure 19.2. The major pathways through which the structure and functioning of boreal forests may influence the climate system.

deciduous forests, and nonforested wetlands and burn scars (Chapter 6). The latter have an albedo of approximately 0.6 when the short-statured vegetation is snow covered.

During summer, the albedo of boreal vegetation is lower than in winter, with deciduous stands and boreal nonforested wetlands having approximately twice the albedo of conifer forests (Chapin et al. 2000b, Chambers and Chapin 2002). This difference in albedo leads to fluxes of sensible heat (i.e., heat that can be sensed) in conifer stands that are two to three times those of deciduous stands, whereas the latent energy fluxes (i.e., evapotranspiration) of deciduous forest stands in the boreal forest are 1.5 to 1.8 times greater than those of conifer forest stands (Chapin et al. 2000b). Because transpiration is tightly linked to photosynthesis, latent heat exchange tends to be dominated by transpiration in boreal forest stands with high productivity (e.g., deciduous forests). In contrast, evaporation plays a more important role than transpiration in the latent energy exchange of forest stands with low productivity (e.g., black spruce forests), where surface evaporation from mosses can account for up to half of total evaporation (Chapin et al. 2000b, Heijmans et al. 2004). The substantial sensible heating over conifer stands leads to thermal convection and may contribute to the frequency of thunderstorms and lightning (Dissing and Verbyla 2003; Chapter 17), which plays an important role in the fire regime of the boreal forest as a source of ignition.

Because there are substantial seasonal and spatial differences in sensible and latent energy exchange in the boreal forest, climate warming could affect regional climate by altering both positive and negative feedbacks. Positive feedbacks associated with climate warming may result from lengthening of the growing season and from the replacement of tundra with boreal forest. A longer growing season that leads to earlier snowmelt and later snow cover effectively reduces annual albedo and should lead to substantial heating of the atmosphere, which may lengthen the growing season. Several lines of evidence indicate that earlier snowmelt is occurring (Dye 2002) and that growing seasons are lengthening both in Alaska (Keyser et al. 2000; Chapter 4) and generally throughout the boreal forest (McGuire et al. 2004).

Another positive feedback is associated with expansion of boreal forest into regions now occupied by tundra. This would decrease albedo in both summer and winter and should cause substantial heating of the atmosphere, a response that could possibly accelerate the replacement of tundra by boreal forest (Chapin et al. 2000a, 2000b). Studies conducted with general circulation models indicate that the position of the northern treeline has a substantial influence on global climate, with effects extending to the tropics (Bonan et al. 1992, Foley et al. 1994). At Alaska's latitudinal treeline on the Seward Peninsula, spruce trees have established progressively farther from the forest limit since the 1880s and have become more dense within existing stands (Lloyd et al. 2002). This has led to a conversion of shrub tundra into low-density forest-tundra within a band extending approximately 10 km from the forest limit. These forests have a lower albedo and greater sensible heat fluxes than the tundra they have replaced, causing a net warming effect on the atmosphere (Chapin et al. 2000b). Field transplants and modeling experiments suggest that fire may play a role in the expansion of treeline by allowing the establishment of tree seedlings in tundra and that large and nearly instantaneous responses to warming are likely at the treeline ecotone (Lloyd and Fastie 2002, Lloyd et al. 2002; Chapter 5).

In contrast to the positive feedback to climate warming caused by forest expansion into tundra, any increase in fire frequency caused by boreal warming (Chapter 17) could lead to negative feedback to warming if fire increases the proportion of deciduous stands on the landscape. Postfire stands are dominated by herbs, shrubs, and deciduous trees (Chapter 7). These stands have a higher albedo (0.14) than do the conifer stands (0.09) that they replace, and they therefore transfer less energy to the atmosphere (Chambers and Chapin 2002), as described earlier. Thus, the degree to which the response of vegetation dynamics to climate warming influences regional climate depends on the interaction of factors that may both enhance and mitigate warming.

Exchange of Radiatively Active Gases

Radiatively active gases are essentially transparent to down-welling short-wave radiation but are effective at absorbing up-welling long-wave radiation and so trap energy close to the surface of the earth. Were it not for the presence of radiatively active gases in the atmosphere, the earth's surface would be extremely cold. Thus, increases in the atmospheric concentrations of radiatively active gases could warm the earth's surface by altering the earth's near-surface energy balance (Ramanswamy et al. 2001). Since 1750, increases in atmospheric carbon dioxide (CO_2) and methane (CH_4) are estimated to be responsible for trapping 1.46 W m^{-2} and 0.48 W m^{-2}, respectively, which is about 80% of the estimated radiative forcing of 2.43 W m^{-2} that has been attributed to greenhouse gases since 1750 (Ramanswamy et al. 2001). The effects of increasing CO_2 and CH_4 are of concern globally because CO_2 is currently increasing by 0.4% per decade and CH_4 is increasing 0.6% per decade. While

CO_2 has been responsible for a majority of the radiative forcing associated with greenhouse gases, the higher rate of increase in atmospheric CH_4 is of concern because CH_4 is 21 times more effective per molecule than CO_2 in absorbing long-wave radiation on a 100–year time scale (Shine et al. 1990). Changes in the boreal forest region that may influence the exchange of CO_2 and CH_4 with the atmosphere include feedbacks in mechanisms that control fluxes of carbon dioxide and methane, climate warming, increases in the atmospheric concentration of CO_2 as well as nitrogen deposition, and land-cover changes.

Feedbacks from CO_2 and CH_4

A complete analysis of trace-gas feedbacks to climate should consider responses of both CO_2 and CH_4 flux in boreal forests (Roulet 2000). Likely changes in these fluxes could both enhance warming (positive feedbacks) and mitigate warming (negative feedbacks; McGuire and Hobbie 1997, Chapin et al. 2000b, Clein et al. 2002). Boreal forests contain approximately 27% of the world's vegetation carbon inventory and 28% of the world's soil carbon inventory (McGuire et al. 1997). Much of the soil carbon is highly labile and has accumulated simply because of cold and/ or anaerobic soils conditions (Chapter 15). Recent high-latitude warming (Chapter 4) could trigger the release of this carbon "time bomb" that has accumulated over millennia.

Responses to Climate Warming

Warming could release carbon as CO_2 from aerobic boreal soils, that is, soils that are not saturated with water, through enhanced decomposition (McGuire et al. 1995b). In anaerobic boreal soils, warming could affect carbon storage by altering soil drainage patterns. Although soil drainage may be especially vulnerable to the response of permafrost to climatic warming, the net effect on radiative forcing of the atmosphere is not clear because drainage can be either enhanced or retarded by permafrost degradation (Chapter 16) and because the response of drainage is likely to affect the release of CO_2 and CH_4 in opposite directions (Roulet 2000). For example, the release of CO_2 from aerobic decomposition is likely to be enhanced if permafrost warming results in a drop of the water table (Oechel et al. 1995, Christensen et al. 1998), but emissions of CH_4 are likely to decrease because methanogenesis is an anaerobic process (Roulet 2000). In contrast, if permafrost thawing results in the expansion of lakes and wetlands, then releases of CH_4 are likely to increase (Zimov et al. 1997) with an associated reduction in CO_2 emissions. In eastern Canadian peatlands, the enhanced CH_4 emissions associated with the creation of wetlands will likely result in a positive feedback to warming for up to 500 years, until the enhanced storage of carbon in the wetlands (i.e., uptake of CO_2 from the atmosphere) offsets the enhanced radiative forcing associated with CH_4 emissions (Roulet 2000).

Although the warming of aerobic soils tends to increase the release of CO_2 from boreal forests, the net effect of warming depends on the balance between production and decomposition. Climate warming is lengthening the growing season in Alaska

(Keyser et al. 2000; Chapter 4) and across the boreal forest region (Dye 2002, McGuire et al. 2004). In temperate forests, annual carbon storage is enhanced by approximately 6 g C m^{-2} for every day that the growing season is lengthened (Baldocchi et al. 2001), suggesting that longer growing seasons in boreal forest should enhance terrestrial carbon storage. In the boreal forest, the start of the growing season, as defined by photosynthetic activity of the canopy, is tightly coupled with the thawing of the soil in conifer stands, because frozen soil prevents transpiration (Frolking et al. 1999, Zhuang et al. 2003). Deciduous stands begin net carbon uptake following leaf-out, which is also sensitive to the timing of snowmelt. Analyses based on satellite data suggest that both production and vegetation carbon storage have generally been enhanced across the boreal forest in recent decades (Randerson et al. 1999, Myneni et al. 2001). One hypothesis for the mechanism of increased production is that warming increases decomposition of soil organic matter to release nitrogen in forms that can be taken up by plants. Since production is often limited by plant nitrogen supply in boreal forests (Van Cleve and Zasada 1976, Van Cleve et al. 1981), an increase in nitrogen availability to plants should increase production. Several boreal warming experiments have provided support for this mechanism (Van Cleve et al. 1990, Bergh et al. 1999, Stromgrem and Linder 2002; Chapter 11).

Although there is substantial evidence that warming can increase production in the boreal forest through the mechanism of enhanced soil nitrogen availability to plants, production may not increase if other factors limit production. For example, warming has reduced the growth in old white spruce trees growing on south-facing aspects in interior Alaska because of drought stress (Barber et al. 2000; Chapter 5). At treeline in Alaska, the growth of trees located in warm dry sites below the forest margin declined in response to recent warming, whereas the growth of trees located at treeline, particularly in wet regions, increased (Lloyd and Fastie 2002). Thus, there appears to be substantial spatial variability in the response of white spruce growth to recent warming in Alaska, and studies conducted elsewhere on other species throughout the boreal forest suggest that growth responses to warming depend on interactions between temperature and the timing and amount of precipitation (Briffa et al. 1998, Jacoby and D'Arrigo 1995).

In situations where warming does enhance production of boreal forests, there is potential to increase soil carbon storage if the transfer of carbon from vegetation to the soil is greater than the enhancement in decomposition from warming. If this condition occurs, then the long-term rate of soil carbon storage depends on whether the carbon that is transferred to the soil decomposes quickly or slowly (Hobbie et al. 2000, Clein et al. 2000). Our understanding of soil carbon and nitrogen transformations in response to warming is incomplete; this gap is key in limiting our ability to make projections of the long-term response of soil carbon to warming (Clein et al. 2000).

Responses to Increases in Atmospheric CO_2 and to Nitrogen Deposition

In general, boreal forests do not respond to elevated CO_2 because their photosynthesis and growth are nitrogen- rather than carbon-limited (McGuire et al. 1995a).

Nonetheless, nitrogen concentrations of plant tissues generally decrease in response to elevated atmospheric CO_2 (McGuire et al. 1995a, Norby et al. 2001). The degree to which nitrogen concentrations decrease in response to elevated atmospheric CO_2 may depend on the degree to which nitrogen is resorbed from plant tissue prior to senescence (McGuire et al. 1997, Norby et al. 2001).

Much of the mass of inorganic nitrogen compounds, primarily NO_Y and NH_X, emitted by the burning of fossil fuels is subsequently deposited onto downwind terrestrial ecosystems (Holland et al. 1999), including a substantial portion of the boreal forest biome (Townsend et al. 1996). Because boreal forests are often nitrogen-limited, this deposition appears to have increased production, resulting in significant carbon storage in the boreal forest region (Townsend et al. 1996, Kicklighter et al. unpublished simulations of the Terrestrial Ecosystem Model). Tundra and boreal forests may account for about one-quarter of global carbon storage associated with anthropogenic nitrogen deposition (Kicklighter et al. unpublished). In Canada, for example, boreal forests may have been responsible for storing approximately 60 Tg C per year in response to anthropogenic nitrogen deposition, which is about equivalent to estimates of carbon losses associated with increased disturbance (Kurz and Apps 1999). Key uncertainties in these estimates include nitrogen retention in these ecosystems (Kicklighter et al. unpublished), the degree to which deposited nitrogen becomes available to plants (Nadelhoffer et al. 1999), and the degree to which enhancements in carbon uptake are allocated to wood—the most persistent form of carbon storage (Townsend et al. 1996). Once these forests become saturated from continued deposition, their rate of carbon sequestration is likely to decrease (Townsend et al. 1996).

Responses to Land-Cover Change

Important changes in land cover that have occurred or could occur in the boreal forest region include changes associated with disturbance (e.g., fire, insect outbreaks, timber harvest, and cropland establishment/abandonment; Chapters 9, 17, 18) and large-scale changes in the distribution of vegetation (e.g., the advance of tree line in regions now occupied by tundra; Chapter 5). Disturbance is generally characterized by a period of ecosystem carbon loss, during which production is less than decomposition, followed by a period of ecosystem carbon gain once production exceeds decomposition. If disturbance regimes are at steady state for a substantial period of time, ecosystem carbon storage does not change because areas of carbon loss are offset by areas of carbon gain. If boreal disturbances become more frequent or more severe, they have the potential to release ecosystem carbon pools (Kasischke et al. 1995, McGuire et al. 2004). Whether changes in the distribution of vegetation result in carbon storage or loss largely depends on the type of transition (McGuire and Hobbie 1997). For example, the replacement of tundra with boreal forest will likely cause a net uptake of carbon from the atmosphere, whereas the transition of boreal forest to grassland will likely cause a net release of carbon to the atmosphere.

As described in Chapter 9, insect outbreaks have substantially increased mortality of larch and white spruce in both Alaska and Canada. Insect infestation is also a major disturbance in the Russian boreal forest, but the extent and timing of these outbreaks are less well known than for similar occurrences in the North American boreal forest. Insect disturbance that causes partial or complete stand mortality causes carbon loss because of lowered production. Subsequent additional loss of carbon may occur if stands are then salvage logged. Stands affected by insect disturbance may also be more vulnerable to fire because tree mortality generally increases the flammability of forest stands.

During the 1970s and 1980s, the fire frequency in northwest Canada increased substantially (Chapter 17; Kurz and Apps 1999, Stocks et al. 2000). Together, both insect disturbance and fire have likely released substantial amounts of carbon into the atmosphere from Canada's forests (Kurz and Apps 1999, Amiro et al. 2001). As with insect disturbance, subsequent loss of carbon may occur if stands are salvage-logged after fire. The degree to which increased fire frequency has the potential to release carbon in the boreal forest depends, in part, on fire severity. Fires in central Russia, which is dominated by Scots pine, tend to be surface fires that burn the ground layer underneath the canopy, and trees survive because of thick bark. In contrast, fires in far eastern Russia and in North America tend to be stand-replacing crown fires. Analyses of the effects of climate change projections on fire-promoting weather suggest that climate change has the potential to increase fire frequency in Canada (Flannigan et al. 2001). Of potential responses to climate change, fire is the disturbance agent that has the greatest potential to release large amounts of carbon from the boreal forest.

Boreal forests represent a wood resource of global significance. In general, forest harvest and management result in lower vegetation and soil carbon stocks than are found in equivalent unmanaged forests. For example, forests in Fennoscandia are so heavily managed for wood production and harvest that both vegetation and soil carbon are lower than in other areas of the boreal forest (McGuire et al. 2002). In Canada, salvage logging following increased incidence of insect disturbance and fire since 1970 have likely led to net carbon losses (Kurz and Apps 1999). Wood harvest could reduce carbon storage in Siberia's boreal forest and may have already done so in far eastern Siberia, where illegal logging has apparently been increasing over the past decade. Carbon loss from this activity, which results in the export of wood to China and other Asian countries, has not offset the drop in legal commercial logging associated with the breakup of the Soviet Union. Thus, it is expected that the change in the degree of forest harvest in the Russian boreal forest will result in net carbon storage over the next few decades unless harvest rates return to previous levels.

The replacement of tundra with boreal forest occurred in earlier warm periods of Holocene in northern Eurasia (MacDonald et al. 2000) and in western Canada (Spear 1993). Over the past half-century, treeline advances into tundra have been documented in Alaska (Lloyd et al. 2002), Canada (Lavoie and Payette 1994), and Russia (Gorchakovsky and Shiyatov 1978). In Fennoscandia, there is evidence at some sites for treeline advance during the first half of the twentieth century (Kullman 1986), while at other sites tree growth form has changed at treeline (e.g., shifts from stunted krummholz trees to upright trees) without substantial changes in position of the tree species' limit (Kullman 1995). Equilibrium modeling studies suggest that the replacement of arctic tundra with boreal forest could substantially increase vegetation carbon storage (gains of 39.2 to 49.2 Pg C among scenarios in McGuire and Hobbie 1997), whereas soil carbon is likely to be less affected (loss of 2.8 Pg C to gain of 9.3 Pg C among scenarios in McGuire and Hobbie 1997). Soil carbon is relatively insensitive to this vegetation change because increases in production associated with forested landscapes, which cause inputs of carbon into the soil to increase, are somewhat offset by increased rates of decomposition associated with increased soil temperature. This potential increase in ecosystem carbon storage by replacing tundra with boreal forest is likely to proceed at a very slow pace because of inertia associated with the ability of boreal forests to migrate into regions of arctic tundra (Chapin and Starfield 1997, Lloyd et al. 2002).

Delivery of Fresh Water to the Arctic Ocean

The delivery of freshwater from the panarctic land mass is of special importance because the Arctic Ocean contains only about 1% of the world's ocean water, yet receives about 11% of world river runoff (Forman et al. 2000). The Arctic Ocean receives freshwater inputs from 4 of the 14 largest river systems on earth (Forman et al. 2000). Additionally, the Arctic Ocean is the most river-influenced and landlocked of all oceans and is the only ocean with a contributing land area greater than its surface area (Vörösmarty et al. 2000). Freshwater inflow contributes as much as 10% to the upper 100 meters of the water column for the entire Arctic Ocean (Barry and Serreze 2000). Changes in freshwater inputs to the Arctic Ocean could alter salinity and sea ice formation, which may have consequences for the global climate system by affecting the strength of the North Atlantic Deep Water Formation (Broecker 1997). Modeling studies suggest that maintenance of the thermohaline circulation is sensitive to freshwater inputs to the North Atlantic (Manabe and Stouffer 1995). Also, freshwater on the Arctic continental shelf more readily forms sea ice in comparison to more saline water (Forman et al. 2000). Changes in freshwater inputs to the Arctic Ocean depend on changes in the amount and timing of precipitation and the responses of permafrost dynamics, vegetation dynamics, and disturbance regimes to global change. For example, evapotranspiration may change with thawing of permafrost or with a shift in vegetation from conifer to deciduous trees, with subsequent consequences for river runoff.

The boreal forest plays a significant role in the hydrology of the circumpolar north because it dominates the landmasses that contribute to the delivery of freshwater to the Arctic Ocean. Over the past 70 years there has been a 7% increase in the delivery of freshwater from the major Russian rivers to the Arctic Ocean (Peterson et al. 2002, Serreze et al. 2002). In the Yenisey and Lena Rivers, this increase has occurred primarily in the cold season (Serreze et al. 2002). The pattern is most pronounced in the Yenisey, where runoff has increased sharply in the spring, decreased in the summer, but increased for the year as a whole. Although the mechanisms responsible for this pattern are not completely clear, the patterns are linked to higher air temperatures, increased winter precipitation, and strong summer drying. It is possible that the changes in runoff patterns for the Yenisey and the Lena are associated with changes in active layer thickness and the thawing of permafrost (Serreze et al. 2002, Romanovsky et al. 2000). A major challenge will be to establish the link between permafrost changes of the boreal forest in response to a warming climate and changes in the discharge of freshwater into the Arctic Ocean.

Conclusions

Although it is clear that ongoing changes in the boreal forest have consequences for the climate system, we do not completely understand whether the net effect of changes will enhance or mitigate warming. Responses of water, energy, and trace-gas exchange may result in either positive or negative feedbacks to both regional and global warming. Of particular concern is whether the net response of boreal forests could lead to positive feedbacks that greatly enhance the rate of regional and global warming. Changes in carbon storage of high-latitude ecosystems have important implications for the rate of CO_2 accumulation in the atmosphere and for international efforts to stabilize the atmospheric concentration of CO_2 (McGuire and Hobbie 1997, Betts 2000). Simultaneous changes in trace-gas exchanges and albedo of boreal forests influence both regional and global energy balance. For example, the reduction in radiative forcing associated with enhanced carbon storage from an expansion of the boreal forest is likely to be offset by the warming effects of lower albedo (Betts 2000). Increased delivery of freshwater from the boreal forest to the Arctic Ocean also has substantial implications for climate if it disrupts thermohaline circulation by weakening the formation of North Atlantic Deep Water, a response to warming that could ironically launch the earth into another ice age (Manabe and Stouffer 1995). Because the exchange of water, energy, and trace gases among boreal forests, the atmosphere, and the ocean are linked, responses of boreal forests to global change will manifest themselves at a spectrum of spatial and temporal scales. Development of an integrated understanding will provide insight as to how climate change is affecting the boreal forest biome and, conversely, how the response of the boreal forest will influence climate change in other regions.

References

Amiro, B. D., J. B. Todd, B. M. Wotton, K. A. Logan, M. D. Flannigan, B. J. Stocks, J. A. Mason, D. L. Martell, and K. B. Hirsch. 2001. Direct carbon emissions from Canadian forest fires, 1959–1999. Canadian Journal of Forest Research 31:512–525.

Baldocchi, D., E. Falge, L. Gu, R. Olson, D. Hollinger, S. Running, P. Anthoni, C. Bernhofer, K. Davis, R. Evans, J. Fuentes, A. Goldstein, G. Katul, B. Law, X. Lee, Y. Malhi, T. Meyers, W. Munger, W. Oechel, K. U. K. Pilegaard, H. Schmid, R. Valentini, S. Verma, T. Vesala, K. Wilson, and S. Wofsy. 2001. FLUXNET: A new tool to study the temporal and spatial variability of ecosystem-scale carbon dioxide, water vapor, and energy flux densities. Bulletin of the American Meteorological Society 82:2415–2434.

Barber, V. A., G. P. Juday, and B. P. Finney. 2000. Reduced growth of Alaska white spruce in the twentieth century from temperature-induced drought stress. Nature 405:668–673.

Barry, R. G., and M. C. Serreze. 2000. Atmospheric components of the Arctic Ocean freshwater balance and their interannual variability. Pages 45–56 *in* E. L. Lewis, editor. The Freshwater Budget of the Arctic Ocean. Kluwer Academic Publishers, Dordrecht, Netherlands.

Bergh, J., S. Linder, T. Lundmark, et al. 1999. The effect of water and nutrient availability on the productivity of Norway spruce in northern and southern Sweden. Forest Ecology and Management 119:51–62.

Betts, R. A. 2000. Offset of the potential carbon sink from boreal forestation by decreases in surface albedo. Nature 408:187–190.

Betts, A. K., and J. H. Ball. 1997. Albedo over the boreal forest. Journal of Geophysical Research 102:28901–28909.

Bonan, G. B., D. Pollard, and S. L. Thompson. 1992. Effects of boreal forest vegetation on global climate. Nature 359:716–718.

Briffa, K. R., F. H. Schweingruber, P. D. Jones, T. J. Osborn, S. G. Shiyatov, and E. A. Vaganov. 1998. Reduced sensitivity of recent northern tree-growth to temperature at northern high latitudes. Nature 391:678–682.

Broecker, W. 1997. Thermohaline circulation, the Achilles heel of our climate system: Will man-made CO_2 upset the current balance? Science 278:1582–1588.

Chambers, S. D., and F. S. Chapin III. 2002. Fire effects on surface-atmosphere energy exchange in Alaskan black spruce ecosystems: Implications for feedbacks to regional climate. Journal of Geophysical Research 108:8145, doi:10.1029/2001JD000530 [printed in 108(D1), 2003].

Chapin, F. S., III, and A. M. Starfield. 1997. Time lags and novel ecosystems in response to transient climatic change in arctic Alaska. Climatic Change 35:449–461.

Chapin, F.S., III, W. Eugster, J. P. McFadden, A. H. Lynch, and D. A. Walker. 2000a. Summer differences among arctic ecosystems in regional climate forcing. Journal of Climate 13:2002–2010.

Chapin, F. S., III, A. D. McGuire, J. Randerson, R. Pielke Sr., D. Baldocchi, S. E. Hobbie, N. Roulet, W. Eugster, E. Kasischke, E. B. Rastetter, S. A. Zimov, W. C. Oechel, and S. W. Running. 2000b. Feedbacks from arctic and boreal ecosystems to climate. Global Change Biology 6:S211–S223.

Christensen, T. R., S. Jonasson, A. Michelsen, T. V. Callaghan, and M. Hastrom. 1998. Environmental controls on soil respiration in the Eurasian and Greenlandic Arctic. Journal of Geophysical Research 103:29,015–29,021.

Clein J., B. Kwiatkowski, A. D. McGuire, J. E. Hobbie, E. B. Rastetter, J. M. Melillo, and D. W. Kicklighter. 2000. Modeling carbon responses of tundra ecosystems to historical and projected climate: A comparison of a plot- and a global-scale ecosystem model to identify process-based uncertainties Global Change Biology 6:S127–S140.

Clein, J. S., A. D. McGuire, X. Zhuang, D. W. Kicklighter, J. M. Melillo, S. C. Wofsy, P. G. Jarvis, and J. M. Massheder. 2002. Historical and projected carbon balance of mature black spruce ecosystems across North America: The role of carbon-nitrogen interactions. Plant and Soil 242:15–32.

Dissing, D., and D. Verbyla. 2003. Spatial patterns of lightning strikes in interior Alaska and their relations to elevation and vegetation. Canadian Journal of Forest Research 33:770–782.

Dye, D.G. 2002. Variability and trends in the annual snow-cover cycle in Northern Hemisphere land areas, 1972–2000. Hydrological Processes 16:3065–3077.

Flannigan, M., I. Campbell, M. Wotton, C. Carcaillet, P. Richard, and Y. Bergeron. 2001. Future fire in Canada's boreal forest: paleoecology results and general circulation model—Regional climate model simulations. Canadian Journal of Forest Research 31:854–864.

Foley, J. A., J. E. Kutzbach, M. T. Coe, and S. Levis. 1994. Feedbacks between climate and boreal forests during the Holocene epoch. Nature 371:52–54.

Forman, S. L., W. Maslowski, J. T. Andrews, D. Lubinski, M. Steele, J. Zhang, R. Lammers, and B. Peterson. 2000. Researchers explore arctic freshwater's role in ocean circulation. Eos Transactions AGU 81(16):169–174.

Frolking, S., K. C. McDonald, J. S. Kimball, J. B. Way, R. Zimmermann, and S. W. Running, 1999. Using the space-borne NASA scatterometer (NSCAT) to determine the frozen and thawed seasons. Journal of Geophysical Research 104:27895–27907.

Gorchakovsky, P. L., and Shiyatov, S. G. 1978. The upper forest limit in the mountains of the boreal zone of the U.S.S.R. Arctic and Alpine Research 10:349–363.

Heijmans, M. M. P. D., W. J. Arp, and F. S. Chapin III. 2004. Controls on moss evaporation in a boreal black spruce forest. Global Biogeochemical Cycles 18:GB2004, doi:10.1029/2003GB002128.

Hobbie, S. E., J. P. Schimel, S. E. Trumbore, and J. R. Randerson. 2000. A mechanistic understanding of carbon storage and turnover in high-latitude soils. Global Change Biology 6:S196–S210.

Holland, E. A., F. J. Dentener, B. H. Braswell, and J. M. Sulzman. 1999. Contemporary and pre-industrial global reactive nitrogen budgets. Biogeochemistry 46:7–43.

Jacoby, G. C., and R. D. D'Arrigo. 1995. Tree ring width and density evidence of climatic and potential forest change in Alaska. Global Biogeochemical Cycles 9:227–234.

Kasischke, E., N. L. Christensen Jr., and B. J. Stocks. 1995. Fire, global warming, and the carbon balance of boreal forests. Ecological Applications 5:437–451.

Keyser, A. R., Kimball, J. S., Nemani, R. R., and Running, S.W. 2000. Simulating the effects of climatic change on the carbon balance of North American high-latitude forests. Global Change Biology 6:S185–S195.

Kullman, L. 1986. Late Holocene reproductive patterns of *Pinus sylvestris* and *Picea abies* at the forest limit in central Sweden. Canadian Journal of Botany 64:1682–1690.

Kullman, L. 1995. New and firm evidence for mid-Holocene appearance of *Picea abies* in the Scandes Mountains, Sweden. Journal of Ecology 83:439–447.

Kurz, W. A., and M. J. Apps. 1999. A 70-year retrospective analysis of carbon fluxes in the Canadian forest sector. Ecological Applications 9:526–547.

Lavoie, C., and S. Payette. 1994. Recent fluctuations of the lichen-spruce forest limit in subarctic Quebec. Journal of Ecology 82:725–734.

Lloyd, A. H., and C. L. Fastie. 2002. Spatial and temporal variability in the growth and climate response of treeline trees in Alaska. Climatic Change 52:481–509.

Lloyd, A. H., T. S. Rupp, C. L. Fastie, and A. M. Starfield. 2002. Patterns and dynamics of treeline advance in the Seward Peninsula, Alaska. Journal of Geophysical Research 107, 8161, doi:10.1029/2001JD000852 [printed in 108(D2), 2003].

MacDonald, G. M., A. A. Velichko, C. V. Kremenetsi, O. K. Borisova, A. A. Goleva, A. A. Andreev, L. C. Cwynar, R. T. Riding, S. L. Forman, T. W. D. Edwards, R. Aravena, D. Hammarlund, J. M. Szeicz, and V. N. Gattaulin. 2000. Holocene treeline history and climate change across northern Eurasia. Quaternary Research 53:302–311.

Manabe, S., and R. J. Stouffer. 1995. Simulation of abrupt climate change induced by freshwater input to the North Atlantic Ocean. Nature 378:165–167.

McGuire, A. D., and J. E. Hobbie. 1997. Global climate change and the equilibrium responses of carbon storage in arctic and subarctic regions. Pages 53–54 *in* Modeling the Arctic

System: A Workshop Report of the Arctic System Science Program. Arctic Research Consortium of the United States, Fairbanks, AK.

McGuire, A. D., J. M. Melillo, and L. A. Joyce. 1995a. The role of nitrogen in the response of forest net primary production to elevated atmospheric carbon dioxide. Annual Review of Ecology and Systematics 26:473–503.

McGuire, A. D., J. M. Melillo, D. W. Kicklighter, and L. A. Joyce. 1995b. Equilibrium responses of soil carbon to climate change: Empirical and process-based estimates. Journal of Biogeography 22:785–796.

McGuire, A. D., J. M. Melillo, D. W. Kicklighter, Y. Pan, X. Xiao, J. Helfrich, B. Moore III, C. J. Vorosmarty, and A. L. Schloss. 1997. Equilibrium responses of global net primary production and carbon storage to doubled atmospheric carbon dioxide: Sensitivity to changes in vegetation nitrogen concentration. Global Biogeochemical Cycles 11:173–189.

McGuire, A. D., C. Wirth, M. Apps, J. Beringer, J. Clein, H. Epstein, D. W. Kicklighter, J. Bhatti, F. S. Chapin III, B. de Groot, D. Efremov, W. Eugster, M. Fukuda, T. Gower, L. Hinzman, B. Huntley, G. J. Jia, E. Kasischke, J. Melillo, V. Romanovsky, A. Shvidenko, E. Vaganov, and D. Walker. 2002. Environmental variation, vegetation distribution, carbon dynamics, and water/energy exchange in high latitudes. Journal of Vegetation Science 13:301–314.

McGuire, A. D., M. Apps, F. S. Chapin III, R. Dargaville, M. D. Flannigan, E. S. Kasischke, D. Kicklighter, J. Kimball, W. Kurz, D. J. McRae, K. McDonald, J. Melillo, R. Myneni, B. J. Stocks, D. L. Verbyla, and Q. Zhuang. 2004. Land cover disturbances and feedbacks to the climate system in Canada and Alaska. Pages 139–161 in G. Gutman, A. C. Janetos, C. O. Justice, E. F. Moran, J. F. Mustard, R. R. Rindfuss, D. Skole, B. L. Turner II, and M. A. Cochrane. Land Change Science: Observing, Monitoring, and Understanding Trajectories of Change on the Earth's Surface. Kluwer Academic Publishers, Dordrecht, Netherlands.

Myneni, R. B, J. Dong, C. J. Tucker, R. K. Kaufmann, P. E. Kauppi, J. Liski, L. Zhou, V. Alexeyev, and M. K. Hughes. 2001. A large carbon sink in the woody biomass of northern forests. Proceedings of the National Academy of Sciences USA 98(26):14784–14789.

Nadelhoffer, K. J., B. A. Emmett, P. Gundersen, O. J. Kjonaas, C. J. Koopmans, P. Schleppi, A. Tietema, and R. G. Wright. 1999. Nitrogen deposition makes a minor contribution to carbon sequestration in temperate forests. Nature 398:145–148.

Norby, R. J., M. F. Cotrufo, P. Ineson, E. G. O'Neill, and J. G. Canadell. 2001. Elevated CO_2, litter chemistry, and decomposition: A synthesis. Oecologia 127:153–165.

Oechel, W. C., G. L. Vourlitis, S. J. Hastings, and S. A. Bochkarev. 1995. Change in arctic CO_2 flux over two decades: Effects of climate change at Barrow, Alaska, Ecological Applications 5:846–855.

Peterson, B. J., R. M. Holmes, J. W. McClelland, C. J. Vorosmarty, R. B. Lammers, A. I. Shiklomanov, I. A. Shiklomanov, and S. Rahmstorf. 2002. Increasing river discharge to the Arctic Ocean. Science 298:2171–2173.

Ramanswamy, V., O. Boucher, J. Haigh, D. Hauglustaine, J. Haywood, G. Myhre, T. Nakajima, G. Y. Shi, and S. Solomon. 2001. Radiative forcing of climate change. Pages 349–416 in J. T. Houghton, Y. Ding, D. J. Griggs, M. Noguer, P. J. Van Der Linden, and D. Xiuaosu, editors. Climate Change 2001—The Scientific Basis: Contribution of Working Group I to the Third Assessment Report of the Intergovernmental Panel on Climate Change (IPCC). Cambridge University Press, Cambridge, UK.

Randerson, J. T., C. B. Field, I. Y. Fung, and P. P. Tans. 1999. Increases in early season net ecosystem uptake explain changes in the seasonal cycle of atmospheric CO_2 at high northern latitudes. Geophysical Research Letters 26:2765–2768.

Romanovsky, V. E., Osterkamp, T. E., Sazonova, T. S., Shender, N. I., and V. T. Balobaev. 2000. Past and future changes in permafrost temperatures along the East Siberian transect and an Alaskan transect. Eos Transactions AGU 81(48):F223–F224.

Roulet, N. T. 2000. Peatlands, carbon storage, greenhouse gases, and the Kyoto Protocol: prospects and significance for Canada. Wetlands 20:605–615.

Serreze, M. C., D. H. Bromwich, M. P. Clark, A. J. Etringer, T. Zhang, and R. Lammers. 2002. Large-scale hydro-climatology of the terrestrial Arctic drainage. Journal of Geophysical Research 107: 8160, doi:10.1029/2001JD000919 [printed 108(D2), 2003].

Shine, K. P., R. G. Derwent, D. J. Wuebbles, and J-J. Morcrette. 1990. Radiative forcing of climate. Pages 41–68 in J. T. Houghton, G. J. Jenkins, and J. J. Ephraums, editors. Climate Change—The IPCC Scientific Assessment. Cambridge University Press, Cambridge, UK.

Spear, R. W. 1993. The palynological record of Late-Quaternary arctic tree-line in northwest Canada. Review of Paleobotany and Palynology 79:99–111.

Stocks, B. J., M. A. Fosberg, M. B. Wotten, T. J. Lynham, and K. C. Ryan. 2000. Climate change and forest fire activity in North American boreal forests. Pages 368–376 in E. S. Kasischke and B. J. Stocks, editors. Fire, Climate Change, and Carbon Cycling in North American Boreal Forest. Springer-Verlag, New York.

Stromgren, M., and S. Linder. 2002. Effects of nutrition and soil warming on stemwood production in a boreal Norway spruce stand. Global Change Biology 8:1195–1204.

Townsend, A. R., B. H. Braswell, E. A. Holland, and J. E. Penner. 1996. Spatial and temporal patterns in terrestrial carbon storage due to deposition of fossil fuel nitrogen. Ecological Applications 6:806–814.

Van Cleve, K., and J. Zasada. 1976. Response of 70-year-old white spruce to thinning and fertilization in interior Alaska. Canadian Journal of Forest Research 6:145–152.

Van Cleve, K., R. Barney, and R. Schlentner. 1981. Evidence of temperature control of production and nutrient cycling in two interior Alaska black spruce ecosystems. Canadian Journal of Forest Research 11:258–273.

Van Cleve, K., W. C. Oechel, and J. L. Hom. 1990. Response of black spruce (*Picea mariana*) ecosystems to soil temperature modification in interior Alaska. Canadian Journal of Forest Research 20:1530–1535.

Viterbo, P., and A. K. Betts, 1999. Impact on ECMWF forecasts of changes to the albedo of the boreal forests in the presence of snow. Journal of Geophysical Research 104(D22):27803–27810.

Vörösmarty, C. J., B. M. Fekete, M. Meybeck, and R. B. Lammers. 2000. Global system of rivers: Its role in organizing continental land mass and defining land-to-ocean linkages. Global Biogeochemical Cycles 14:599–621.

Zhuang, Q., A. D. McGuire, J. M. Melillo, J. S. Clein, R. J. Dargaville, D. W. Kicklighter, R. B. Myneni, J. Dong, V. E. Romanovsky, J. Harden, and J. E. Hobbie. 2003. Carbon cycling in extratropical terrestrial ecosystems of the Northern Hemisphere during the 20th century: A modeling analysis of the influences of soil thermal dynamics. Tellus 55B:751–776.

Zimov, S. A., Y. V. Voropaev, I. P. Semiletov, S. P. Davidov, S. F. Prosiannikov, F. S. Chapin III, M. C. Chapin, S. Trumbore, and S. Tyler. 1997. North Siberian lakes: A methane source fueled by Pleistocene carbon. Science 277:800–802.

20

Communication of Alaskan Boreal Science with Broader Communities

Elena B. Sparrow
Janice C. Dawe
F. Stuart Chapin III

Introduction

An important responsibility of all researchers is to communicate effectively with the rest of the scientific community, students, and the general public. Communication is "a process by which information is exchanged between individuals through a common system of symbols, signs or behavior" (Merriam-Webster 1988). It is a two-way process that requires collaborations, best-information exchange practices, and effective formal and informal education. Communication of this knowledge and understanding about the boreal forest is important because it benefits scientists, policymakers, program managers, teachers, students, and other community members. Good data and a firm knowledge base are needed for improving understanding of the functioning of the boreal forest, implementing best-management practices regarding forests and other resources, making personal and communal decisions regarding livelihoods and quality of life, coping with changes in the environment, and preparing future cadres of science-informed decision makers.

Communication among Scientists

Communication among scientists is an essential step in the research process because it informs researchers about important ideas and observations elsewhere in the world and allows boreal researchers to contribute to general scientific understanding. For example, the Bonanza Creek LTER has developed its research program by incorporating many important concepts developed elsewhere, including ecosystem dynamics (Tansley 1935), succession (Clements 1916), state factors (Jenny 1941), predator interactions (Elton 1958), and landscape dynamics (Turner et al. 2001). Through active research and regular communication and collaboration with the international scientific community, these "imported" ideas have been adapted to the boreal forest and new ideas and insights have been developed or communicated to the scientific community, as described in detail throughout this book. New ideas have originated among boreal researchers, and their "export" has sparked research elsewhere in the world (Chapter 21).

The pathways of communication are changing. Alaskan boreal researchers have participated actively in traditional modes of communication, including hundreds of peer-reviewed publications, several books, reports intended for managers, and participation in meetings and workshops. However, some of the greatest benefits of long-term research reside in the records of changes that occur. These long-term data are now available to the rest of the world through internet Web sites that house databases, publications, photographs, and other information (http://www.lter.uaf.edu). Web-based publications and data archives provide new opportunities, such as the rapid sharing of information, but also new challenges such as protection of sensitive information (e.g., locations of rare and endangered species or game animals) and long-term availability of Web sites.

Communication and Science Education

Science education plays a crucial role in the communication of science. It involves not only knowledge (data and principles) but also the insights and the procedures involved in acquiring the knowledge. It begins with kindergarten and continues through the rest of life. In the lifelong learning community, teachers can be students and students can be teachers. Science education is a shared responsibility among schools, teachers, scientists, parents, and other community members. The goals of science education are to develop and improve science literacy, develop and enhance critical thinking and personal decision-making abilities, and enable the public to give informed input to policymakers and researchers.

Some aspects of science education come easily to research communities. Most Alaskan boreal researchers live and teach in the boreal zone, so they automatically incorporate boreal research findings in lectures, discussions, and laboratories. The link between ecological research and university education is particularly strong in Alaska because the state attracts those people who enjoy living and breathing the ecology of Alaska. Most boreal biologists, whether molecular, physiological, or ecological by training, are ecologists at heart. Second, having classrooms on the doorstep of relatively undisturbed forests facilitates the engagement of students in research through both laboratory instruction and individual research by undergraduate and graduate students.

The linkage between boreal research and K-12 education also happens relatively easily. Boreal researchers, as parents in a small community, take an active interest in the schools. Moreover, K-12 teachers, who were drawn to Alaska by its wilderness qualities, typically have a keen interest in boreal ecology, as do students, many of who come from families with outdoor interests. Given these natural affinities among teachers, students, and boreal researchers and the availability of established scientific environmental measurements (Butler and MacGregor 2003) for K-12 student use in the classroom, it is not surprising that the Bonanza Creek "Schoolyard LTER" (SYLTER) was an instant success.

Bonanza Creek SYLTER developed a miniresearch program in which K-12 students are the researchers who, with the help of teachers or LTER researchers, design their own long-term research program, collect data, and archive them in Web-accessible databases. Scientist involvement and teacher buy-in are essential in linking research and K-12 science education and in engaging students in the science process. Fourth graders at University Park Elementary School, for example, conducted a year-long investigation of the ecology and history of their backyard slough. On their school Web site, students published field reports and pictures through the seasons. Fifth graders at the same school and sixth graders at Joy Elementary have been collecting and entering weather data on the Internet, as they gather phenological data on birch trees. Similarly, West Valley High School students (in studies related to ongoing LTER research) have been recording weather data in relation to effects of climate on tree growth and patterns of green-up and green-down of deciduous shrubs and trees. Other investigations have been conducted by individual high school students mentored by boreal scientists.

Engaging K-12 students in boreal research is more challenging in remote rural villages where the main connections to the rest of the world are by air, riverboat, or snow machine. The challenge of distance also represents a unique opportunity. Boreal researchers are hampered in their efforts to generalize their results to broad geographic areas by the limited road network in Alaska. Researchers at the University of Alaska have teamed up with K-12 teachers and their students in remote villages in a unique program to document the geographic pattern of seasonal vegetation development that may be related to climate change. University researchers provide satellite images, and village students document the seasonal changes in leaf development and senescence. The village and university researchers communicate about their findings by Internet and maintain a database that can be compared with similar observations made by K-12 students throughout the world. The challenges of remoteness in Alaska led to the development of a K-12 research program that expanded globally. This plant phenology investigation developed by a team of UAF boreal researchers and an educator is part of the Global Learning and Observations to Benefit the Environment (GLOBE) program (http://www.globe.gov) and has succeeded in converting an impediment to education and research into an asset. The GLOBE Program is an international environmental science and education program that connects K-12 students, teachers, and scientists around the world for research collaboration and cross-cultural enrichment (Sparrow 2001). GLOBE engages precollege students in environmental and earth systems investigations at or close to their schools as a way of teaching and learning science.

Perhaps the most exciting potential link between research and education is a challenge that remains a "work in progress," that is, the integration of Western science and traditional Native knowledge. Many remote Alaskan villages (and even towns and cities) have a large indigenous population whose cultural ties to the land are still strong. These include Inupiaq, Yupik, Aleut, Tlingit, and Haida people in coastal areas and Athabascans in interior Alaska (Aigner et al. 1986). For centuries, these people have depended on understanding their links to the land for food and survival. This traditional knowledge, that is, knowledge based on long-term observations and tradition, is transmitted orally from generation to generation. It is still essential where subsistence hunting is an important source of food and where winter travel requires a keen understanding of snow, weather, and winter hydrology. There are many important commonalities, in spite of differences, between traditional Native knowledge and Western science (Fig. 20.1; Stephens 2000a).

Traditional ecological knowledge (TEK), because it is based on the long-term relationship of people with the land, is a useful basis for ecological understanding, scientific research, impact assessment, and resource management (Berkes 1999, Huntington 2000, Usher 2000). However, strong barriers prevent wider application of TEK in research and resource management; these include a preference for established scientific methodologies, the difficulty of accessing TEK (rarely written), the need to access and describe TEK in Western scientific ways and terms, the reluctance of researchers and managers to work with nonscientists (indigenous or not), the reluctance of holders of TEK to share what they know, and issues of ownership and control over TEK (Huntington 2000). Some of these obstacles also apply to use of Native knowledge in science education. However, the common ground between traditional knowledge and Western science needs to be cultivated to maximize the benefits of using more than one knowledge base and approach in both research and education (Fig. 20.1).

Global change and its northern manifestations create some new opportunities for transfer of knowledge between TEK and western science (Riedlinger and Berkes 2001). Traditional knowledge has always viewed people as an integral component of ecological systems, whereas Western science typically studies ecological and social systems separately (Berkes and Folke 1998). The large anthropogenic changes in the global environment have made it increasingly evident that ecological and social processes can no longer be studied in isolation. Adopting the TEK perspective of linked social-ecological systems would provide a more realistic context for future studies of regional processes (Chapin and Whiteman 1998, Berkes et al. 2003).

Conversely, Western science can contribute to the sustainability of traditional lifestyles. Village elders now risk being "strangers in their own land" (Berkes 2002), because the climate and ecological patterns that traditionally enabled them to predict conditions for safe and successful travel and hunting are changing and are no longer predictable from the cues traditionally used by village elders. Western science provides enough understanding to improve prediction of some of the emerging climatic and ecological patterns in ways that may prove useful to local elders. Despite the potential mutual benefits provided by traditional knowledge and Western science, the lack of mutual understanding between these two epistemologies makes their integration a slow process.

Traditional knowledge can contribute to science education in several potentially important ways. Science education often neglects to connect students' "knowledge into a coherent picture of how the world works and how we come to know it" (Nelson 2002). This holistic perspective is central to traditional knowledge and provides a powerful educational approach that could enhance science learning. Thus, scaffold-

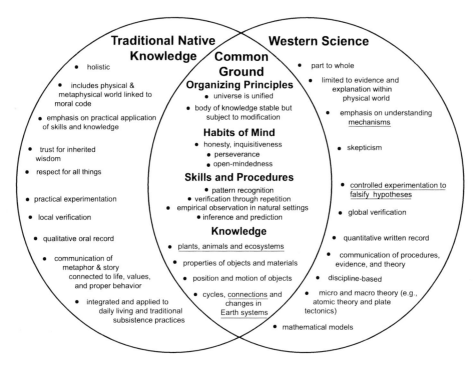

Figure 20.1. Similarities and differences between traditional Native knowledge and Western science. Modified with permission from the Alaska Rural Systemic Initiative (Stephens 2000a), with the modifications underlined.

ing environmental studies with traditional knowledge, especially in rural schools, would be very useful (Sparrow 2002).

Storytelling is a feature of Alaskan and other Native cultures that is a particularly important educational tool (Schneider 2002). Stories play a key role in the oral transfer of traditional knowledge (Cruikshank 1998) because facts presented in stories are easier to remember than facts in lists (Wilson 2002). Metaphors and analogies that strike a familiar chord help us to make sense of information by connecting the story to our prior experience and knowledge. Thus, Wilson (2002) advocates the use in teaching of stories that reveal the trial-and-error processes of science. Stories have proven to be effective teaching tools for subject matters ranging from natural history to atomic physics (Hakim 2002) but are underutilized in science education. Educational programs in Alaska have begun to bridge the gap between traditional knowledge (TK) and Western science. One such program is the "Observing Locally Connecting Globally" (OLCG) program (http://www.uaf.edu/olcg) at the University of Alaska Fairbanks. OLCG engages K-12 students in global change research by building environmental studies based on local community knowledge and by developing skills and knowledge common to both TK and Western science (Sparrow et al. 2000). Teachers are supported throughout the year as they work with students and are prepared during a two-week OLCG Institute, in which Native elders share observation skills, long-term environmental observations, holistic knowledge, and a systems approach (Stephens 2000b). Teacher participants also learn and practice scientific measurements used in GLOBE and other ongoing climate change research projects. They apply the research and teaching methods they learned to develop and conduct a research investigation relevant to their schools and communities (Gordon 2001). This integration of traditional and Western approaches to science yields positive effects in the villages in terms of student learning and attitude toward science; it also increases teacher comfort with inquiry-based pedagogies (Gordon et al. 2005).

Education that integrates social and ecological processes is equally relevant at the graduate level. This is being implemented at the University of Alaska through an integrative graduate education and research training (IGERT) program called "Resilience and Adaptation." This program combines the ecological roots of regional sustainability, which come largely from the LTER program, with economic and cultural issues, which are equally critical in developing a sustainable future. This program has found it necessary to stretch beyond the academic community to managers, Native communities, and other community and regional groups to understand how these components of sustainability might be combined.

Communication about Research with the General Public

Scientists are notoriously reticent and ineffective in communicating about research results and conclusions to the general public. The public frequently wants to know the certainty of findings. In contrast, Western science is built around rejection of hypotheses (a paring away of possibilities), rather than absolute proof. Traditional knowledge built around ecological understandings of how an ecosystem functions may someday provide a useful bridge between the public's need to know and scientists' reluctance to make pronouncements in the face of uncertainty.

Given this communication gulf between scientists and the public, it is perhaps not surprising that the major initiative to enhance communication about boreal research has come from the public. In 1993, Governor Walter Hickel, of Alaska, proposed significant changes to the management of interior Alaska's state forested lands and asked for public input. Community members spoke at length about the diverse ways in which the forest supports their livelihoods and quality of life. They identified three barriers to their ability to give informed feedback on the governor's proposals: insufficient access to forest management information, inadequate information with which to judge the likely consequences of the proposed actions, and poor understanding of the available technical data. In the absence of sound information, the forest development debate became increasingly divisive, and trust in professional forest management plummeted.

Concerns over these developments led to the formation of the Alaska Boreal Forest Council as a grassroots response to the need to dissolve the barriers to informed public participation in natural resource policy and decision-making processes. The council's first two actions were (1) to disseminate responses of community members to the governor's proposals to the general public (Dawe et al. 1994) and (2) to commission an independent public opinion poll to gather additional information about the public's desires for forest development. The polling results led the council to initiate a nine-month-long consensus-building process among scientists, resource managers, legislators, recreationists, commercial interests, and other stakeholders. This process allowed the public to communicate concerns to resource managers and the broader scientific community in a neutral setting. It also gave the resource managers a community forum in which to explain management practices to a more diverse audience than ordinarily showed up at agency functions.

The consensus-building process was very successful. The freedom to express strong feelings and opinions in a safe environment proved to be the necessary first step in building trust and consensus for all involved. Two features were especially important: each round table was chaired by a professional facilitator to keep the discussions fair and on track, and the consensus output from each round table was accepted exactly as written; people who had not been direct participants at a round table could make no edits or changes to it (Alaska Boreal Forest Council 1995).

The increased communication and respect generated among round-table and workshop participants led to the identification of common interests regarding forest development and fruitful avenues for collaboration among diverse stakeholders. The Alaska Boreal Forest Council began a weekly brown-bag discussion group so that community members and technical experts could continue the dialogue in an informal setting. Interests and concerns identified during the course of the brown baggers led the council to spearhead two additional workshops: a technical session entitled "Managing for Changing Forest Conditions: What Are Our Options?" (Alaska Boreal Forest Council 1998) and a follow-on community session, "People and the Forest: Creating a Partnership."

The consensus that emerged from these two workshops was that ecosystem management is the paradigm of choice for managing the interior Alaska boreal forest ecosystem. Three features of ecosystem management—its emphasis on adaptive management, its explicit recognition of humans as part of the ecosystem, and its insistence that the public participate in the establishment of management goals—were seen as essential to developing responsible forest management policy and practices. Accordingly, the council's board resolved to make ecosystem management the central tenet of its approach to developing sustainable communities and economies.

Today, when we consider the community dialogue and consensus-building process that began in 1993, we realize that it was difficult and halting at first. Although pockets of distrust still exist, the community dialogue has become increasingly productive over time. This, we believe, is the result of the open communication that was established early on among diverse boreal forest stakeholders. In addition to the more positive tone of the forest development debate and specific collaborative community projects that have been initiated as a result of this community consensus-building effort, an LTER researcher established an important precedent at one of the first council workshops: simultaneous reporting of significant research results to the scientific community and to the local community of concern (Alaska

Boreal Forest Council 1996). The importance of this communication to the broader community cannot be overemphasized. The public is turning increasingly to an integration of science, economics, and social values as the foundation for responsible resource development.

References

Aigner, J. S., R. D. Guthrie, M. L. Guthrie, R. K. Nelson, W. S. Schneider, and R. M. Thorson, editors. 1986. Interior Alaska: A Journey through Time. Alaska Geographic Society, Anchorage, AK.

Alaska Boreal Forest Council. 1995. Community Round Tables: Fall 1995 Consensus Statements.

Alaska Boreal Forest Council. 1996. Coming together: weaving a sustainable future in the Tanana river basin. An interior forest sustainability workshop. Available at http://www.akborealforest.org.

Alaska Boreal Forest Council. 1998. Managing for forest condition in interior Alaska: What are our options? Technical workshop, 9–10 April, Fairbanks, AK.

Berkes, F. 1999. Sacred Ecology: Traditional Ecological Knowledge and Resource Management. Taylor and Francis, London and Philadelphia.

Berkes, F. 2002. Epilogue: making sense of arctic environmental change? Pages 335–349 *in* I. Krupnik and D. Jolly, editors. The Earth Is Faster Now. Arctic Research Consortium of the United States, Fairbanks, AK.

Berkes, F., and C. Folke. 1998. Linking social and ecological systems for resilience and sustainability. Pages 1–25 *in* F. Berkes and C. Folke, editors. Linking Social and Ecological Systems: Management Practices and Social Mechanisms for Building Resilience. Cambridge University Press, Cambridge, UK.

Berkes, F., J. Colding, and C. Folke, editors. 2003. Navigating Social-Ecological Systems: Building Resilience for Complexity and Change. Cambridge University Press, Cambridge, UK.

Butler, D. M., and I. D. MacGregor. 2003. GLOBE: Science and education. Journal of Geoscience Education 51:9–20.

Chapin, F. S., III, and G. Whiteman. 1998. Sustainable development of the boreal forest: Interaction of ecological, social and consumer feedbacks. Conservation Ecology [online] 2:12. Available at http://www.consecol.org/vol2/iss2/art12.

Clements, F. E. 1916. Plant Succession: An Analysis of the Development of Vegetation. Publication 242. Carnegie Institution of Washington, Washington, DC.

Cruikshank, J. 1998. The Social Life of Stories: Narrative and Knowledge in the Yukon Territory. University of Nebraska Press, Lincoln.

Dawe, J. C., A. N. Whitworth, R. J. McCaffrey, and D. A. Yates, editors. 1994. Voices of the Forest: Public Testimony on the Future of the Tanana Valley State Forest: Issues and Opportunities Related to Land Use Management in Interior Alaska. ABFC Misc. Publication No. 1.

Elton, C. S. 1958. The ecology of invasions by animals and plants. Methuen, London.

Gordon, L. S. 2001. Breaking new ground in Alaska. ENC Focus 8:38–39.

Gordon, L. S., S. Stephens, and E. B. Sparrow. 2005. Applying the national science education standards to Alaska: weaving Native knowledge into teaching and learning environmental science through inquiry. Pages 85–98 *in* R. Yager, editor. Exemplary Science: Best Practices in Professional Development. National Sciences Teachers Association Press, Arlington, VA.

Hakim, J. 2002. The story of the atom. American Educator 26:12–25.
Huntington, H. 2000. Using traditional ecological knowledge in science: Methods and applications. Ecological Applications 10:1270–1274.
Jenny, H. 1941. Factors of Soil Formation. McGraw-Hill, New York.
Merriam-Webster. 1988. Webster's Ninth New Collegiate Dictionary. Merriam-Webster, Inc., Springfield, MA.
Nelson, G. D. 2002 (spring). Statement quoted by Editors of American Educator. 26:8.
Riedlinger, D., and F. Berkes. 2001. Contributions of traditional knowledge to understanding climate change in the Canadian Arctic. Polar Record 37:315–328.
Schneider, W. 2002. So They Understand. Cultural Issues in Oral History. Utah State University Press, Logan.
Sparrow, E. B. 2001. GLOBE: A new model in K–12 science education. Global Glimpses. Center for Global Change and Arctic System Research 9(1):1–4.
Sparrow, E. B. 2002. Community collaborations: Cultivating the common ground. Education and Implementation Panel Report: Impact of GLOBE on Your Community. Pages 175–178 *in* Proceedings of the Seventh GLOBE Annual Meeting, Chicago.
Sparrow, E. B., S. Stephens, and L. S. Gordon. 2000. Global Change Education Using Western Science and Native Knowledge. Abstract published in the Proceedings of the Fifty-first American Association for the Advancement of Science Arctic Science Conference. Whitehorse, Canada.
Stephens, S. 2000a. Handbook for Culturally Responsive Science Curriculum. Alaska Rural Systemic Initiative. Available at http://www.ankn.uaf.edu/handbook/.
Stephens, S. 2000b. Observing Locally, Connecting Globally. Sharing Our Pathways. Newsletter of the Alaska Rural Systemic Initiative 5:6–7.
Tansley, A. G. 1935. The use and abuse of vegetational concepts and terms. Ecology 16:284–307.
Turner, M. G., R. H. Gardner, and R. V. O'Neill. 2001. Landscape Ecology in Theory and Practice: Pattern and Process. Springer-Verlag, New York.
Usher, P. J. 2000. Traditional ecological knowledge in environmental assessment and management. Arctic 53:183–193.
Wilson, E. O. 2002. The power of story. American Educator 26:8–11.

21

Summary and Synthesis
Past and Future Changes in the Alaskan Boreal Forest

F. Stuart Chapin III
David McGuire
Roger W. Ruess
Marilyn W. Walker
Richard D. Boone
Mary E. Edwards
Bruce P. Finney
Larry D. Hinzman
Jeremy B. Jones
Glenn P. Juday
Eric S. Kasischke
Knut Kielland
Andrea H. Lloyd
Mark W. Oswood

Chien-Lu Ping
Eric Rexstad
Vladimir E. Romanovsky
Joshua P. Schimel
Elena B. Sparrow
Bjartmar Sveinbjörnsson
David W. Valentine
Keith Van Cleve
David L. Verbyla
Leslie A. Viereck
Richard A. Werner
Tricia L. Wurtz
John Yarie

Introduction

Historically the boreal forest has experienced major changes, and it remains a highly dynamic biome today. During cold phases of Quaternary climate cycles, forests were virtually absent from Alaska, and since the postglacial re-establishment of forests ca 13,000 years ago, there have been periods of both relative stability and rapid change (Chapter 5). Today, the Alaskan boreal forest appears to be on the brink of further significant change in composition and function triggered by recent changes that include climatic warming (Chapter 4). In this chapter, we summarize the major conclusions from earlier chapters as a basis for anticipating future trends.

Forest Dynamics

Alaska warmed rapidly at the end of the last glacial period, ca 15,000–13,000 years ago. Broadly speaking, climate was warmest and driest in the late glacial and early Holocene; subsequently, moisture increased, and the climate gradually cooled. These changes were associated with shifts in vegetation dominance from deciduous woodland and shrubland to white spruce and then to black spruce. The establishment of stands of fire-prone black spruce over large areas of the boreal forest 5000–6000 years ago is linked to an apparent increase in fire frequency, despite the climatic trend to cooler and moister conditions. This suggests that long-term features of the Holocene fire regime are more strongly driven by vegetation characteristics than directly by climate (Chapter 5).

White spruce forests show *decreased* growth in response to recent warming, because warming-induced drought stress is more limiting to growth than is temperature per se (Chapters 5, 11). If these environmental controls persist, projections suggest that continued climate warming will lead to zero net annual growth and perhaps the movement of white spruce to cooler upland forest sites before the end of the twenty-first century. At the southern limit of the Alaskan boreal forest, spruce bark beetle outbreaks have decimated extensive areas of spruce forest, because warmer temperatures have reduced tree resistance to bark beetles and shortened the life cycle of the beetle from two years to one, shifting the tree-beetle interaction in favor of the insect (Chapter 9). At its altitudinal and latitudinal limits, the boreal forest is expanding into tundra because of tree recruitment beyond treeline during recent warm decades. At arctic treeline, spruce establishment in tundra depends at least partially on thawing permafrost and other disturbances. In summary, current trends show a slow expansion of forest into tundra in the north, retreat of boreal forest in the south, and impending major compositional changes in central portions of Alaska's boreal forest.

Some major Holocene vegetation transitions occurred over a time scale of a few centuries (e.g., deciduous to white spruce) to perhaps 1000 years (e.g., white to black spruce). Future changes in fire regime (Chapter 17) could trigger further widespread change by creating the potential for altered successional trajectories. Although self-replacement, in which the prefire tree species returns to dominance after fire, occurs in the coldest and warmest environments dominated by black spruce and aspen, respectively, successional sequences with multiple stages are more common in intermediate sites. Here, changes in a number of processes could alter vegetation composition and successional trajectory. Late-successional conifers establish during the first and second decades after fire, but their establishment success is sensitive to the depth of the organic mat remaining after fire, understory species composition, and seed availability from on-site serotinous cones (black spruce) or off-site seed sources (white spruce). Insect outbreaks can disrupt or redirect succession by reducing abundance of preferred hosts (Chapters 7, 9). Mammalian herbivores speed succession by eliminating early successional species and by shifting the competitive balance in favor of the less palatable, more slowly growing late-successional species (Chapters 8, 13). Changes in any of these processes could alter vegetation composition and successional trajectory.

After initial establishment, competition, facilitation, and herbivory interact to drive successional change (Chapter 7). Ecosystem controls change at key *turning points* (thresholds), where a shift in dominance of plant functional types radically alters the physical and chemical environments that govern ecosystem processes and disturbance probability (Chapter 1). In the floodplain, intense herbivory by moose initially constrains canopy development, creating an ecosystem dominated by physical controls over soil water movement, driving surface evaporation and gypsum accumulation at the soil surface (Chapters 7, 13). Colonization by thinleaf alder shifts the system from physical to biological control, adds 60–70% of the nitrogen that accumulates during succession, and causes herbivory to change from a deterrent to an accelerator of succession by eliminating palatable early successional species (Chapters 7, 13, 15). Other key turning points include (1) a shift to balsam poplar dominance, where changes in productive potential and litter chemistry enhance NPP and nitrogen cycling rates (Chapters 11, 14), and (2) a shift to white spruce dominance, where mosses grow rapidly in the absence of smothering broadleaved litter, nutrient cycling rates are reduced by low temperature and the sequestering of nutrients in low-quality litter, and fire probability increases because fuels dry quickly and support fire spread (Chapters 6, 7, 14, 17).

Species diversity is low in the boreal forest (Chapters 6, 13) and varies dramatically through succession, with peaks in early succession (e.g., fire-specialist plants, herbivorous insects, neotropical migrant birds, and mammals) and late succession (nonvascular plants and saprophagous insects).

Many boreal animals exhibit large population fluctuations. Moose and hare densities appear to be sensitive to food availability and predation, whereas vole densities correlate more strongly with climate (Chapter 8). Two native insects have changed from decadal outbreaks to consistently low populations (large aspen tortrix since 1985; spear-marked blackmoth since 1975), whereas other species that had negligible populations before 1990 now show large outbreaks (eastern spruce budworm, spruce coneworm, larch sawfly, and aspen leaf miner; Chapter 9).

Long-term forest harvest studies permit an assessment of potential future human impacts on Alaska's boreal forest. Low-intensity forest harvest (no scarification) reduces initial seedling establishment but maximizes long-term growth of tree seedlings (Chapter 18). Overstory retention treatments have no long-term effect on tree recruitment and growth. These studies suggest that low-intensity management after clear-cutting, an approach that mimics *certain* aspects of natural fire cycles, may maximize ecological recovery. Economics currently preclude widespread commercial forestry in interior Alaska, and future developments of this industry would require substantial changes in world markets and local infrastructure.

Biogeochemistry

Although most surfaces of interior Alaska have not been glaciated during the Quaternary, they exhibit properties of young soils, with minimal profile development being widespread on the landscape (Chapter 3). This occurs because weathering rates are slow because of the cold, and frequently anaerobic, environment and because surface loess deposition, frost-heaving, and/or erosion counteract these soil-forming processes (Chapters 3, 16).

Aboveground production varies by more than an order of magnitude among major forest types in interior Alaska, primarily due to variation in topography and successional age (Chapters 6, 11). It is greatest in midsuccessional stands on floodplains, where soil temperature and moisture are relatively high. On south-facing slopes, production is constrained by moisture and on north-facing slopes, by soil temperature. On temperature-limited sites, mosses account for half of aboveground production. Except on water-limited south-facing slopes, nitrogen appears to be the factor that directly limits growth, with temperature exerting indirect effects through its impacts on nitrogen cycling (Chapters 11, 12, 14, 15).

The critical controls over the productivity and nutrient cycling of Alaskan boreal forests occur belowground (Chapters 11, 12). Carbon and nutrient cycling rates in fine roots are several orders of magnitude faster than in aboveground tissues because of high belowground allocation and turnover (Chapter 12). Cold temperatures confine fine-root production to zones close to the soil surface, and the progressive extension of fine-root production into deeper soil horizons occurs slowly, as the soil warms through the season. Alaskan trees and shrubs have roots with morphological, phenological, and physiological traits that are similar to those of most forest biomes. Alaskan trees are unusual primarily in their large root allocation and in the large quantity and rapid turnover of fine roots (Chapter 12).

Ratios of aboveground litterfall to soil respiration in interior Alaskan forests are among the lowest recorded in North American forests, suggesting that a large proportion of boreal soil respiration originates in root respiration and the rapid turnover of fine roots (Chapter 12). For example, fine-root respiration constitutes about 60% of total soil respiration in black spruce forests. The soil carbon derived from fine roots is quite labile, so soil carbon stocks can decline rapidly in the absence of continued inputs, such as might occur after fire.

Mammalian herbivores play a key role in the biogeochemistry of the boreal forest. In the floodplain willow communities, they consume 40% of aboveground NPP (Chapter 13). When herbivores are experimentally excluded, biogeochemistry changes rapidly from a system dominated by inorganic C cycling and solubility equilibria to a biologically controlled pattern of cycling dominated by NPP and decomposition.

N_2 fixation inputs by green alder in the uplands and thinleaf alder in the floodplains account for the largest percentage of total N accumulated during succession (Chapters 14, 15). Fixation inputs appear to exceed plant N demand, and significant amounts of fixed N may be lost via leaching or denitrification, particularly in midsuccessional stages, where nitrification potential is high and soil microbial biomass is more C- than N-limited (Chapters 14, 16).

Despite the large quantities of organic N that accumulate in boreal soils, the vegetation is strongly N-limited, and conversion of insoluble organic N to plant-available forms appears to be the rate-limiting step (Chapters 11, 15). Once this organic N becomes soluble, it is quite dynamic. Amino acids turn over more rapidly than inorganic N and are a major source of N absorbed by both plants and microbes (Chapters 14, 15). The natural abundance of ^{15}N in vegetation suggests that uptake of organic N plays a significant role in the N nutrition of the boreal forest.

Landscape Processes and Disturbance

Presence or absence of permafrost is probably the single most important threshold regulating the structure and functioning of Alaska's boreal forest (Chapters 4, 16). Permafrost is discontinuous in most of interior Alaska, being generally present on north-facing slopes and in valley bottoms, where it leads to cold, water-logged soils, and generally absent on the upper parts of south-facing slopes, where soils drain freely. Permafrost temperatures are now typically warmer than -2°C and have warmed about 0.7°C per decade since 1970 in response to regional warming and changes in insulation by snow and vegetation (Chapter 4). Continued warming will likely lead to extensive permafrost degradation within 10 to 25 years. Currently 38% of the research watersheds that have been intensively studied by the Bonanza Creek LTER have unstable or thawing permafrost. Thus, permafrost is likely to be lost much more rapidly in the coming decades than during the past 90 years, over which span a 2.1% loss occurred.

Permafrost response to climate warming involves multiple ecosystem feedbacks that involve changes in insulation by snow, moss, the surface organic mat, and soil drainage (Chapter 16). Insulation declines dramatically after fire, increasing the depth of thawed soil from about 50 cm to 2–4 m. As permafrost recovers during postfire succession, an unfrozen layer (talik) forms between a seasonally frozen or newly developed upper permafrost layer and the lowered surface of original permafrost. In sloping terrain, water drains laterally through the talik, drying surface soils, whereas on flat terrain water may accumulate at the surface, forming ponds or waterlogged soil. In cases where this permafrost contains large volumes of ice, thawing may cause subsidence of the ground surface (thermokarst). Thus, the impact of climate warming on soil moisture in permafrost terrain depends strongly on topography, fire history, and other factors that control talik formation and drainage conditions.

Low-permafrost watersheds or watersheds with well-developed taliks have greater base flow (80% of discharge) and are less flashy (i.e., less likely to cause floods) than high-permafrost watersheds, in which base flow increases from 50% to 60% of discharge in early summer to values similar to those of low-permafrost watersheds in late summer, when mineral soils have thawed (Chapter 16). Areas with abundant groundwater flow also generate aufeis (areas of thick winter ice that forms when groundwater is forced to the surface by freezing soil) that kills most woody vegetation, substantially altering riparian dynamics. Groundwater flow also generates higher concentrations of base cations, inorganic nitrogen, and dissolved CO_2 and less dissolved organic carbon and nitrogen than in permafrost-dominated watersheds, where most water flows through the organic mat. Thus, permafrost and talik distribution strongly influence soil moisture, land-water interactions, and stream discharge and chemistry. In contrast to temperate ecosystems, nitrate losses in Alaskan boreal streams are four to five times greater than deposition inputs, a result that we cannot currently explain in light of the strong nitrogen limitation of watershed vegetation.

The running waters of the boreal forest derive most of their physical and biological characteristics from a cold climate syndrome. Extreme seasonality induces formation of several kinds of ice in streams and drainage basins. Long-lasting ice cover can limit gaseous exchanges between water and the atmosphere, frazil and anchor ice can disturb substrates, and permafrost in the drainage basin increases transport of dissolved organic carbon and makes streams hydrologically flashy in response to runoff and precipitation (Chapters 10, 16). Glaciers in headwaters transport cold water and sediments during summer melt, reducing substrate diversity and stability and so limiting biotic diversity. As is the case elsewhere, autumnal leaf litter is the major source of food to stream foodwebs and a connection between riparian and aquatic ecosystems. The cascading effects of herbivory on soil systems (changes in species composition of vegetation and chemical characteristics of leaf litter; Chapter 13) affect stream foodwebs via autumnal leaf litter. High-latitude limitations on primary production by both periphytic microbes (attached to streambed sediments) and riparian vegetation make running waters nutrient-poor. Therefore, the uphill transport of marine-derived nutrients by spawning salmon provides a key source of nutrients for steam and riparian consumers. Even so, limited food resources and cold waters slow growth rates and maturation of stream consumers, so many major taxa are absent or in low abundance in Alaskan running waters, compared to their presence in temperate streams. Extreme seasonality renders some habitats uninhabitable in winter, forcing vertical migrations of benthic invertebrates and longitudinal migrations of fishes. The arctic grayling, a ubiquitous fish in boreal forest streams, has a complex life history that maps the seasonal availability of habitats for spawning, juvenile rearing, adult feeding, and overwintering.

Fire is the dominant disturbance agent in interior Alaska but is highly variable in space and time. In the average year, the fire season lasts only three weeks, sandwiched between early summer, when soils are wet from snowmelt, and late summer, when precipitation increases (Chapter 17). However, 55% of the total area burned between 1961 and 2000 burned in just 6 years, when the fire season lasted much longer, giving rise to very large fires. Thus, the fire regime is dominated by unusual years, rather than by average conditions. The 7% increase in area burned in Alaska in the past 40 years is much less than the doubling reported for western Canada. Fire return time varies regionally from <50 years to more than 100 years. It correlates positively with temperature and vegetation cover and negatively with precipitation. Lightning, which accounts for 90% of the area burned, is controlled by both synoptic processes related to El Niño and by local factors such as topography and presence of forest vegetation. Human ignitions, which account for 60% of the fires in Alaska, generally produce small fires because they are lit at times and places where fire does not readily spread.

Recent and projected changes in the boreal forest could feed back to the climate system. The lower albedo and greater sensible heat flux of spruce compared to deciduous forests or nonforested wetlands (Chapter 19) suggest that northward forest expansion could be a positive feedback to regional warming but that loss of forests to the south or net conversion from conifer to deciduous forests resulting from fire could have a net cooling effect, one of the few negative feedbacks to high-latitude warming that has been identified. Boreal forests contain approximately 27% of the world's vegetation carbon inventory and 28% of the world's soil carbon inventory (equivalent to 75% of the total atmospheric carbon), so warming effects on net ecosystem production (NPP minus respiration) or on fire regime could substantially alter the global climate system. Warming appears to enhance carbon release in dry areas, enhance uptake in wet areas, and enhance methane release in wet areas. The net effect of fire depends on fire severity and on changes in fire frequency. All of these effects on trace-gas feedbacks hinge on permafrost and hydrological changes, which are poorly known. The recent shrinkage of lakes and wetlands in interior Alaska suggests, however, that the CO_2 efflux is increasing and methane efflux is decreasing.

Preparing for the Future

Although the boreal forest is the northernmost outpost of forested biomes, whose current distribution is clearly linked to temperature, its immediate future appears to be more sensitive to projected changes in moisture than to temperature. If warming leads to surface drying, as suggested by recent trends and many climate projections, drought could enhance soil drying and cause more frequent insect outbreaks and fires, triggering rapid changes in vegetation, permafrost, soils, and streams. Alternatively, a more ice-free ocean upwind of Alaska could increase precipitation (Chapters 2, 4), with quite different ecosystem consequences. Although our crystal ball does not allow a clear choice between these alternative futures, it seems quite likely that the Alaskan boreal forest will change dramatically from its current state and that future planning must account for this eventuality (Chapter 20). Only with long-term research, such as that conducted by the Bonanza Creek LTER program, can we understand the controls and patterns of long-term change adequately to prepare society for the choices that it must face.

Index

abiotic factors
  insect responses to, 138, 140, 142
  successional controls over, 102–104
aboveground forest production
  evaporation effects on, 177, 178
  precipitation effects on, 177, 178
  temperature effects on, 176–177
access, to timber sale areas, 306
acidic black spruce/lichen forest, 95
acidic bogs. *See* bogs
active layer, 27, 42, 50–51, 57, 242, 269, 270, 271, 278, 280, 281, 318
  fire effects on, 28
aerobic conditions
  decomposition responses to, 230
  soil responses to, 23, 280
aerosols, anthropogenic, 273, 274
age distribution
  and mammal population turnover, 129
  of plant species, 213–214, 286–297
A horizon, 22, 26, 27
  formation of, 22, 31
Alaska Boreal Forest Council, 328, 329
Alaska Fire Service, 287, 290, 296
Alaska Primary Manufacturing Law, 302
Alaska Range, 25
albedo, 46, 56, 271, 273, 310–312, 318
  soil environment effects of, 111

alder, insect use of, 134. *See also* green alder; thinleaf alder
algae, 148–149, 158–160
allocation, to fine roots, 201–203, 245–256, 335
alluvial deposits, 24, 25, 92, 102, 103, 172, 269
alluvial terrace. *See* terrace
alpine, 88
  treeline, 70
amino acids, 252–254, 257, 261
  turnover of, 237, 253–254, 335
ammonium
  deposition of, 273, 275
  in soils, 251–254, 257–258
  in streams, 276
anaerobic conditions
  decomposition responses to, 230, 250–251
  soil responses to, 23, 29, 275, 280
anchor ice, 337
anions, 273–280
anthropogenic burning, 70, 73. *See also* ignition, human
anthropogenic changes, 326
arctic grayling. *See* grayling
arctic lamprey. *See* lamprey
arctic treeline, 70

arthropods, 133, 141
Asia, export of timber to, 303
aspect, 21
   fire frequency responses to, 28
   soil formation responses to, 29
   soil property responses to, 29
aspen
   diameter of timber, 306
   environmental requirements of, 89
   insect use of, 134–135, 138
aspen forest
   decomposition in, 231, 246
   distribution of, 88, 90
   fire return time in, 71, 287
   harvest of, 303–304
   Holocene history of, 67
   post-fire establishment in, 293
   production of, 91
aufeis, 273, 274, 281, 336
autogenic mechanisms, of succession, 115

backswamp, 72
bacteria, 149, 233
balsam poplar. *See* poplar
bark beetles, 134–135, 137–142
   competition among, 139
   development of, 137, 138, 140
   diversity of, 136
   flight of, 141
   Holocene record of, 64
   life cycle of, 139
   outbreaks of, 333
   phenology effects on, 139
   progeny of, 140
basalt, 274
base flow, permafrost effects on, 277–278, 336
base saturation, 32
bears
   insect responses to, 140
   predation, 220
   stream responses to, 152
beavers, 121
   diet of, 212
   successional role of, 108
bedload, 102
beetles, interactions with hosts, 138, 141.
         See also bark beetles; buprestid beetles; cerambycid beetles
Beringia, 63, 64, 69, 85

B horizon, 26, 30
   formation of, 22
biofilm, 148–149, 158–160
biogeochemistry, 241–261, 334–335
   herbivore effects on, 213–220
   models of, 232
   successional changes in, 217, 218
   See also ecosystem, dynamics of
biomass
   of fine roots, 190
   of insects, 133, 135
   of microbes, 233, 235
   of plants, 105
biotic factors
   insect responses to, 138, 140, 142
   successional changes in, 102
birch
   distribution of, 88, 90–91
   environmental requirements of, 89
   insects of, 134, 138
birch forests
   decomposition in, 229, 231, 233, 242, 246, 257
   diameter of timber, 306
   fire return time in, 287
   harvest of, 303–304
   post-fire establishment in, 293
   production of, 91
   stream responses to, 149
black bears. *See* bears
black carbon, 28
black spruce
   environmental requirements of, 89
   fine-root processes of, 190–203
   insect use of, 134, 135, 136
black spruce forests
   all-season roads in, 306
   areal extent of, 113
   community classification of, 95–96
   community composition of, 94–95
   decomposition in, 229, 234, 242, 244–246, 249, 251, 253, 256–260
   distribution of, 88, 92
   fire effects in, 246, 286–287, 291, 293
   fire return time in, 287
   in floodplain succession, 108–109
   Holocene history of, 64–66, 68, 69, 71, 72, 74
   production of, 91
   in upland succession, 114–115

blackwater streams, 93, 158, 161
blotch miners, 133, 134
bluejoint, successional changes in, 111
bluffs (south-facing), vegetation of, 89–90
bog-fed streams. *See* blackwater streams
bogs
  distribution of, 88
  species composition of, 92
Bonanza Creek Experimental Forest, 18
boreal forest, 329
broadleafed forest, 64, 66–69
broadleafed trees. *See* deciduous species
Brooks Range, 25, 65, 75
browsing
  community responses to, 221–223
  decomposition responses to, 149, 218
  geochemistry responses to, 221
  soil process responses to, 217, 221
  successional effects of, 105, 108
  vegetation responses to, 214–216, 217
brunification, 30
budworm. *See* spruce budworm
bulk density, 23
buprestid beetles, 140, 141, 142
burbot, 152, 158, 161
buried organic horizon, successional role of, 107
burned area, 287–290

C. *See* carbon
caddisflies (Trichoptera), 149–151, 156, 160
calcium, 103, 273
cambisol, 32
Canada, 286, 293–294, 295
canopy structure
  and community classification, 94
  insect effects on, 138
capillary rise, 103, 221
carbon, 312–318
carbon cycling
  carbon balance, 243, 245–250, 261
  fine-root contributions to, 200–203, 245, 256
  herbivore effects on, 217–218
  successional increases in, 103, 235
carbon dioxide, 312–318
carbon stocks
  climatic effects on, 247
  successional increases in, 250

Caribou-Poker Creek Research Watershed (CPCRW), 18
cation exchange capacity (CEC), 22, 26
cations
  in deposition and streams, 273–274
  successional changes in, 107
cerambycid beetles, 140, 141, 142
charcoal, 28, 286
  ecosystem effects of, 256
  Holocene record of, 64–73, 75, 76
  influx of, 66, 72
checkered beetle, 39
chemistry
  of groundwater, 274, 276, 278, 280
  of soils, 274, 276, 278, 280
  of streams, 274, 276, 277, 278, 279, 280
chinook salmon, 152, 158
chironomid, 231
chloride, 273
C horizon, 22
chronosequence, 31, 261
chum salmon, 152, 158, 161
cisco, 152, 158, 161
clay, 26
clear-cutting, 303. *See also* forest harvest
climate
  C balance response to, 241, 244, 246–248, 261
  feedbacks to, 309–318, 337–338
  forest production responses to, 185–186
  gradients of, 16, 93
  Holocene changes in, 64, 68, 69
  hydrologic responses to, 281–282
  insect responses to, 138, 139
  as state factor, 5, 27
  system, 309–318
  threshold, 74
  warming, 3, 50–55
closed shrub stage, in floodplain succession, 105–107
coarse particulate organic matter (CPOM), 148
coarse woody debris, and forest harvest methods, 305
collector-gatherers, 148, 150–151, 160
colluvium, 24
colonization
  by insects, 138–141
  successional effects of, 101
common names of plants, 82–83

communication
  among scientists, 323–324
  benefits, definition of, 323
  consensus building, 329
  with general public, 328–330
  information exchange, 323
  and science education, 324–328
community composition
  of insects, 133–141
  of mammals, 122–125
  of plants, 86–97
  of soil microbes, 228
  of stream invertebrates, 148–151
competition
  among insects, 134–136, 139, 141
  among plants, insect effects on, 133
  successional changes in, 115, 334
coneworm. See spruce coneworm
conifers, insect use of, 134. See also black spruce; larch; white spruce
contributing area, 277–279
Copper River Basin, 22, 24
Cordillera
  Northern, 12
  Southern, 13
cranberry, distribution of, 90
cropland establishment/abandonment, 315
crowberry, 256
crown fire, 285, 295–296
cryosol, 32
cryoturbation, 23, 24, 27
  and soil classification, 32
  and soil morphology, 27
cryptogams. See lichens; mosses
cumulative degree-days. See degree-days

deciduous forests
  distribution of, 87
  harvest of, 306
  Holocene record of, 64, 66, 69
  successional role of, 112
deciduous species
  environmental requirements of, 89
  insect use of, 134
decomposition
  carbon cycle responses to, 313–315
  controls over, 228, 230, 231, 234, 237, 245
  of fine roots, 199–200, 245–246, 261
  of forest floor, 202–203, 241–246, 260–261
  groundwater chemistry responses to, 270, 280
  of leaf litter in streams, 148–149
  nutrient effects on, 245
  plant species effects on, 244–246
  of soil organic matter, 244
  successional changes in, 108, 242
defensive chemicals. See plant chemistry
defoliators, 133, 134, 140
degree-days, 42
  carbon balance response to, 247–248
  freezing degree-days, 42–44, 49
  growing degree-days, 42–44, 247–248
  insect responses to, 134
  soil degree-days, 42–43, 52
  stream water degree-days, 154
  thawing degree-days, 42–44
denitrification, 275, 280
density of stands, insect effects on, 133
deposition, 275
  successional responses to, 102
diagnostic species, 95
DIN. See dissolved inorganic nitrogen
Discharge, permafrost effects on, 336
Diseases, insect-associated, 142
dissolved inorganic nitrogen (DIN)
  in groundwater, 280
  in soils, 228
dissolved organic carbon (DOC)
  in groundwater, 278–281
  in soils, 236, 237
dissolved organic matter (DOM), 148
dissolved organic nitrogen (DON)
  in groundwater, 280–281
  in soils, 237
disturbance
  climate feedbacks from, 315–317
  Holocene record of, 62–65, 69, 70, 73, 74
  hydrologic responses to, 270, 271, 280
  insect interactions with, 138, 141
  regime, 62, 70, 73, 74
  See also fire; flooding; insect outbreaks; snow
diversity of plants, 81–99
  forest harvest effects on, 86–87, 305
  insect responses to, 133, 141
  patterns of, 85–86, 334
DOC. See dissolved organic carbon

DOM. *See* dissolved organic matter
DON. *See* dissolved organic nitrogen
drainage class, 33
   and community type, 95
drought
   decomposition responses to, 230, 233
   experimental manipulation of, 179–180
   fire responses to, 52
   Holocene record of, 67, 73, 75
   insect responses to, 138
   seedling responses to, 105
   white spruce responses to, 314, 333
duff, 285. *See also* soils, organic horizon of
Dune Lake, 65–68, 72

earth hummocks, 27
earthquakes, 14
eastern larch beetle, 135, 136, 137, 140
ecoclimatic province, 39–40
   Boreal Ecoclimatic Province, 39
   Interior Highland Ecoregion, 39
   Subarctic Cordilleran Ecoclimatic Province, 39
   Subarctic Ecoclimatic Province, 39
ecological cascade, 221–222
ecoregion, 17, 21
   Interior Bottomlands, 22, 23
   Interior Highlands, 22
ecosystem
   engineers, 211
   health of, 142
   higher-order interactions in, 222
   insect effects on, 138, 141
   temporal and spatial scales of, 13
education
   formal and informal, 323
   graduate, 328
   K–12, 324
   and research, 325
   science, 324, 326
   tools, 324
   western, 326
effective moisture, 63, 65, 68, 69
electricity, in early Fairbanks, 302
element cycling. *See* carbon; nutrient cycling
energy exchange, 309–312, 318
enzyme. *See* extracellular enzyme

eolian deposits, 24–25, 269
epistemologies, 328
equivalent latitude, 46–47
erosion, 274
   of banks, 281
   successional responses to, 104, 108
establishment, successional effects of, 111
evaporation
   herbivore effects on, 221
   salt crust responses to, 103
   soil moisture responses to, 229
evapotranspiration, 43, 53, 311
   at equilibrium, 271
   hydrologic role of, 271, 273, 277
exotic species, 82
extracellular enzyme, 228, 236, 237

facultative diapause, 139
feathermoss
   distribution of, 91–92
   successional role of, 107, 242
   thermal effects of, 242
feedbacks, 73, 74, 114, 228
Fennoscandia, 69
Fens, distribution of, 88
fertilization experiments, 180–183
   decomposition in, 245
filter feeders, 148, 150, 160
fine particulate organic matter (FPOM), 148
fine roots
   decomposition of, 199–200
   longevity of, 197–199
   production and phenology of, 191–197, 199, 202
   temperate forest, comparisons of, 190–195, 197–198
fire
   climatic effects on, 52–53, 285–288, 291, 294, 296
   climate system responses to, 315–316
   cycle, 249, 285, 287, 292–293, 296
   damage to timber, 303
   databases of, 287–290, 292–294, 296, 298
   decomposition response to, 249
   as disturbance agent, 337
   frequency of, 67–73, 286–288, 293, 296, 312, 316
   future changes in, 333

fire (*continued*)
    Holocene history of, 62, 64–74
    hydrologic effects of, 271
    ignition, 287, 290–293, 295–296
    insect interactions with, 138, 141–142
    interannual variation in, 285, 288–290, 293–296
    interval, 70, 71
    locations of, 287–288, 297
    logging as substitute for, 307
    management, 291, 292, 295, 296, 297
    map accuracy, 287, 294, 297
    microbial responses to, 227
    permafrost responses to, 53
    records, 287–288
    regime, 64–66, 68–73, 285–287, 292–297
    scar, 71–73, 286–287, 296
    season, 288, 291, 293, 295
    size and distribution, 287–290
    soil carbon response to, 249
    soil properties affected by, 28
    soil respiration response to, 249
    soil temperature effects of, 43, 56
    spread, 293
    successional responses to, 56, 110, 113
    suppression or protection, 292–293
    temporal patterns of, 286, 296
    type, 285, 287, 295–296
    weathering responses to, 28
fire severity
    climate feedbacks from, 315–316
    permafrost responses to, 57
    successional responses to, 57, 110
fireweed, successional changes in, 111
fish
    food habits of, 151–152
    life histories of, 160–161
    seasonal migration of, 162
    species composition of, 152, 158
flammability, 71–72, 316
flies (Diptera), 150–151, 155, 156, 158–160
flooding
    geomorphologic responses to, 101–102, 281
    successional responses to, 100–109
    See also precipitation
floodplain, 25, 31
    decomposition in, 235, 245
    fine-root production in, 190–203

forest distribution on, 88, 172
long-term forest history of, 70–72
organic nitrogen in, 252
phosphorus in, 258–259
flora
    composition of, 81–99
    geographic affinities of, 84–85
    state factor influence on, 81
floristic classification, 94–95
flow path, 280
fluctuations in animal populations, 334
flumes, 302
fluvial processes, 101–102, 281
folivorous insects, 133, 135, 136, 137, 138, 141, 142
    leaf chewers, 133
    miners, 133
foodwebs, stream
    litter decomposition in, 148–149
    primary production in, 149, 158–160
    terrestrial herbivore effects on, 149
foraging
    soil responses to, 213
    strategies of, 212
forbs, species names of, 82–93
forest dynamics, 333–334
    Holocene history of, 69
forest floor
    decomposition of, 229, 231, 232
    and germination after forest harvest, 304
forest growth, 184–185
forest harvest, 302–307, 315–316, 334
    insect responses to, 141
    plant diversity responses to, 86–87
forest management, 302–307
forest planning, public participation in, 306
forest-steppe, 67
forest-tundra, 69
fragmentation, of litter, 230
frazil ice, 337
freeze-thaw, 22, 27, 29
freshwater, 309, 317–318
fuel
    consumption, 285, 286, 296, 297
    flammability, 286, 291, 293
    forest harvest effects on, 305
    moisture or condition, 285, 291, 293, 297
    type, 285, 291, 292, 293
    wood as a source of, 302–303

functional types
  of consumers in streams, 148–151
  as an interactive control, 9
fungi
  decomposition by, 149, 230, 233
  insect-associated, 139, 141

gap formation
  insect effects on, 138
  successional responses to, 112
gelifluction, 29, 31
gelisol, 27
general circulation model (GCM), 64
genetics, of insects, 139
geographic information system (GIS), 286–287
geologic processes, 12–17
geologic time periods, 8
geology, 274–277
geomorphology, successional responses to, 101–102
germination, successional consequences of, 111
glacial deposits, 25–26
glaciation, 15–17
  stream responses to, 157–162
glaciology, 274–276
gley soil, 23
GLOBE program, 325, 328
gold mining, 70, 73, 302
grass species, 82–83
grayling, 151–152, 158–159, 161–162
grazers, 148
green alder
  distribution of, 90, 92
  streams responses to, 149, 159
  successional responses to, 112
greenhouse gases, 312–318
ground ice, 27
groundwater
  discharge, 273
  permafrost effects on, 269–271, 336
  recharge, 45, 269
growing season, and climate feedbacks, 311, 313–314. *See also* fire, season
gypsum, 103

hardwoods. *See* deciduous species
hares. *See* snowshoe hares
headwater stream, 274–275

heat flux, 291
hemipterans, 134
herbicides, forest management uses of, 304
herbivores
  fine root responses to, 194–197, 215
  microclimatic effects of, 220–221
  nutrient-cycling responses to, 213–220, 335
  plant allocation responses to, 215–216
  plant chemical responses to, 149, 216–218
  plant production responses to, 219
  successional responses to, 105, 115, 333–334
herb-resprout successional stage
  in permafrost dominated sites, 113–114
  in permafrost free sites, 110–112
histisol, 32
history of forest harvest, 302–303
Holocene, 25, 27, 31
  boreal events, 8
  fire patterns during, 286
  transition to, 85
  vegetation history during, 62–74
homopterans, 134
horsetail, distribution of, 91
host
  distribution, 138
  preference, 134
  physiology, 140
  suitability, 138, 139, 140, 141
  susceptibility, 135, 136, 139, 140, 141, 142
human activity, 70, 72, 73
  soil formation in responses to, 27
humpback whitefish. *See* whitefish
humus, 22, 26
hunting, and moose management, 305
hybrid spruce, 140
hydrology
  of glacial streams, 275, 277
  of non-permafrost-dominated watersheds, 275–277
  of permafrost-dominated watersheds, 275–277
hymenopterans, 134

ice breakage, 53
ice bridges, 306

ice content. *See* soils, ice content of
ice lens, 27
ice storms
  insect responses to, 141
  successional responses to, 108
icing. *See* aufeis
IGERT program, 328
ignition, 337
  human, 287, 290–293, 295
  lightning, 287, 290–291, 293, 295, 296, 337
immobilization, 231, 236, 253–256
Indigenous Peoples
  Aleut, Haida, Inupiaq, Tlingit, Yupik, 325
  Athabascan, 326
  fire use by, 293
infiltration, 45, 269, 270, 276–278, 281
  capacity, 276
informed feedback, 328
initial floristics, 116
inquiry-based pedagogies, 328
insect outbreaks, 138–139, 141–142
  climate feedbacks caused by, 315–316
  as disturbance agent, 53
  fire effects of, 56
  successional responses to, 333
insects
  biomass of, 133, 134, 135
  competition among, 134, 139
  damage to timber by, 303
  densities of 133
  distribution and range of, 133
  diversity of, 133, 136, 141
  flight of, 141
  outbreaks of (*see* insect outbreaks)
  population dynamics of, 139, 140, 141
  reproduction of, 135
  species composition of, 133, 134
  tree interactions with, 134–135, 138–141
  See also macroinvertebrates
interactive controls
  definition of, 7
  successional changes in, 115
interior Alaska, 17
interspecific competition, 135, 141
inversion, 48–49
invertebrates
  decomposition role of, 231, 237
  See also macroinvertebrates, stream

Ips beetle. *See* bark beetles
iron, 30
  oxide of, 31
isotope, 63
isotopic ratio, 65
  foliar nitrogen, 258–259

Kanuti Flats, 22
knowledge
  holistic, 328
  Native, 326
  traditional, 326, 327, 328
  traditional ecological, 327
  traditional Native, 325, 326, 327
Koyukuk Innoko Lowland, 22
Kuskokwim Highlands, 22

Labrador tea, 256
lacustrine deposits, 22, 24, 26
lake-bottom sediments, charcoal record of, 286
lake level, 63, 65, 67
lamprey, 151–152, 158
land bridge, 85
landform, 33
land ownership
  and forest planning, 306
  and timber harvest, 303
landscapes
  decomposition patterns across, 230, 237
  insect effects on, 138, 139, 141, 142
  soil distribution across, 24–26, 30
  vegetation distribution across, 81–99
larch
  distribution of, 92
  insect interactions with, 134, 135, 136, 138
larch budmoth, 135, 136, 137, 140
larch forests
  fire in, 293
  Holocene history of, 69
larch sawfly, 136, 137, 138, 140
large aspen tortrix, 137, 138
Late Glacial Maximum (LGM), 85
latent heat, 273, 276, 310–312
leaching, 22, 148, 280
  base status responses to 30
leaf eating insects. *See* folivorous insects
leaf miner, 133, 134, 137
least cisco. *See* cisco

lepidopterans, 134
lichens, species names of, 83
life history traits and succession, 115
life-long learning community, 324
light, successional changes in, 106
lightning
    fires caused by, 110, 287, 290–291, 293, 295, 296, 337
    insect responses to, 139
lignin, 235, 245, 252
litter
    decomposition of, 228, 229, 230, 232–234, 243–246
    fine root contributions to, 201–203, 245
    fire consumption of, 296
    quality of, 234, 235, 237
    seedling recruitment on, 108, 111
    in streams, 148–149
litterfall
    foliar, 174
    woody, 174
loam, Fairbanks silt, 277
loess, 15–17, 24–25, 172, 274
logging. See forest harvest
longnose sucker. See sucker
lowlands
    soil distribution on, 21, 27, 30
    vegetation distribution on, 92, 172
    See also floodplain
lumber production, 302, 304
Lutz spruce, 140
lynx, 126

macroinvertebrates, stream
    consumption of salmon carcasses by, 152
    functional feeding groups of, 149–151, 158–160
    life histories of, 155–156
    role as fish food, 151
    taxonomic composition of, 150, 158–160
    winter freezing responses of, 156
magnesium, 274
mammals
    herbivorous, 121, 211
    large, 122
    small, 124
management
    of ecosystems, 329
    of forests, 323, 328, 329

    of moose, 305
    of resources, 326
mass attack, 141
mayflies (Ephemeroptera), 149–150, 158, 160
meadows, 230
meander belt, 72
metamorphic rocks, 274
methane, 234, 250–251, 312–317
    consumption of, 251
    production of, 250–251
methanogenesis, 313
microbes
    biomass of, 233, 235
    community composition of, 231, 235
    respiration of 229, 246, 250
    seasonal variation in, 227
    soil-formation effects of, 26
    stream decomposition of leaves by, 148–149
    See also bacteria; fungi
microbial turnover, 257
microclimate
    herbivory effects on, 220–221
    as interactive control, 9, 116
microorganisms. See microbes
microtines
    abundance of, 124–126
    climate effects on, 126–128
mineralization, 229, 231, 236, 252, 254–257
    seasonal changes in, 255–256
mineral soil. See soils
mineral weathering, 23, 24
Miocene
    boreal events during, 8
    tree flora of, 82, 84
mixing model, end member, 279
models, 232
moisture, insect responses to, 138, 140, 141. See also drought; soil moisture
moose
    activity budgets of, 212, 219
    Alaska population of, 122
    biomass of, 126
    food consumption by, 219
    foraging by, 219–220
    habitat of, 219
    hunting for, 305
    life span of, 220

moose (*continued*)
　management of, 305
　population density of, 122, 305
　population turnover of, 220
　predation on, 220
　stream responses to, 149
mortality
　fire effects on, 285, 296, 297
　insect effects on, 134–135
　of seedlings, 107
　of trees, 173
mosaic, successional, 110
mosses
　fine-root growth responses to, 193
　production of, 202
　species names of, 83
　See also feathermoss; *Sphagnum*
mottles, 31
mountain barriers
　to air masses, 15, 17
　to rivers, 13
multiplier effects, on energy and nutrient flow, 220
muskeg, 234
mycorrhizae, and fine-root processes, 199, 204–205
mycetophilid, 231

N. *See* nitrogen
natural enemies, 139
Nenana Valley, 25
net primary production, 52, 91, 202
　aboveground (*see* aboveground forest production)
　and climate feedbacks, 313–315
　of cryptogams, 202
　of fine roots, 202, 335, 261
　fluvial effects on, 281
　soil respiration relationship to, 235, 247
　in streams, 148–149, 158–160
nitrate, 273, 274, 275, 279, 280
　in soils, 252–253, 258
nitrification, 235, 237, 254
nitrogen, 251–258
　microbial uptake of, 253
　plant chemistry effects on, 256
　plant uptake of, 256–258
　successional increases in, 103, 107
　See also nitrogen fixation

nitrogen cycling
　and climate feedbacks, 314–315
　fine-root contributions to, 192, 200–205
　microbial effects on, 231, 232, 235
nitrogen fixation, 252, 258
　nitrogen cycling responses to, 236
　successional changes in, 107, 335
nitrogen immobilization. *See* immobilization
nitrogen limitation, 175, 335
nitrogen mineralization. *See* mineralization
nitrogen productivity, 184–185
nonacidic black spruce/rose/horsetail forest, 95
northern engraver beetle, 135, 137
northern pike. *See* pike
Northwest Territories, 293
NPP. *See* net primary production
nutrient cycling
　above vs. belowground, 335
　climate effects on, 241–242, 244, 246–247, 261
　insect effects on, 138
　mammal effects on, 215, 217
　open vs. closed, 104, 107–108
　parent material effects on, 242, 259–260
　successional changes in, 108
　topographic effects on, 242
　vegetation effects on, 242, 249, 251–252, 255, 257, 261
　See also nitrogen; phosphorus
nutrient uptake, soil fertility effects on, 258

ochric, 22
O horizon
　carbon content of, 244
　decomposition in, 232, 242, 245–246, 249
　description of, 22
　mean residence time of, 246
　soil properties of, 26
　thermal effects of, 242
OLCG ("Observing Locally Connecting Globally") program, 327
oleoresin flow, 141
open willow stage, in floodplain succession, 104–105
organic carbon, 26

organic matter, 21
  accumulation of, 30
  CEC responses to, 30
  soil formation in responses to, 26, 28
organic nitrogen, 335
organic soil. *See* soils, organic horizon of
organisms as state factors, 26
osmotic agents, 233
overland flow, 276–277

P. *See* phosphorus
Paleosol, 23, 31
paludification, 260
paper birch. *See* birch
parasites, 133
parent material
  and soil properties, 23, 24
  as a state factor, 6, 21, 24–26, 172
particle size, 25
pathogenic stain fungi, 139
peak flow, 275, 277
peatland, 285, 315
pedogenesis, 22–24
periphyton, 148–149, 158–160
permafrost
  climate change and, 313, 336
  commercial timber in relationship to, 306
  continuous, 39, 49
  decomposition responses to, 229, 230, 234, 242, 250, 260
  discontinuous, 26, 27, 39, 49, 53, 269, 278, 281
  distribution of, 27, 41, 42, 48, 49, 53, 55, 92, 269–271, 276–279
  ecological effects of, 3
  fire effects on, 285, 286
  ice-rich, 269–270
  isolated, 49
  soil formation in response to, 27
  sporadic, 49
  stability of, 50, 53
  sub-, 269
  successional changes in, 114
  thermal regime of, 57, 242, 250, 260
  thickness of, 49
  See also active layer
pH
  community type in relation to, 95
  of deposition and runoff, 273, 280
phenolics, 216, 236

pheromones, 134, 139
phloem tissue, 139, 141
phloeophagous insects, 133, 134, 136, 137, 138, 141, 142
  cambium feeders, 133
  sapwood borers, 133
phosphorus, 258–260
  availability of, 259–260
  fire effects on, 111
  pH effects on, 259
  total, 258
physiognomy, and community classification, 94
physiographic regions, 12–14
phytophagous insects, 133, 134, 142
pike, 152, 158, 161
pine, 69, 256
pipe flow, 278
plant chemistry
  decomposition responses to, 149, 216, 256
  as defense, 216–217
  herbivore effects on, 216–218
  insect responses to, 138, 141
  nitrogen mineralization response to, 256
  phenolics, 216, 256
  terpenes, 216
  toxic compounds, 218
plant community. *See* vegetation
plant-soil feedbacks, 229
plate tectonics. *See* tectonic geology
Pleistocene
  floristic changes during, 8, 84
  soil formation during, 24–26, 31
Pliocene
  boreal events during, 8
  cooling during, 84
pollen, 62, 63, 67, 68
polyphenolics, 236
poplar
  environmental requirements of, 89
  fine root processes of, 190–203
  insect use of, 134, 135, 138
poplar forests
  decomposition in, 229, 231, 235, 242, 252, 256
  harvest of, 303–304
  Holocene history of, 64, 67
  production of, 91
  successional stage, 107–108

population, human, 18
population dynamics. *See* insects;
    mammals; succession
porcupine, 121, 213
potassium, 274
potential biota
  as a state factor, 6, 93
precipitation
  chemistry of, 273, 275
  decomposition responses to, 227, 234
  Holocene history of, 15, 65, 66, 68, 69
  hydrologic budgets in relation to, 276–277
  insect responses to, 141
  mean annual, 40, 43, 45, 54
  as rainfall, 41, 43, 45
  as snow, 41, 43–45, 52
predation
  insect responses to, 133, 139
  mammal responses to, 121, 126
  stream effects of, 148, 151
primary production. *See* net primary production
primary succession
  definition of, 100
  herbivore effects on, 212
  pathways of, 101
  See also succession
productivity. *See* net primary production
profile, soil
  carbon distribution in, 249
  development of, 22–24, 334
  fine-root exploration of, 194–197
proxy records, 62–64, 75
psychrophiles, 229

quaking aspen. *See* aspen
Quaternary, 8, 35, 63
  floristic changes during, 84

radiation, 46
  and climate feedbacks, 310–312, 318
  and daylength, 40, 46
  longwave, 47
  net, 47
  shortwave, 47
radiatively active gases, 309, 310–312, 318
radiocarbon dating, 63
rainfall. *See* precipitation

recreation and forest harvest, 306
recruitment of seedlings
  in floodplain succession, 108
  in uplands, 111
red squirrel, 128
reduced condition, 27, 30, 31, 32
reforestation, 304
refugium during glaciation, 12
regeneration, 67
  after forest harvest, 304
  successional variation in, 102–115
region. *See* physiographic regions
relay floristics, 109, 115
remote sensing, 286
  of insects, 135
residual material, 22
residuum, 24
resources, as an interactive control, 9, 107
respiration
  of fine roots, 192, 231, 235
  fire effects on, 286
  of microbes, 229, 231–237
  of soil, 244
response time, 276, 278
resprouting, and succession, 111
R horizon, 23
riverboats, stern-wheeled, 302, 303
  fuel for, 302
roads
  extent of, 306
  winter, 306
roots
  herbivore effects on, 194–197, 215
  pathogens of, 140
  production of, 335, 261
  turnover of, 335
  See also fine roots
rose
  distribution of, 90
  insect use of, 134
round-logs, 302–303
round whitefish. *See* whitefish
runoff, 317
Russia, 293–296

salmon
  anadromy and habitat template of, 161
  management of, 163
  marine-derived nutrients from, 151–152
  trophic importance of, 152–154

salt crust, successional changes in, 103
salvage logging, 303
sand dunes, 15, 25
sap feeders, 134
saprophages, 133, 138
sapstain fungi, 138, 141
sawlogs, definition of, 303
scarification, 304–305
scarified surfaces, regeneration on, 304
scavengers, 133
schoolyard LTER, 325
scientific names, 82–83
scolytid beetles. *See* bark beetles
sculpin, 152, 158, 161
secondary chemicals. *See* plant chemistry
secondary succession. *See* succession
sediments
  burial of seedlings, 107
  charcoal in, 70–73
seed rain, successional effects of, 111, 281
self-replacement, 109, 115
sensible heat, 273, 311–312
sheefish, 152, 158, 161
shelterwood harvest. *See* forest harvest
shoulder season, nutrient cycling during, 256
shredding, 148–149, 231
shrub, species names, 82–83
shrub-sapling stage, of upland succession, 112
silt. *See* loess
silviculture, and research, 304–305
simultaneous reporting, to local and scientific communities, 329
Sitka spruce, 140
slimy sculpin. *See* sculpin
slope, 21, 23
  permafrost effects of, 27
  soil formation effects of, 29, 30
  and soil properties, 29
smoldering fire, 285
snags, retained during timber harvest, 305
snow
  disturbance effects of, 53
  See also precipitation
snowmelt, 276, 277, 311, 314
snowpack, 44–46
  duration of, 50, 273
  thermal properties of, 45, 50, 256

snowshoe hares
  biomass of, 126
  density of, 122–124
  lifespan of, 128, 129
  population dynamics of, 121–124
  predation on, 126, 128
  stream responses to, 149
  successional effects of, 107
sodium, 273, 274
soil moisture
  decomposition responses to, 230–231
  deficit, 277
  fire responses to, 286, 293
  holding capacity, 270
  patterns of, 29–31, 179–180, 231, 272, 277, 281
  regime, 26, 28
soil organic nitrogen, 252–254, 257, 261
  turnover of, 252–254
soils
  age of, 24, 25, 31, 334
  carbon accumulation in, 217, 229
  carbon chemistry of, 221
  carbon dynamics of, 244–250
  carbon stocks of, 243–244
  chemistry effects on decomposition, 244–246
  classification of, 22, 32, 33, 34
  color of, 26, 30, 31, 32, 33
  depth of, 24, 25
  development of, 25
  drainage in, 27, 28, 30
  fauna of, 205, 231–232
  formation of, 21–23, 30
  horizons of, 22, 23, 30, 31
    nomenclature of, 23
  ice content of, 276
  mineral horizon of, 272, 278, 280, 281
  morphology of, 27
  nutrients in, 180–184
  organic horizon of, 269, 270, 272, 276, 278
    consumption by fire, 285–286, 296–297
  organic matter (SOM) in, 26, 70, 228, 241–244, 246, 252, 254, 259–260
  pH of, 23, 30, 251, 254, 259–260
  profile (*see* profile, soil)
  reaction of, 28
  riparian, 280

soils (*continued*)
  saturated, 270, 277
  structure of, 23, 26, 27, 29
  texture of, 26, 29, 33, 34
  warming of, 178–179
soil temperature, 32, 46, 48, 50–51, 269, 273
  fine-root responses to, 193–196, 201
  fire effects on, 286
  successional changes in, 106
solutes, 273, 275
spear-marked black moth, 134, 137, 138
species effects, successional changes in, 115
species richness, 87, 334
specific discharge, 277
Sphagnum, distribution of, 92
Spruce, Holocene changes in, 63–72. *See also* black spruce; white spruce
spruce beetle, 137–142
spruce budworm, 135, 137, 138
spruce coneworm, 135, 138
spruce-fir forests, 293
squirrel, predation on hares, 128
stand age, 70–71
state factors
  biogeochemical responses to, 242
  definition of, 4
  and ecosystems, 115, 171–177, 184
steppe, 67
  distribution of, 88–90
stoneflies (Plecoptera), 149–151, 156, 158, 160
stone pine, 69
streamflow, 277, 278
streams
  chemistry of (*see* chemistry, of streams)
  glacial influences on, 157–162
  high-latitude constraints on, 154–155
  ice and winter habitats in, 156–157, 162
  landscape characteristics of, 147–148, 163
  phenology of, 154–156
  types of, 156–160, 275–276
subsoil, 22, 26, 27
substrate quality, 232
  decomposition responses to, 244–246
  nitrogen cycling responses to, 251–252, 254–255

subsurface water, and fine-root depth distribution, 197
succession
  definition of, 100
  facilitation of, 115, 334
  fine-root production and longevity in, 194, 196, 198, 202
  on floodplains, 100–109, 174–175, 281
  insect effects on, 134, 138, 141
  mammalian effects on, 212, 217
  microbial activity in, 227, 229, 231, 235, 237
  nutrient cycling in, 242, 246, 252–253, 255, 258–261
  in permafrost-dominated sites, 72, 113–115, 175–176
  in permafrost-free uplands, 109–112, 175–176
  plant populations in, 214
  processes controlling, 100–120
  state factor effects on, 9
  topographic effects on, 110
  See also primary succession
successional trajectory
  definition of, 100
  fire effects on, 114, 285, 286, 293
  variations in, 109
sucker, 151–152, 158, 161
sulfate, 273–275
surface energy balance, 56, 58, 271
surface fire, 285, 287, 296
surface runoff. *See* overland flow
sustainability, and forest harvest, 305
symbionts, 133
syngenesis, 23, 25
synoptic storm, 281

talik, 57, 271, 281, 336
Tanana Basin/Tanana Valley
  area of productive forest in, 305
  distance to timber markets from, 306
Tanana-Kuskokwim Lowlands, 22
Tanana River, 15, 101, 161–162
tannins, 245
  decomposition responses to, 149, 235–236
  successional responses to, 107
tectonic geology, 13–15
TEK (traditional ecological knowledge), 326

temperature
  of air, 42, 247
  decomposition responses to, 229–230, 242, 244, 246–248
  ecosystem responses to, 3
  insect responses to, 134, 138, 139
  maximum air, 41, 42
  mean annual air, 40, 41, 42, 46, 50, 54, 269
  minimum air, 41, 42
  range air, 40, 42
  of soil, 247
  stream invertebrate responses to, 154–156
  white spruce responses to, 91
  See also soil temperature
tephra, 25
terpene, 216
terrace, 15, 25
Tertiary
  boreal events during, 8
  boreal species distribution in, 84
texture, 24, 25
thermal offset, 50–51
thermal processes, 271
thermal regime. *See* temperature
thermohaline circulation, 4, 317–318
thermokarst, 27, 51, 56
thinleaf alder
  distribution of, 92
  litter production ratio of, 214–215
  litter quality of, 229, 235
  relative production of, 217
  soil nitrogen response to, 258
  stem age distribution of, 214
  successional changes in, 105
thinning, commercial and pre-commercial, 304
thunderstorms, 291
timber
  allowable cut in interior Alaska, 305
  diameter of logs, 306
  distribution of commercial, 306
  export to Pacific Rim, 303
  harvest (*see* forest harvest)
  industry in interior Alaska, 305–306
  large processing facilities, 306–307
  markets, 303, 306
  partial-harvest methods, 304
  resources in interior Alaska, 305
  salvage, 303

timber harvest. *See* forest harvest
time, as a state factor, 7, 31, 172
topography
  community responses to, 95
  hydrologic effects of, 270–271, 276
  microclimatic effects of, 40, 46–48
  soil responses to, 24
  as a state factor, 7, 31, 172
  vegetation responses to, 86–88
toxicity, 236
trace gas feedback, 318, 338
  fire effects on, 286, 297
trajectory. *See* successional trajectory
translocation, 141
transpiration, 311, 314
  insect effects on, 141
transportation network, 19
trap trees, 134
treeline, 312, 314–316
  movement of, 67, 70, 333
treeline black spruce woodland, 95
tree planting, 304
tree rings, 286–287
tree species names, 82
tundra, 291, 312, 316–317
turning point, 334
  abiotic to biotic control, 105
  definition of, 9, 115
  in floodplain succession, 102
  in upland succession, 112, 114
twinflower, distribution of, 90

unscarified surfaces, regeneration on, 304–305
uplands, 21
  north-facing, 92
  south-facing, 89–91
  successional patterns in, 109–115, 172

vegetation
  distribution of, 81–99
  diversity of, 85–86, 334
  Holocene changes in, 62–64, 68, 69, 73, 74
  insect interactions with, 133, 134, 137, 141
  mosaic, 81
  soil interactions with, 26
  successional changes in, 100–120, 312

vertebrates, 211
volatilization of nitrogen, 111

water
  balance, 271
  deficit, 74
  exchange, 309–312, 318
  holding capacity, 25, 29
  -logged, 68
  potential, 233
  -shed, 269–282
  subsurface, 197
  table, 230, 313
  See hydrology; moisture; transpiration
weathering, 22–24
  climate effects on, 334
  during primary succession, 103
  oxidative, 28
  and phosphorus availability, 258, 260
  rate of, 31
  stream chemistry responses to, 274, 279, 280
weevil, 140
western science, 326, 327, 328
wetland, 92, 234, 313
whitefish, 152, 158, 161
white spruce
  environmental requirements of, 89
  germination of, 304
  insect interactions with, 134, 138, 140, 141
  temperature effects on, 91
white spruce forests
  amount of, 305
  cone crops in, 304
  decomposition in, 229, 242, 246, 252, 253, 255, 256, 259
  diameter of logs, 306
  distribution of, 88, 91
  fine-root processes of, 190–203
  fire effects in, 286, 293
  fire return time in, 287
  floodplain successional stage, 108
  harvest methods in, 303
  Holocene changes in, 64–69, 71, 74
  production of, 91
  sawmill use of, 302
  seed rain in, 303
  upland successional stage, 112
wildfire. *See* fire
wildlife
  consumptive use of, 211, 224
  services to society, 223–224
  timber harvest effects on, 305
  and tourism, 223
willow
  biomass of, 219
  compensatory growth of, 219
  decomposition of leaf litter of, 216–219
  insect interactions with, 134, 135, 138
  stream responses to, 149, 159
  successional changes in, 105
willow shrub stage, 114
  decomposition in, 229, 235
  distribution of, 90
  litter biomass ratio, 213
wind, 45, 48
  insect responses to, 141
winter
  and fine-root longevity, 198–199
  stream organisms during, 156–157, 160–162
Wisconsin glaciation, 25
wolf predation, 220
wood-boring beetles. *See* buprestid beetles; cerambycid beetles
Wrangell Mountains, 25

xylem, 141

Yukon Flats Ecoregion, 17
Yukon-Tanana Uplands, 24
Yukon Territory, 293